CLAIMS TO FAME
THE
Lancaster

CLAIMS TO FAME
THE
Lancaster

NORMAN FRANKS

ARMS AND
ARMOUR

Arms and Armour Press
A Cassell Imprint
Villiers House, 41/47 Strand, London WC2N 5JE.

Distributed in the USA by Sterling Publishing Co. Inc.,
387 Park Avenue South, New York, NY 10016-8810.

Distributed in Australia by Capricorn Link (Australia) Pty. Ltd,
2/13 Carrington Road, Castle Hill, NSW 2154.

British Library Cataloguing-in-Publication Data:
a catalogue record for this book is available from the British Library

ISBN 1-85409-220-0

Designed and edited by DAG Publications Ltd.
Designed by David Gibbons; layout by Anthony A. Evans;
edited by Jonathan Falconer; printed and bound in Great Britain
by Hartnolls Ltd, Bodmin, Cornwall.

Front of jacket: Lancaster ED860 Nuts
flew 130 operations. Painting by Air Vice Marshal N. E. Hoad,
CVO, CBE, AFC

Contents

Acknowledgements

I have been pleased and grateful to receive generous help in producing this book, both from individuals and members of a number of squadron associations and their secretaries. All have given recollections and made photographs available for which I thank the following:

G. M. Bailey (100 Sqn), C. R. Baird (103 Sqn), S. Baker DSO DFC (635 Sqn), J. B. Bell DFC (103 Sqn), E. N. Bickley (101 Sqn), G. E. Blackler DFC (550 Sqn), N. D. Bryon (576 Sqn), S. Burrows DFC (44 Sqn), W. F. Caldow DSO AFC DFM (550 Sqn), E. H. Cantwell DFM (101 Sqn), J. Chatterton DFC (44 Sqn), J. R. Clark DFC (100 Sqn), C. Cunnington (115 Sqn), E. A. Davidson DFM (61 Sqn), K. T. Dorsett DFM (15 Sqn), A. W. Downs DFM (166Sqn), J. F. Dunlop DFC (166 Sqn), B. C. Fitch DFC (61 Sqn), J. S. Griffiths DFC (103 Sqn), J. Harris OBE DFC (550 Sqn), B. R. W. Holmes (50 Sqn), L. A. Humphries DFM (75 NZ Sqn), S. A. Jennings DFC (61 Sqn), E. Jones (156 Sqn), H. E. Jones (156 Sqn), J. H. Kemp DFM (101 Sqn), H. W. Langford (153Sqn), R. R. Leeder DFC (101 Sqn), I. Lucas (115 Sqn), G. Luckraft (166/153 Sqns), A. G. Manuel DFM (166 Sqn), R. D. Mayhill DFC (75 NZ Sqn), D. P. McElligott (75 NZ Sqn), P. R. Mellor DFC (635 Sqn), J. D. Melrose DFC (9 Sqn), G. J. H. Murphy (101 Sqn), J. Nicholson (550 Sqn), J. R. O'Hanlon (576 Sqn), G. P. Pickering DFC (115 Sqn), G. H. Rodwell DFM (166 Sqn), A. D. Simpson DFC (75 NZ Sqn), L. W. Sparvell (166/153 Sqns), J. R. Stedman DFC (576 Sqn), C. Straw (166 Sqn), G. E. Tabner (576 Sqn), F. W. Thompson CBE DSO DFC AFC (44 Sqn), D. E. Till DFC (576 Sqn), Mrs S. Westrup (103 & 576 Sqns), J. Wright (166 Sqn), V. C. Viggers DFC (101 Sqn).

I should also like to thank friends and fellow historians who have given help, and in particular photographs from their own collections. They are: Chaz Bowyer, Martyn Ford-Jones, Stuart Howe, Bruce Robertson, Andy Thomas, Patrick Otter and Peter Wright; also photos via Ron Durran, Vic Redfern, G. Robinson, Bob Walkinton, Mrs Barbara Linacre, and Mr R. Pritchard of No 101 Squadron.

Thanks are due to Air Vice-Marshal Norman Hoad CVO CBE AFC, not only for his help but also for making available his painting for the dust jacket; the staff of the Air Historical Branch (MoD), Imperial War Museum, Royal Air Force Museum and the Public Record Office, Kew.

And to Heather...

Introduction

The mere name of the Avro Lancaster, the RAF's most famous four-engined heavy bomber of World War 2, is enough to conjure up all sorts of mental images, not only to the men who flew them, flew in them, to the men and women who helped to build them, service them, or who had friends or relatives who may have died operating them, but to people of other generations who have an interest in any famous aeroplane. The Lancaster is so well known that sometimes it is easy to forget or ignore the statistics of this sturdy machine that helped take Bomber Command's war to targets all over Germany and its occupied territories between 1942 and 1945. Because this book deals with just a handful of these aeroplanes, the statistics become a highlight.

A total of 7,366 Lancasters were built, of which over 3,400 were lost on operations and a further 200 plus were destroyed or written-off in crashes.[1] They carried out approximately 156,000 operational sorties and carried over 600,000 tons of bombs.

Some 125,000 aircrew served in Bomber Command in World War 2, of whom 73,700 became casualties – a staggering 60 per cent. Of this total 63,750 occurred on operations, the others being casualties during training or associated flying duties. In all, 55,500 died, over 47,000 of them on operational sorties; over 9,800 others were taken prisoner. The other 8,400 or so were those who returned wounded or were injured in accidents, divided almost exactly by those 'on ops' and those on other duties.

Of those who died, nearly 38,500 were members of the Royal Air Force (RAF), almost 10,000 Royal Canadian Air Force (RCAF), over 4,000 Royal Australian Air Force (RAAF), while some 1,700 were Royal New Zealand Air Force (RNZAF); men from Poland, France, South Africa and other European or Dominion countries made up the balance.

For a bomber crew to complete a tour of operations (a tour of duty) they had to carry out a set number of sorties against hostile targets. The number varied from time to time, but was generally 30, although sometimes it was 25 and at other times it went up to 35. Different squadron commanders occasionally varied these again, so that if a crew completed a particularly hazardous few trips towards the end of their tour he may allow them to finish on 29 or even 28. By the same token, if casualties had been heavy and a need to mount a 'Maximum Effort' was called for, a crew might be asked to do an extra one. Some keen 'press-on types' may even have gone on to do 32 or 33, especially if a crew member had missed one or two and the others wanted them all to finish together.

However, with each raid having the full potential of turning out to be their last, few really chanced their arm, and as soon as they were allowed to finish, finish they did. But the aeroplanes went on. As long that is, as they performed well, didn't get too badly damaged and passed the regular maintenance service.

Having read the statistics above one can appreciate that neither the crew nor the aeroplane had a particularly good chance of completing a set number of bombing missions, and a crew could just as easily go missing on their first trip as on their 30th - and often did. The chances of any Avro Lancaster completing a large number of raids was even less likely, for as one crew either completed a tour (or were lost in another aeroplane), their usual machine had to keep on going until either it too 'failed to return' or was damaged beyond repair or simply became worn out.

One cannot say with any real accuracy just how many Lancasters flew on operations, but it had to be around 6,500. Yet of this supposed total, only 34 – a mere 1.9 per cent – managed to complete 100 or more ops. The highest number of operations completed by a Lancaster was a miraculous 140.

To list these 34 in any sort of sequence, either by serial number or by ops flown is to miss the point. To know how many operations a particular aeroplane flew is one thing, to know what those operations were, when they were flown, and just as importantly, who flew them, is the reason for this book. So, whether you flew them, flew in them, helped to build or service them or just have an interest in them, the record of these 34 very gallant 'ladies' is here for all to see. It is their claim to fame.

[1] Figures do vary depending on which source is used so I have used approximate figures which still make the points required.

The Avro Lancaster

Developed from the disappointing Avro Manchester, the Lancaster began to equip Bomber Command squadrons in 1942, the first being No 44 (Rhodesia) Squadron at Waddington, Lincolnshire, followed by No 97 Squadron at nearby Woodhall Spa. By 1945 there were 56 squadrons flying Lancasters and over 1,000 of the type were either in these squadrons or on the strength of training or operational conversion units (OCUs).

The Lancaster achieved many firsts: first to fly Pathfinder missions in August 1942; first to carry 8,000lb bombs in April 1943; first to carry a 12,000lb bomb, in September 1943, followed by the 12,000lb deep penetration bomb – the Tallboy – in June 1944; then finally the mammoth 22,000lb Grand Slam bomb in March 1945.

It achieved fame in other ways: the famous raid on the Ruhr Dams by the specially formed No 617 Squadron in May 1943; the sinking of the Tirpitz by Nos 9 and 617 Squadrons in November 1944. It played a major part in the important raid on the German experimental rocket base at Peenemunde in August 1943 which resulted in severe delays to the V1 and V2 flying bomb and rocket programmes. Of the 32 awards of the Victoria Cross awarded to the men of the RAF in World War 2, 10 were given to men who flew in the Avro Lancaster.

Some more facts. Empty, the Lancaster weighed 36,900lb (16.5 tons), while fully loaded it was 68,000lb (30 tons). Bomb loads could and did vary, but the average was 14,000lb (6.25 tons). With this figure it had a range of 1,660 miles and a service ceiling of 24,500ft. Four 1,460hp Merlin 20 or 22 (or 1,640hp Merlin 24) engines hauled this vast load into the sky, together with a seven-man crew (sometimes eight).

For those unfamiliar with fuel loads the following figures will seem even more fantastic. These are actual figures for actual raids with actual bomb loads in 1943:

Target	Date	Fuel load	Bomb load
Essen	25/26 Jul	1,500gals	1 x 4,000lb bomb, 3 x 1,000lb bombs, 540 x 4lb incendiaries
Hamburg	24/25 Jul	1,500gals	– ditto –
Berlin	23/24 Aug	1,743gals	1 x 4,000lb, 2 x 1,000lb

Target	Date	Fuel load	Bomb load
			690 x 4lb incendiaries, 48 x 30lb incendiaries
Nurnberg	10/11 Aug	1,800gals	1 x 4,000lb, 1 x 1,000lb 1 x 500lb,690 x 4lb
Berlin	31 Aug/1 Sep	1,900gals	1 x 4,000lb, 1 x 1,000lb 90 x 4lb
Berlin	3/4 Sept	2,062gals	1 x 4,000lb, 630 x 4lb 48 x 30lb
Milan	12/13 Aug	2,062gals	1 x 4,000lb, 540 x 4lb 48 x 30lb

Anyone who fills their car up each week with, say, 15 gallons of petrol will easily calculate that the bomber that went to Milan carried enough fuel to fill his car up for over 137 weeks – two and a half years! Multiply that by the number of aircraft that went to Milan that August night – 504 – and the total comes to 1,039,248 gallons (the family car will be supplied for 1,330 years). And that was just one night, with another 152 bombers going to Turin as well.

The pure logistics of the amount of aviation fuel needed to sustain not only Bomber Command, but Coastal and Fighter Commands too, plus training, transport, communications aircraft et al, not to mention the massive US 8th Air Force operating in Britain, is almost incomprehensible.

The Lancaster crew probably didn't bother themselves with this mind-boggling statistic. One of them could be trying to work out how he might get hold of just two gallons of petrol for his own car in order to see his girlfriend the next night – if he got back from Milan. What concentrated the mind of the Lancaster pilot, of course, was that his skill was needed to get this powerful but heavily laden beast off the ground. He and his six crewmen were fully aware of the potential death-trap they were flying; a huge metal tomb stuffed with high-explosives, incendiaries, 2,062 gallons of high octane fuel not to mention several thousand rounds of .303 ammunition for use in the eight machine-guns it carried. That was the 30 tons mentioned earlier.

Bomb loads varied of course, depending on the type of target and the range. Less fuel meant a higher bomb load, longer range (ie: more fuel), less bomb load. By 1943 when the Lancaster was really into its stride, the usual bomb load would be one 4,000lb 'cookie', another 2,000lb of High Explosive (HE) – either 2 x 1,000-pounders or 4 x 500-pounders, plus 4lb and/or 30lb incendiaries in varying quantities according to room and load capacity.

Researching this Book

When researching for this book it was necessary to plough through the squadron diaries (Forms 540 and 541) to list each occasion a particular Lancaster flew an operational sortie. If originally I thought this was just a time-consuming chore I was soon to discover that the chore was barbed with all sorts of difficulties.

Anyone reading through the old wartime diaries will know that they are as accurate as the men - or perhaps women - who compiled them. They range from the excellent to the very poor. If a squadron had, say, a former journalist as its diary compiler then the diary could be not only interesting but well kept and well documented as well as accurate. If, on the other hand, someone who could type with at least one finger was 'volunteered' and was far more interested in getting down to the NAAFI, or trying to get away with the minumum amount of information he knew his CO would be happy with – and to sign, if the CO signed at all and didn't leave the formality to his adjutant – those records could be poor and perhaps, occasionally, untrustworthy.

With bomber operations happening almost every night, the orderly room staff's daily chore of typing up anything from 12 to 22 Lancaster serial numbers with seven crew names, plus flying times and comments has to be riddled with typing errors as well as errors of fact. On several occasions I found the same serial flying at the same time with two different crews – an obvious error. What was less obvious of course, was when the aircraft I was researching flew and the serial was typed out incorrectly or listed with a fictitious serial number. On occasions the copy page I was reading was a very poor carbon copy – almost illegible. Crews might be listed wrongly, mis-spelt, or last minute crew changes, where say a crew member went sick and was replaced by someone else, wasn't noted in the records. That would show a man taking part in a sortie when in reality he was in bed with two aspirins.

What is recorded here is the best of the surviving evidence, but it has to be fairly accurate, although I would not be able to vouch that it is 100 per cent so. However, what is produced will give a pretty accurate picture of what each of these 34 Lancasters did and who flew them.

Another problem I found was that where publicity had occurred when a particular Lancaster reached its 100th op, it often proved very difficult to reconcile this to the records. All sorts of things could have happened, of course, and one can picture the scene.

A flight commander happens to mention to the CO that A-Able is nearing its 100th op according to the bombs painted on the side of the cockpit. The CO appears interested and asks for a full breakdown of the sorties flown. The flight commander in turn tells the adjutant, say, who gives the task to a corporal clerk who is about as interested in the job as he is of being away from his bride of two months who lives 300 miles away from where he is based. So he swiftly runs through the Form 541 pages and quickly lists the ops; this list eventually goes back up the chain to the CO.

Thus on such-and-such a night A-Able flies its 100th op. However, what the corporal clerk failed to see, was that one sortie was aborted right after take-off and not counted, and that another was listed as being flown when in fact it was being repaired. Thus its 100th op was in reality only its 98th! That it eventually went on to fly 109 sorties, is one thing, but because there was now a record of its 100th op date, counting forward from that, the record shows that A-Able flew 111.

In any event, it was the aircraft's groundcrew who painted the bomb symbols on the nose – it had little to do with the aircrew other than being of interest if it was 'their' particular aircraft. Moreover, it was the groundcrew's aircraft! They looked after it, checked it, refuelled it, repaired minor battle damage, changed the engines and so on. Few if any of the flyers took the trouble to check-out the totals, especially if they returned from a short leave, saw that some more bombs had gone up and assumed the aircraft had done some additional trips. Their main preoccupation was in staying alive and in finishing their tour.

The difficulty was that what the groundcrew interpreted as a sortie may not have been taken as such by the record keepers. That is not to say the groundcrews were taking liberties, far from it, they were just proud of their machine and if they understood it had flown an op, then up went another bomb. The only problem comes later, when the actual bomb symbols do not easily tally with surviving records, and someone assumes that a photograph of the aircraft in question must be accurate in terms of ops flown.

As in all things, once a fact, however inaccurate, becomes established it is difficult to reconcile or change it. Therefore the reader will find in some of the aircraft-biographies that follow, an occasional reference to total sorties being different – in dispute if you like – to some long-assumed established 'fact'.

Mentioning aborted missions, these too caused problems, in that sometimes they were counted as ops, sometimes not. Obviously if a Lancaster lost an engine right after take-off and the pilot went round and landed again then the mission was not counted. If on another raid, the Master Bomber called off the raid near the target or even over it, the raid was abandoned but the op counted – quite naturally. The grey area is somewhere in between.

Ron Clark, who flew EE139 with No 100 Squadron in 1943 (see The Tour) was asked about aborts and the like. He says:

'The issue of aborts in 1943 was a grey area and sometimes these were credited or half credited or not credited, depending on the circumstances.

'The decision to abort was not easy, it was a terrific anti-climax and a lot of trouble had gone for nothing. One had the feeling of letting the side down. We aborted our trip to Oberhausen on 14 June because of rear turret failure before reaching the coast. The CO criticised my decision and probably thought a pep-talk was in order as I was a new boy. The night-fighter usually approached from the rear and below and Geoff Green in the defective turret would still have had the chance to detect it. Of course, the enemy had tremendous advantage in fire-power and the ability to pick up the glare of the bomber's exhausts. It might have been impossible for Geoff to bale out in an emergency though.

'After having an engine fail on the Nurnburg trip on 27 August, we decided to press-on but we gradually lost height and rather than stress the remaining engines for such a long flight we dropped the bombs in the Channel and returned to base. At the debriefing I was taken aback when the ground engineer officer threatened me with a court-martial after I said that I wanted a new engine instead of another re-conditioned one. As I walked away I glanced at "Dinger" Bell, the Flight Commander, who seemed equally surprised. Neither trip was credited of course.'

Bomber Command set up a bomb-line later in the war, a line at which by flying past it the sortie counted towards a crew's tour; but aborting on the near side of it, it didn't. For obvious reasons the 'line' was not advertised and the line in any event changed from time to time. It has not been clear anywhere in the Form 541s whether each and every raid was made official or not when aborts occurred. The reader will have to pick his own way through that minefield as he reads the raid lists. At this distance I am certainly not going to make decisions of that nature. Ron Clark remarks on this bomb-line:

'My rear gunner, Geoff Green, who operated his second tour in 1944 tells me that the 'Bombing Line' was instituted after the Invasion of Europe to protect the Allied forces from our own bombs. This line fluctuated with the rapid advance and was the cause of some mis-understandings and casualties on the ground. Geoff recalls some awkward moments with Canadian soldiers in a bar in Paddington. They had taken casualties in this way but they ended up by buying him a pint.'

As author of this book I have not, by the same token, set out to prove any points, merely to record the details of these 34 'above average' Lancasters - centenarians - whose rise to fame was in the number of ops they flew. In some cases, where in the past a certain number of ops has been credited to them, the total figure is a problem. Others are not so problematic, but all went to, or over, the 100. Some squadrons claimed credits for sorties flown during the food-dropping sorties to Holland in 1945

– Operation 'Manna', while others also gave credit to trips flown during Operation 'Exodus' – the return of released Allied prisoners of war by air.

Where these occurred they have been noted, but it will be up to the reader if he wishes to think whether a Lancaster which flew 109 bombing raids plus two 'Manna' and two 'Exodus' trips made a total of 113 sorties or not. That is not germane to the aims of this book. All I might say is that what is good for one Lanc should be good for another!

By and large a pattern does emerge when looking at the life of almost any Lancaster during World War 2. It arrives on a squadron, is checked over by the groundcrew and assigned to a Flight. Unless it is specifically a replacement for a crew's Lancaster which has been seriously damaged a night or so before, then the new bomber is available to any crew who needs an aeroplane. As there were usually more crews than available/serviceable machines, one or two crews regularly shared a Lancaster. From time to time it then becomes the more or less permanent aeroplane for one specific crew. After a few ops the Lanc is taken from the Availability State for a couple of days while it is looked over by the maintenance section – something like a family car's first 1,000-mile service.

Once back in action, its usual crew will continue with it, but then for a couple of nights another crew will take it, this occurring as the regular crew takes a 48-hour leave or some such break period. They return and take over the machine again until they finish their tour – or fail to return while flying another aircraft. In between times, of course, anything can happen, from slight damage to an engine failure which will take the bomber out of the line for a few days. A Lancaster might have as many as eight or nine engine changes on a squadron. The RAF's accident damage categories referred to in the text need a little explanation: Cat AC = Minor damage; Cat E = Write-off.

Then the regular crew may be lucky and finish their tour. The Lancaster then seems to be in a sort of limbo while two or three other crews – probably new crews – start their tours. Suddenly one finds that once again a more or less regular crew is flying it on most ops. This continues until the bomber has another major service or, as the ops mount up, it may go away from the squadron – even back to the makers for a complete refit and service. If it returns to the squadron it will continue although it may just as easily be reassigned to another squadron.

One might also imagine there were many aspects to flying these Lancs which had now become veterans, with 60, 70 or 80 missions completed. Not everyone was keen to fly them. On the one hand they were getting old and had perhaps lost some of their spriteliness. Why would a crew choose to fly an old crock when it may have the chance to fly a brand new machine?

Then again, as superstition and luck – real or imagined – played a very large part in any air force crew's flying, one would obviously start thinking that this particular Lancaster with its 80-plus-ops was living on borrowed time. Its luck must surely be running out soon and do I want to be flying it over Germany when it does? A few Lancasters got into the 90's before they were lost, one even going down on its 99th sortie.

What follows now is a brief account of each of the 34 known Lancs that became centenarians, and continues their story until either they were 'retired', the war finished or they were shot down – or in the language of the time, simply 'failed to return'.

Understanding the Lists

By and large the operational sortie lists are self explanatory, with date, target, the times the aircraft took off and landed back, and the crew as listed in the Form 541.

Where a crew's subsequent sortie in an aircraft was made, then merely the name of the captain is recorded. Where there has been a crew change, then the new name is noted as 'in' and the missing man as 'out'. These will not be completely accurate due to human error; what is recorded here is what is recorded in the Form 541.

The addition of a second pilot (2P) is generally a new pilot who was required to fly at least one, sometimes two, trips with a seasoned pilot and crew, to get the feel of operations, prior to taking his own crew on ops. Sometimes the second pilot will be a senior officer, who decides to go along for the ride. In some squadrons it became the custom to have navigators, bomb aimers and even air gunners do a 'practice' trip too, hence occasionally these will be noted, eg: (2BA), (2N) etc. However, on some sorties and in some squadrons, a second navigator would be part of the crew to operate and monitor the H2S radar. On No 101 Squadron special operators made up the eighth crew position and are recorded as SO in the lists.

Crew positions/duty has not always been possible to record accurately. Some records simply do not show what duty a man performed. However, most squadrons did note the seven men down in their crew positions. That is to say that if they recorded them as, pilot (P), navigator (N), bomb-aimer (BA), flight engineer (FE), wireless-operator (WOP), mid-upper (MU) and rear-gunner (RG), then they always (mostly) listed them in that order. In the majority of these cases, I have listed the crew in the way it was generally done, so one can assume that subsequent crews show the same positions.

In any event, the pilot was always first, and in the vast majority of cases, the two gunners were listed last and the wireless-operator was listed third from last. The variation almost always came between the other three members, navigator, bomb-aimer and flight engineer, sometimes being listed, second, third or fourth or in any combination thereof. Hopefully, at some stage, the squadron's way of listing them has been shown.

Occasionally, different squadrons going to the same target will record a different name. The main reason for this was that one unit might record the general target area, while another its specific aiming point. I have chosen, generally, to record the target as it appears in the Form 541. Any reader wishing to tie-up these targets should cross-check them with either Martin Middlebrook's excellent reference book *The Bomber Command War Diaries* or the list of Bomber Command targets in Air-Britain's *The Lancaster File* compiled by J. J. Halley.

GLOSSARY OF RANK ABBREVIATIONS USED IN THE LISTS

Sgt	Sergeant	FO	Flying Officer
FS	Flight Sergeant	FL	Flight Lieutenant
WO	Warrant Officer	SL	Squadron Leader
PO	Pilot Officer	WC	Wing Commander

The ranks of officers on detachment from the US 8th Air Force to RAF Bomber Command are abbreviated in the lists as follows:

| T/Sgt | Technical Sergeant | 1/Lt | First Lieutenant |
| Flt Off | Flight Officer | 2/Lt | Second Lieutenant |

The Tour

When researching this book I was in contact with Mr Ron Clark DFC who had completed a tour of operations in 1943, flying for the most part Lancaster EE139 'Phantom of the Ruhr'. He and his crew were with No 100 Squadron, although their last operational sortie was with No 625 Squadron which had been formed from C Flight of No 100 Squadron on 1 October 1943.

Although all I really needed was a couple of stories about EE139, what he produced was so good and rich in detail that it really could not be cut down so it has been reproduced here in its entirety. In this way the reader, provided he was not himself part of wartime RAF Bomber Command, can hopefully have some insight into what it meant to fly operations during World War 2 and be a part of a bomber crew.

In reading his words one will begin to appreciate the stories of the 34 Lancasters that are featured in this book and remember that whatever the aircraft itself achieved, there were seven, sometimes eight, men taking it on operations.

'At the end of May 1943 after two years training in my case, my new crew and I were posted to No 100 Lancaster Squadron based at Waltham, near Grimsby. This was it! I had a total of 437 hours 45 minutes flying time in my log book.

'RAF Waltham seemed strangely quiet after the bustle of RAF Lindholme where we had completed our crew training and we spent the next 10 days doing local flying. On 2 June we became acquainted with EE139 for the first time. After some local flying details and a X-country flight we began to realise that EE139 was ours.

'Settling into the squadron routine we visited the aircraft every morning, then sat in the Flight Office – always a snappy salute when entering – got to know the other crews and then took tea and toast with the two charming ladies who kept the village Post Office over the hedge from the aircraft dispersal. As a crew we had a Nissen hut to ourselves and sometimes the groundcrew dropped in. Occasionally we carried out a 'low-level' on Grimsby, some more occasionally than others.

'We were a seven-man crew, all NCOs. Harold Bennett the flight engineer was a Lancastrian from Preston and a regular, an ex-brat from Halton. He had joined the crew at the Heavy Conversion Unit at Lindholme, where we had flown Halifaxes, then finishing on Lancasters. His expertise stood us in good stead.

'Jim Siddall, navigator, was a strong minded Yorkshireman whose job was to keep us on track although he had little to go on when the Gee navigation system was jammed. The Station routine had little attraction for him and his Flight Office was not often graced with his presence, for which I had some explaining to do. Jim, as the only married man in the crew, was shot down and killed in a Mosquito aircraft over Holland later in the war.

'Lishman "Lish" Easby was the wireless-operator. Some of the crew were tickled by his North Riding accent, now considerably modified after postwar years in the Civil Service.

'At 29 years-old, Les Simpson, the mid-upper gunner, was the Daddy of the crew. He had recourse to the sick-bay once or twice, one occasion being after falling off the top of the fuselage after giving his turret perspex an extra polish. By then we were all a bit touchy about these details, especially if the aircraft had been used by another crew whilst we were on leave. Occasionally, on our early trips, Les, no doubt half frozen, had inadvertently fired a burst from his twin Brownings and sometimes saw things which weren't there. Gleefully, and I suspect mainly by his fellow Londoner, Doug Wheeler our bomb-aimer, Les was dubbed "Trigger" after these unwelcome interruptions. Les secretly revelled in this soubriquet as it marked him out as the veteran gunner he was becoming.

'If there was any escapade in the offing, Geoff Green our rear-gunner from King's Lynn, was sure to be involved. On one occasion after we had paid a visit to the "Marquis of Granby", we saw Geoff hobbling down the street on some-body's crutches. He was a sartorial example to the rest of the crew and at our first meeting we noticed with approval his custom-made uniform. Geoff went on to complete a second tour of Ops on Lancasters, and was awarded a DFC for shooting down a jet fighter. After a permanent commission, he retired years later as a squadron leader.

'Finally on 11 June we were posted for operations for that coming night. The Station became more active, the heavy chain was locked around the telephone kiosk and we waited in anticipation for the briefing. Our EE139 was at the bottom of the Battle Order but all eyes were on the tapes pinned to the map of Europe. "Gentleman, the target for tonight is Dusseldorf," said the Squadron Commander. "Happy Valley again," muttered the older hands as the Station Commander, the CO, and finally the Intelligence Officer had their say.

'A welter of crews with their equipment and parachutes boarded the crew buses which dropped us off at the aircraft. The long bomb-trollies and belts of ammunition of the armourers were on hand as we greeted the ground crew and checked over EE139. No tea at the Post Office tonight!

'Finally settled we could hear Merlin engines spluttering into life as we awaited our turn. The gunners were wedged into their turrets with thick electrically-heated suits and fur-lined flying boots. The rest of us wore normal bat-tledress on the adequately heated flightdeck. In my case I had on army boots and neck scarf, with emergency rations and shaving kit stuffed into my battledress pockets. Later, one had to decide when to consume the bar of chocolate and the can of orange juice which was an aircrew concession.

'Strapped into the pilot's seat with the Mae West and parachute harness, one became conscious that the Elsan toilet was located right at the back of the aircraft, so I could see a problem looming. Ben Bennett solved the problem by attaching a tube to an empty can with an empty orange juice tin on the other end. We passed this contraption round the crew like the port at a Mess dinner. This was rarely used after the first couple of trips, probably because on those early sorties the added tension had resulted in more frequent calls.

'I settled into my seat. I had already noticed that on the square of armour plate situated behind the pilot's head (later removed to save weight) somebody in the Avro factory had pencilled, "May good luck follow you everywhere". We would certainly need it.

'The aircraft in the first wave were taxying out of their dispersal points and we signalled the groundcrew to stand-by for start-up. We always did this in the same sequence to bring in certain services powered by the engines. With strict radio silence being maintained, we waved the chocks away and taxied out to join the queue. The line of well-groomed purposeful Lancasters were a brave sight with their big props flicking over. Although each aircraft had six machine-guns facing aft, there was a solitary Bren gun mounted on a pick-up truck guarding the approach to the runway in case an intruder showed up.

'One by one the heavily laden Lancasters swept down the runway to climb lazily over the bungalows of Waltham village and head for the rendezvous point on the coast above Mablethorpe. Finally it was our turn. The Aldis lamp winked green and we taxied onto the main runway. A final check with the crew and I opened the throttles to full power. This had to be done asymmetrically along with coarse use of the rudder to keep the aircraft straight, cancelling out the effect of the torque from the propellers and any cross-wind until the tail came up as the speed increased. With the speed now rapidly increasing and with the high pitched crescendo of sound from the Merlins, ears alert for engine failure before the critical speed, the end of the runway was coming up fast.

'The ASI needle flickered to flying speed and I heaved back on the control column. As the rumble of the wheels ceased, EE139 was airborne for her first operation. Landing gear up quickly, increase speed to prevent stalling as the flaps came up and then reduce power when they were retracted. Turn onto course and do the after take-off checks.

'We arrived at the coast and circled to gain height until it was time to set course across the North Sea. A rate-one 360 degree turn took two minutes and the aim was to depart right on time and maintain the density of the stream of aircraft. Darkness rapidly fell as we set course, flying away from the sun, but that was no signal to switch on navigation lights.

'We kept a good look-out but there was the occasional sickening lurch as we encountered the slipstream of an unseen bomber. One thought of the people at home who would read tomorrow about the raid and also the casualties in the morning newspapers, then we were on our way.

'As we approached the Dutch coast the tension increased and the crew kept in contact by intercom, especially with Geoff in the rear turret. I soon abandoned the gentle weaving technique so as not to disorientate the gunners. As we were in the last wave, the target came into view early as the attack had started. It reminded me of the glow of the steelworks near my home-town in West Cumberland. Soon one could see what looked like sparklers high above the target. Then with the final turn-in, we were exposed to the inferno below.

'The sparklers had now become large puffs of smoke from the flak barrage. The widespread fires augmented by the flashes of the guns, exploding bombs, photographic flares dropped from aircraft, coupled with the probing searchlights, and an occasional stream of tracer bullets created a terrifying picture. We were unprepared for this but Doug, lying on his stomach, with his parachute pack beneath him, was already guiding us to the target markers dropped by the Pathfinders.

'An aircraft not far away was coned in the searchlights. It appeared as a bright object in the middle of a colossal web of light. The flak turned its attention to this visual target and the unfortunate bomber (and crew) were soon enveloped in the shell bursts. Before long one could see the bobbles of tracer fire from the night fighters and then the aircraft exploded. As the searchlights were looking for further prey, Doug was lining up his bomb-sight and ordered, "Bomb doors open!".

'This was the most testing part of the operation as it meant holding steady for 15 seconds whilst following his instructions minutely. "Left...left...right a bit...steady..." Could we avoid that column of smoke, often a feature for the last wave? Another aircraft was reported above us, its bomb-doors open.

'"Bombs gone!" What relief! A quick turn onto the course out as the bomb-doors closed just as the 4,000lb 'cookie' from the aircraft above plunged downwards. Lightened of its load EE139 was now much livelier and gradually the target receded and we climbed to take advantage of favourable winds.

'We were not sure of our position as we neared the English coast and called "Darkie". Do we need any help? No thanks, just confirm our position. We landed safely at base. Over the egg and bacon after de-briefing we pondered momentarily over our future; 29 more ops?

'The next night it was operations to Bochum, again the Ruhr - the Happy Valley. Over the target it was our turn to be Buggins as we were coned in the searchlights. I immediately kicked the Lancaster into a steep diving turn - no time to warn the crew; they knew what to expect. The Elsan would shoot its contents onto the roof. We slid into comparative darkness and breathed again with only slight flak damage.

'The Battle of the Ruhr wore on but on 12 July we were surprised to see the tape leading to Turin - something new! There was a long detour over the Bay of Biscay on the way back in order to avoid flying over enemy territory in daylight. We were promised fighter escort at dawn and a diversionary airfield in Cornwall was nominated in case we ran short of fuel.

'Compared to the German targets there was only token resistance over Turin and we headed for home. As dawn broke over the Bay of Biscay there was no fighter escort to be seen. In fact there was nobody to be seen at all and we plotted a solitary course to Cornwall. Any marauding Ju88 would have got a hot reception but we were glad to see the coast and Ben said we had just enough petrol to reach Waltham. People were going to work as we skimmed over Cleethorpes.

'I picked out the runway through bleary eyes and suddenly – bang! Everything in the aircraft rattled and what a bounce. With the Lancaster, one bounce usually meant more – very undignified in daylight – so I opened the throttles and went round again. Ben squeezed the last drops of fuel out of the tanks and I was wide awake enough to make a normal landing after being 11 hours in the air strapped to the pilot's seat.

'On all RAF Stations there was a sprinkling of Commonwealth aircrews. The Australians in dark blue, the New Zealanders and Canadians in RAF blue but different; the South Africans with their army ranks and khaki uniforms plus a few Americans in the RCAF who usually changed into US uniforms which improved their pay prospects. To serve with them was one of the privileges of being in the RAF and many of them stayed on to fly in the airlines after the war.

'Although we were an all-English crew, some of the gunners in the squadron were French-Canadian and at about this time they all moved from the sergeants' mess to the officers' quarters through their government's policy. The RAF sergeant pilots were given accelerated promotion and within a few months I moved from sergeant through flight sergeant to warrant officer. The other crew members remained as sergeants.

'I was not offered a commission but when I applied, it seemed a formality and it was promulgated just before my last operation of the tour. The others were commissioned after they finished their tour except for Ben who became a warrant officer. The Uxbridge Syndrome was apt to break out now and then. The CO asked me how my relations with the crew would be affected if I was commissioned. I suppose he had to ask something and what could I say? Not much!

'On 24 July the assault on Hamburg began and during the next nine nights we carried out, with EE139, four operations, interspersed with one to Essen and another to Remscheid where our good photograph of the aiming point showed a town of roofless buildings.

'The weather on our trip of 2 August was very bad and after avoiding thunderstorms we found ourselves off track in the middle of a heavy flak barrage over what was probably Bremen. We dropped the bombs and returned, being credited with half a trip for our pains. Fortunately aborts were infrequent, with our well-maintained aircraft.

'By the third trip to Hamburg the Germans had considerably strengthened the defences and a veritable forest of searchlights stretched along the Elbe. Fortunately with the introduction of "Window", the radar element was neutralised. Lish Easby spent a lot of his time stuffing packets of "window" – aluminium foil strips – down the flare-chute.

'Normally, if the pilot was incapacitated, the flight engineer would do his best to fly the aircraft, there being no co-pilot of course. In our case I thought it best to nominate Doug Wheeler for the job as he had received some pilot training and he sometimes flew the aircraft when we carried out local exercises. Ben kept his counsel about this but in an emergency he would have had his hands full in any case without flying the aircraft.

'Ben had already painted the Phantom of the Ruhr on the nose of the aircraft. As the skeleton in a shroud, clutching a bomb in each bony fist it had rather a grim appearance despite the sardonic grin on its face. There was also a line of small bombs depicting the German targets and an ice-cream cornet for Turin shortly to be followed by one for Milan.

'After Ops to Nurnberg and Milan we entered the briefing room on 15 August to see the tape running along the Baltic coast but couldn't make out the target. A terse message read out from the C-in-C, universally known amongst the crews as "Butch" Harris, said that if we did not destroy the missile plant at Peenemunde this night we would have to repeat the process until we did. As the bombing height was only 8,000ft it all sounded pretty interesting.

'We had moved up in the Battle Order and being now in the first wave we could see the usual flak over Flensburg on the way in and arrived over the target just as the markers were going down. The people on the ground must have been surprised and the night-fighters had been lured to Berlin as the likely destination. However, there was the usual spirited defence with a flak-ship just off the coast taking part on this occasion.

'In the brilliant moonlight I thought it would be a good idea to have a closer look at the enemy and our substitute mid-upper, Flying Officer Wilson, was enthusiastic, so after bombing I pushed the nose down and we were soon belting over the flat Pomeranian countryside at about 200ft. Geoff quibbled a bit about opening up on civilian targets and I said without a lot of conviction, "Oh, they're all Germans". He scored hits on a large building to which I alerted him as we flashed past while the other gunner was enjoying himself. There was not a light to be seen and no retaliation so we climbed and headed for home.

'The "Master of Ceremonies" Squadron Leader John Searby, was the hero of that night as he circled the target seven times redirecting the attack by radio as the markers were tending to drift off. But who knows what dramas were played out in the 40 aircraft that were lost after the night-fighters redeployed.

'On 25 August 1943 a frisson ran through the briefing room when the Big City appeared on the map, although this was not totally unexpected with the longer nights having arrived. This was the first of three trips we carried out to Berlin with EE139. The Station Commander was present at debriefing. His smart attire was in stark contrast to the dishevelled state of the crews, without a trouser crease between us. I always got the impression that he disliked my scarf; but perhaps he liked it.

'I was detailed to take Sergeant Cook and his crew in EE139 on their first operation on 22 September, which was to Hannover. They were a good crew and they reminded me of our first operation. The next night we went to Mannheim – a well defended target judging from our previous visit. Over the target we were coned and this time they held on and the aircraft was badly damaged by flak and a night-fighter. Ben Bennett was awarded the DFM for his cool action under heavy fire.

This operation put EE139 into the workshops. We said a fond farewell to Phantom of the Ruhr the next day as she stood stricken but defiant in the big hangar. She was to haunt the skies of Europe again and complete 121 operations before being retired for crew training.

'A week later we were "top of the bill" to Hagen and we completed our tour in JA714 after taking her to help form No 625 Squadron at Kelstern. The Flight Commander asked if we would like to space-out our last few trips but the others wanted to finish quickly. The odds were lengthening against us and now always top of the Battle Order, usually we were first off and first back. It was preferable to be ahead of about 20 aircraft approaching the same airfield at night with navigation lights off.

'With thumb hovering over the transmitter button to get the first call in, the WAAF operator in the control tower came straight back almost before I had finished - "Join circuit R for Roger". Was that a sound of relief in her voice? She would know us by now.

'Our first and last operation with No 625 Squadron was yet again to Hannover and we had a brand new pilot on his first sortie. He must have thought he was pushing his luck. Arriving at dispersal everything was behind time. After all the others had taken off I stopped the bombing-up, cleared all the equipment away and hurriedly started the engines. We broke radio silence to request permission to taxi down the deserted runway to save time after pondering a down-wind take-off. We thundered down the runway into the empty sky and 625 had achieved the maximum effort.

'A few mornings later the crew parted at the local railway station to go our separate ways. We were grey-faced but a load had been lifted. I saw Doug Wheeler several times at Lindholme where I was a flying instructor and once met him by chance in the Strand after the war. I also met Ben one day on a train when he was with an aeronautical company. Les worked at Heathrow airport for a while and I met him occasionally.

'Years after the war, Geoff, Lish and I regained contact and later Ben too, through a magazine article he had written. I expect that in our retirement our thoughts turn back to those heady days and the bond that has been forged between us.'

THE AIRCRAFT

R5868
QUEENIE/SUGAR

Perhaps it is fitting that we start with this Lancaster as it has become famous as being the one credited with the highest number of raids to survive to the present day. Although due to battle damage, repairs and maintenance, much of what one sees of the aeroplane today cannot possibly be the same as was seen during World War 2, the machine as a whole is preserved and on permanent display at the Royal Air Force Museum at Hendon, North London.

It was a Mk I built by Metro Vickers at Trafford Park, Manchester as part of Contract 982866, R5868 being the 27th off the Lancaster production line. Once built and equipped with four Merlin XX engines the aircraft was delivered to Avro's at Woodford for final assembly and tests on 20 June 1942.

Nine days later R5868 was delivered to No 83 Squadron at Scampton, Lincolnshire, assigned to B Flight and given the squadron code of OL and the individual letter Q-Queenie. After a few days in the hands of No 83's ground personnel to bring the new bomber up to squadron specification, it was put on the operational strength and deemed ready for ops.

Its first captain and crew were not freshers but an experienced bomber man and the B Flight Commander, Squadron Leader Ray Hilton DFC and bar, who had already flown with No 214 Squadron in 1941 before becoming a Flight Commander with No 83 Squadron. His first tour covered 34 ops, 27 being flown with No 83. Now he was back with No 83 Squadron on his second tour.

Ray Hilton took Queenie to Wilhelmshaven on the night of 8/9 July 1942, the aircraft's load comprising 1,260 x 4lb incendiairies, which the bomb-aimer, Sergeant C. H. Crawley dropped from 16,000ft, all falling in the target area in the early hours of the morning.

Hilton went on to fly Queenie 18 times during his second tour which covered 27 ops, completing his tour on 19 February 1943. By this time he had been promoted to Wing Commander.

Queenie was to have several different painted artworks on its nose, at first a nude female kneeling in front of a bomb, just aft of the front turret, on the port side. This was later painted out and replaced by a red devil (Mephistopheles – to whom Faust sold his soul in German legend), thumbing its nose, dancing in flames with the motto 'Devils of the Air' beneath it. That was painted on by Sergeant A. W. Martin, the Flight Engineer in Pilot Officer Neale M. McClelland's crew when the bomber went to No 467 (RAAF) Squadron in September 1943. In February 1944, Flight

29

Sergeant Dan Smith and LAC Ted Willoughby painted on the words 'No Enemy Plane will fly over the Reich Territory–Hermann Goering', with an arrow pointing to the ever-growing number of bomb markings on Queenie's nose. This of course was the famous boast by the Luftwaffe Commander, Reichsmarschall Hermann Goering made in the early days of the war. Other items of nose-art were a DFC ribbon after Queenie had completed 30 ops and then a DSO after 60. The Lanc was later 'awarded' a bar to the DFC.

Although Bomber Command generally only flew night sorties in the mid-war period, Queenie's fourth mission was a daylight op to the Krupps Works at Essen. Only 10 Lancs (four from 83) took part and it was a cloudy day, so the bombers flew with the aid of Gee and the three that reached the target bombed through cloud. Queenie was one of them and Hilton received the DSO for it and Pilot Officer A. F. MacQueen, his mid-upper gunner, the DFC. Flying through one clear patch a couple of FW190 fighters were seen approaching but they did not attack.

On a night raid to Duisberg on 25/26 July night fighters did attack but Hilton and crew evaded two attempts by FW190s and a Me110 to get at them. On 5/6 August the Squadron CO, Wing Commander D. Crichton-Biggie, took Queenie on a mining trip to the Gironde river where the Lanc was slightly damaged by flak. Then on 18/19 August, Hilton took it to Flensburg on the first-ever Pathfinder (PFF) sortie, Queenie carrying 14 x 4in flares, but the night proved too dark to locate the target and the flares had to be brought back.

The squadron now began to fly PFF sorties, often carrying both bombs and flares – later PFF markers. Flares and bombs went to Wilhemshaven on 14/15 September but Queenie was attacked by a 'four-engined night fighter' on the way home – that is, another Lanc. Flight Sergeant H. Kitto the WOP was wounded.

Queenie went to Italy for the first time on 6/7 November 1942, the target Genoa. Italy was always a long haul often with the Alps to cross. Fifteen crews of 83 took part, two going down over the target, and Queenie had to land at Mildenhall on its return. One of the other crews flying Queenie on occasions was that of Flight Lieutenant J. E. Partridge DFC, but they were to go missing in another aircraft in March 1943.

Flight Lieutenant J.Hodgson DFC took Queenie on its 30th trip on 21/22 December to Munich, and on its next mission, 16/17 January 1943, Hilton took the Lanc to Berlin for the first time, taking also for the first time Target Indicator (TI) markers and flares. Trying to ensure the target was marked accurately, Hilton spent 25 minutes over the Big City that night, but finally haze and broken cloud beat them and they had to bring their markers back.

Hilton finished his second tour on 19 February, strangely enough with a raid on Wilhemshaven where he had taken Queenie on its first mission. Flying Officer F. J. Garvey and crew began to fly the aircraft that month and soon became its regular crew. Over Cologne on 26 February flak damaged the bomb doors and they had to fly home with them partially open. More excitement followed over Berlin on 29/30 March when for 12 minutes they were chased by a FW190 then a Me110 night-fighter. Rick Garvey flew Queenie to Cologne on 3/4 July, recorded as its 60th op. On the 8/9th, Pilot Officer Hugh Ashton, Garvey's rear gunner completed his 30th sortie and received the DFC at the end of this his first tour.

Squadron Leader R. J. Manton flew Queenie to Hamburg on 24/25 July, the first time aircraft carried 'Window' – bundles of thin metal strips to confuse German radar – on this the start of the Battle of Hamburg. Garvey took Brigadier General F. L. Anderson of the US 8th Air Force to Essen on the night of 25/26 July as an observer, together with the Group Navigation Officer, Squadron Leader A. Price. Anderson also went to Hamburg two nights later.

Queenie's period with No 83 Squadron came to an end after a trip to Milan on 14/15 August 1943, with Garvey as skipper. This was the aircraft's 67th op and Rick Garvey had flown on 20 of them, with Hilton a close second with 18. Awards to crewmen who had flown Queenie with 83, apart from Hilton's DSO, were DFCs to Garvey, Hugh Ashton, Pilot Officer S. Sukthanker, Garvey's navigator and the DFM to Garvey's bomb-aimer, Sergeant John A. Cook, whose last sortie in Queenie was his own 35th op. Of Hilton's crew, Pilot Officers O. R. Waterbury (navigator) and A. F. MacQueen (mid-upper) received the DFC while Flight Sergeant Roy Beaven (flight engineer) won the DFM. All were killed in 1943. Ray Hilton later went to No 467 (RAAF) Squadron and was killed in action over Berlin on 23/24 November 1943.

One new crew to fly Queenie was that of Flight Sergeant M. K. Cummings, to Cologne on 16/17 June. On their return from the trip, Cummings received his commission but they all went missing the next night in another Lanc. Almost the same thing happened to Pilot Officer E. Mappin and crew, who took Queenie to Montchanin on 19/20 June and all but one of the crew were lost over Krefeld on 21/22 June.

Queenie left No 83 Squadron on 16 August, having run-up a total of nearly 368 operational flying hours. After a major service it was assigned to B Flight, No 467 (RAAF) Squadron at Bottesford, Leicestershire, moving to Waddington, Lincolnshire, in November. With 467 the code letters were changed to PO and R5868 became S-Sugar. The aircraft replaced the previous Sugar (JA981) which had become 'unreliable for ops'. Its first trip with 467 was to Hannover on 27/28 September, Pilot Officer A. M. Finch as captain, then to Bochum on the night of 29/30 September 1943, flown by Pilot Officer Neale McClelland RAAF, taking a newly arrived pilot with him - Flying Officer J. A. Colpus – on his 'Dickie' trip. Flight Lieutenant Harry Locke took Sugar to Munich on 2/3 October, even though the squadron records show him as flying the previous 'S' – JA981 – but Locke had written R5868 in his log book. Locke received the DFC this same month and later flew with No 97 Squadron.

Colpus flew Sugar himself on 3/4 October (Kassel) although once more the squadron clerk got the serial number wrong in the Form 541. In fact JA981 turns up a few times in early October which is very confusing, as the Lancaster with this serial, while with No 467 (RAAF) Squadron for a brief period, was then with No 617 Squadron and was lost at sea on 15/16 September 1943!

November still saw serial JA981 being recorded, but McClelland took Sugar to Berlin on 18/19th although the Lanc was hit by flak over Bonn and damaged. McClelland received the DFC for this trip. A few nights later – 26/27th – again attacking Berlin, this time with Colpus as skipper, Sugar was coned by searchlights after bombing. Colpus took evasive action but another Lanc came from the port quarter, losing height, and the two bombers collided. Sugar fell away to port in a dive, Colpus applying full rudder and aileron to control the aircraft but then found he could only get 140mph out of it, so Colpus lightened the load on the port side by ordering the engineer to run all four engines off the port outer tank. They got back but landed at Tholthorpe, Yorkshire.

This put Sugar out of action until early in 1944, its next raid being yet another to Berlin on 15/16 February, with a new crew skippered by Pilot Officer J. W. McManus RAAF. It was an eventful night. Twice Sugar swung on take-off but McManus got the Lanc off at his third attempt. Then over the target Sugar lost an engine. In the rear turret sat Sergeant Cliff K. Fudge RAF from Bristol, flying his eighth trip, and over the German capital he celebrated his 21st birthday, an event which was featured in the *Daily Sketch* newspaper on the 17th.

The following two raids, Leipzig on the 19/20th and Stuttgart on 20/21 February both had problems. On the former all the crew suffered sickness due to bad oxygen, while on the latter the port outer engine cut at 13,000ft so McManus had to jettison the load and abandon the mission. A trip to Schweinfurt on 24/25th went alright but over Augsburg the next night they were coned on the bomb-run and came under heavy flak fire for five minutes.

Obviously Sugar was beginning to show signs of age. Coming back from Frankfurt on 22/23 March all four engines began to splutter and the Lanc lost height to 10,000ft. However, some switch jiggling caused them to pick up again. Just as well for the WOP, Sergeant M. Williams, had forgotten his parachute! On landing the tailwheel-tyre burst.

Sugar's next trip was to Berlin again, but the op was not completed as the port outer failed and there was an oil leak from the port inner, causing them to turn back. A night fighter trailed them for 15 minutes but was finally lost in cloud and by evasion. The next night the port outer failed on the bomb-run over Aulnoye and coming back at 5,000ft only just made Tangmere, the RAF fighter base on the Sussex coast. Fortunately it had been a French target.

Above: Lancaster R5868 'Q' in May 1943 while with No 83 Squadron, showing 58 bomb symbols and the 'Devils of the Air' insignia. Rick Garvey is in the cockpit; the others l to r are: Bill Webster, Len Thomas, C. Turner, Jimmy Sukthanker, Jack Cooke and Hugh Ashton.

Right: Neale McClelland and crew in front of R5868 with No 467 (RAAF) Squadron. The original bomb log has been painted out and started again with a kneeling nude holding a bomb insignia.

Above: 'Sugar' in flight, with JO-A of No 463 (RAAF) Squadron in the foreground.

Below: Flying Officer T. N. Scholefield after 'Sugar's' reported 100th op, 12 May 1944 — in fact its 91st. From l to r: Scholefield, I. Hamilton, R. T. Hillas, F. E. Hughes, R. H. Burges, K. E. Stewart and J. D. Wells.

Above: Pilot Officer J. W. McManus RAAF in R5868's cockpit, showing Goering's famous boast. Note, too, the DSO and two DFC ribbons marked under the bomb tally.

Below: 'Sugar' on 7 May 1945 at Kitzingen airfield, Germany, with 125 ops showing.

Top left: No 9 Squadron's 'Johnny Walker', W4964, showing 104 ops. Note the kangaroo insignia beneath the navigator's window.

Left: James Melrose and crew. The large 100th bomb was painted on after hitting the Tirpitz with a 12,000lb Tallboy bomb on 15 September 1944. From l to r: Melrose, S. A. Morris, E. C. Selfe, J. W. Moore, E. E. Staley, E. Hoyle, and R. G. Woolf.

Above: DV245 SR-S of No 101 Squadron over Cap Gris Nez on 26 September 1944. The bomb log and saint insignia can just be discerned below the cockpit.

Right: 'The Saint' insignia, with 40 sorties recorded.

Left: Roy Leeder (DFC) took DV245 on its first sortie, 7/8 October 1943.

Below: Three of Leeder's crew, Geoff Smith, Roy Brown and Percy Drought. All three received the DFM.

Right: Eric Bickley was Roy Leeder's rear-gunner and also won the DFM.

Below: Stan Bowater (front) and crew. From l to r: Gerry Murphy, Ted Reeves, Hugh Dickie, Ken Dickinson, Andy Oliver and Freddie Campbell.

Above: DV302 of No 101 Squadron showing bomb log and long service ribbon.

Left: Flying Officer D. H. Todd RNZAF (right) took DV302 on its first op, to Berlin, on 18/19 October 1943. His wireless-operator was Vic Viggers, also RNZAF.

Opposite page, top: A. E. Netting and crew. From l to r: M. A. Ordway, C. H. N. Complin, E. H. Cantwell, Netting, P. R. Gunter, V. R. Burrill and T. C. Brown.

Right: Sergeant M. A. 'Basil' Ordway, mid-upper gunner in Netting's crew with DV302. Note the ABC aerials.

Above: ED588 VN-G showing 125 bomb symbols. The aircraft failed to return on 30 August 1944, its 128th op.

Below: ED588 and its rival, ED860 QR-N, being waved off in the summer of 1944. N went on to complete 130 trips.

Above: 'Uncle Joe', ED611 of No 463 (RAAF) Squadron, with stars for ops rather than bomb symbols. This photograph was taken on the occasion of the 100th op.

Below: ED611 a few days later with a new paint scheme; note the earlier block of stars painted out and the original block increased.

Above: Bernard Fitch and crew operated in ED860 seven times in their tour. From l to r: A. Lyons, Len Whitehead, Les Cromerty, Fitch, Sid Jennings, Johnnie Taylor and C. Kershaw.

Below: Up goes the 100th bomb on ED860, accomplished on 29 June 1944.

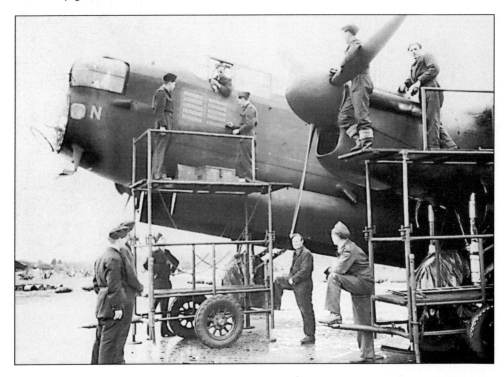

Top: Norman Hoad and crew. Front: Hoppy Boyd and Hoad; rear: Bill Ball, Moosh Embury, Bill Pullin, Norman England, Wilson and Lucky Webb.

Above: Preparing to go to Bordeaux on 11 August 1944 — Op No 120. Norman Hoad in the cockpit of ED860. The crew, l to r: Ball, Pullin, Embury, England, Boyd and Webb.

Above: 'Mike'/'Mother of them All' — ED888 of No 103 Squadron in 1945 showing 140 bomb symbols, the DSO, DFC and two swastikas.

Left: ED888's first regular crew. From l to r: Denny Rudge, Jack Baird, Chiefy Catton, Jack Fitzpatrick, Trevor Greenwood, George Lancaster and Syd Robinson.

Top right: Bert Booth (ED888's fitter) and Jack Fitzpatrick, Rudge's Australian rear-gunner, sit on a 4,000lb 'cookie'.

Right: Jim Bell and Jimmy Griffiths, who took ED888 on its 99th and 100th ops, watch Group Captain W. C. Sheen 'award' the DSO to their charge during July 1944.

Above: 'Mike-Squared' — 140 not out!

Below: The early days of ED905, showing crossed British and Belgian flags and 21 bombs, mostly flown by Belgian, Flying Officer F. V. P. van Rolleghem of No 103 Squadron. John Lamming who painted on the insignia, is third from the left, sitting on a 'cookie'.

Following a few days of service, Sugar was taken over by Pilot Officer A. B. L. Tottenham RAAF, taking it on seven straight raids without a hitch. In early May Pilot Officer T. N. Scholefield RAAF took over as skipper and about this time Goering's famous words were painted on the nose. With the invasion coming up, targets were now mostly against French targets which included V1 flying bomb sites, gun positions, and all manner of transport centres, both rail and road.

On the night of 11/12 May, Bourg Leopold was the target for Sugar's 100th raid. Officially that is, but due to various mix-ups in the records it was only the Lanc's 91st. In any event it was not an auspicious occasion. Haze made it impossible for the target to be identified so the attack was called off. Then two Ju88 night-fighters made several determined attacks over a 10-minute period, the gunners claiming one Ju88 damaged. The bombs were jettisoned over the sea. However, having returned, there were celebrations outside the watch office and later LAC Poone painted on the 100th bomb symbol.

Scholefield, who was later to fly with No 97 Squadron and become a squadron leader, took his crew to visit the Metro Vickers factory and during a lunch break, addressed the workers who had built Sugar.

Flying Officer I. Fotheringham now took over Sugar, taking it to St Pierre Du Mont on the night of D-Day, 5/6 June 1944, although they had a few moments of anxiety when another Lanc was spotted right above them during the bomb-run. This crew took Sugar on four of its next five sorties but they were lost on 29 July.

Sugar's actual 100th operation came on 7/8 July when Pilot Officer M. G. Johnson took it to St Leu d'Esserent, the target being some quarries where the Germans were storing FZG 76s – otherwise known as V1 flying bombs. Mid-July saw Sugar's last sortie with No 467 (RAAF) Squadron with Johnson taking the Lanc to Revigny. After this the squadron said a sad farewell to their old favourite, S-Sugar, which now had 114 bomb symbols on its nose (actually about 103 true ops).

On 3 August Sugar went to the Repair Inspection Works (RIW) but after an extensive overhaul and being converted to take H2S, it returned to No 467 (RAAF) Squadron on 3 December 1944. Various crews operated in the aircraft during that month, including Johnson, now a flight lieutenant. Even in January 1945 there was no set crew flying the Lanc although Flying Officer L. W. Baker RAAF took it on most of its final trips, including one to Czechoslovakia on 18/19 April and the final raid to the U-Boat pens at Flensburg on 23 April – Laurie Baker and crew's 21st operation.

This made 22 raids since Sugar's return to 467, thus 136 according to records, 127 otherwise, plus four 'Exodus' trips, but count them as you will. The aircraft left No 467 (RAAF) Squadron on 23 August 1945 bound for No 15 Maintenance Unit (MU), where it became non-effective in August 1947. Struck off Charge as an 'Exhibition Aircraft' on 22 February 1956 R5868 became part of the Historic Aircraft Collection at No 13 MU, Wroughton, on 16 March 1956. On 24 November 1970, after some years as

Gate Guard at Scampton, commencing in 1959, the Lanc went to Bicester's MU for refurbishment before going to the RAF Museum on 12 March 1972.

1942
No 83 Squadron
8/9 Jul
WILHELMSHAVEN
0007-0420
SL R Hilton DFC
PO O R Waterbury (N)
Sgt R Beavan (FE)
FS H Kitto (WOP)
Sgt C H Crawley (BA)
PO A F MacQueen (MU)
S*g*t A Lavey (RG)
11/12 Jul
DANZIG (Mining)
1652-0257
SL R Hilton and crew
FO H Barber (2P)
14/15 Jul
BORDEAUX (Mining)
2306-0644
PO J E Partridge
FS L W Kleeman (N)
Sgt R Scott (FE)
Sgt J H Ridd (WOP)
Sgt J Allen (BA)
Sgt A Ireland (MU)
Sgt E Mills (RG)
18 Jul
ESSEN
1050-1450
SL R Hilton and crew
Sgt D R Gilchrist in, Sgt Crawley out
19/20 Jul
VEGESAK
2357-0557
FS C D Calvert
Sgt D Dunmore
PO R M Rees
Sgt H A Hafield
Sgt L A Connett
Sgt W D Henderson
Sgt T W Strong
21/22 Jul
DUISBURG
2356-0337
SL R Hilton and crew
23/24 Jul
DUISBURG
0053-0450
FS L T Goodfellow
PO E H Penfree
Sgt J J Mathieson
Sgt E Webster
Sgt F M Tutton
Sgt C H J Byrd
Sgt D L Howe
25/26 Jul
DUISBURG
0034-0406
SL R Hilton and crew
26/27 Jul
HAMBURG
2303-0411
PO J E Partridge and crew
Sgt H Lavey in Sgt Ridd out
5/6 Aug
GIRONDE (Mining)
2227-0541
WC D Crichton-Biggie
PO H L Mazengarb
FS D J Calderwood

Sgt Chaster
Sgt P J Musk
Sgt J Rogers
Sgt C J Millard
6/7 Aug
DUISBURG
0053-0456
PO J Marchant
PO F C Oldmeadow
Sgt L Edwards
Sgt F C Milton
Sgt J Phipps
Sgt R L Romig
Sgt J E Smith
9/10 Aug
OSNABRUCK
0012-0409
SL R Hilton and crew
Sgt D G Lovell in, Sgt Gilchrist out
10/11 Aug
MAINZ
2246-0424
PO J Hodgson
Sgt W A Hamilton
Sgt D N McCartney
Sgt A B Smart
FS G R Pheby
Sgt T Williams
Sgt S A Hathaway
18/19 Aug
FLENSBURG
2110-0215
SL R Hilton and crew
24/25 Aug
FRANKFURT
2050-0235
SL R Hilton and crew
8/9 Sep
FRANKFURT
2106-0230
FS L T Jackson
Sgt D Smith
PO J McMillan
Sgt K C Taylor
FS B E Hargrove
Sgt L R Brettle
Sgt D Crossthwaite
13/14 Sep
BREMEN
2303-0326
SL R Hilton and crew
14/15 Sep
WILHELMSHAVEN
2011-0017
SL R Hilton and crew
2/3 Oct
KREFELD
1925-2305
SL R Hilton and crew
PO B Becker and FS P J Musk in,
Sgts Kitto and Beaven out
5/6 Oct
AACHEN
1915-0100
FL J E Partridge and crew
Sgt D L Coen in, Sgt Allen out
6/7 Oct
OSNABRUCK
1915-2340
SL R Hilton and crew

WC D Crichton-Biggie (2P)
Sgt A MacFarlane in, Sgt Becker out
13/14 Oct
KIEL
1815-2340
SL R Hilton and crew
6/7 Nov
GENOA
2125-0645
SL R Hilton and crew
7/8 Nov
GENOA
1800-0110
SL J K M Cooke DFC
Sgt T C Milton
FL H D M Ransome
Sgt W L Gibbs
Sgt T R Cairns
Sgt S A Hathaway
Sgt H Plant
9/10 Nov
HAMBURG
1725-2245
SL R Hilton and crew
FL E R Simpson DFM in, Sgt Warren out
13/14 Nov
GENOA
1810-0230
SL R Hilton and crew
Warren back
15/16 Nov
GENOA
1810-0130
PO R N H Williams DFM
FS T R Armstrong
FS G L Davies
FS C H Crawley
FS G H Bishop
FS J R Bushby
FS C C Y Lambert
29/30 Nov
TURIN
0050-0815
FS H A Partridge
Sgt R O Felton
PO L W Sprackling
Sgt J M Freshwater
Sgt H Fell
Sgt A D Finnie
Sgt A D Organ
2/3 Dec
FRANKFURT
0205-0800
PO J Marchant DFC RAAF and crew
Sgt F C Milton in, Sgt Powell out
21/22 Dec
MUNICH
1756-0103
FL J Hodgson DFC and crew
Sgt C W Candlin and FS C F J Sprack in,
Sgts McCartney and Hathaway out

1943
16/17 Jan
BERLIN
1655-0015

SL R Hilton and crew
11/12 Feb
WILHELMSHAVEN
1747-2310
FS H A Partridge and crew
13/14 Feb
LORIENT
1830-2310
FS H A Partridge and crew
14/15 Feb
MILAN
1911-0246
SL J K M Cooke DFC and crew
FS A B Smart in, Sgt Milton out
16/17 Feb
LORIENT
1916-2338
SL S Robinson DFM
FS A B Smart
PO D Norrington DFM
FS J H Henderson
Sgt M B W Hambrook
FS C F J Sprack
Sgt W A Rings
(all but Smart were killed in 1943, five on 26 February, Sprack DFM on 13 June)
18/19 Feb
WILHELMSHAVEN
1846-2308
FO F J Garvey
Sgt R B Hicks (FE)
SL R Anderson (N)
Sgt C E Turner (WOP)
Sgt J A Cook (BA)
FS R O Fulton (MU)
Sgt H A Ashton (RG)
19 Feb
WILHELMSHAVEN
1811-2218
WC R Hilton DFC and crew
FS J Rodgers in, Warren out
25/26 Feb
NURNBERG
2032-0256
FO F J Garvey and crew
Sgt S Sukthanker and Sgt D L J Turner in,
SL Anderson and FS Fulton out
26 Feb
COLOGNE
1940-2304
FO F J Garvey and crew
Sgt D B Bourne, Sgt D L J Turner,
and Sgt J Goldie DFM in Sgts Hicks, Turner
and Ashton out
28 Feb
ST NAZAIRE
1838-2312
PO U S Moore DFM
Sgt J H Wright
FO G H Wilson
FO D L Giggory
Sgt D L Coen
Sgt G K Finnie
FO D H Luck

1/2 Mar
BERLIN
1910-0125
PO U S Moore and crew
8/9 Mar
NURNBERG
1957-0244
FO F J Garvey and crew
FS H Lavey in, Sgt Goldie out
11/12 Mar
STUTTGART
2010-0147
FO F J Garvey and crew
FO N C Johnson and Sgt H A
Ashton in
Sgt Bourne and FS Lavey out
12/13 Mar
ESSEN
1928-2336
FO F J Garvey and crew
Sgt D B Bourne and Sgt J A
Cook in
FO Johnson and Sgt Coen
out
27/28 Mar
BERLIN
2009-0305
FO F J Garvey and crew
29/30 Mar
BERLIN
2126-0434
FO F J Garvey and crew
Sgt W L Webster in, Sgt
Bourne out
2/3 Apr
ST NAZAIRE
2009-0045
FS G A McNichol
Sgt G C Mott
PO H H F Beaupre
Sgt G S MacFarlane
PO T W Lewis
Sgt H R Willis
FS C E Hobbs
(PO McNichol was killed in
action, 17 April 1943)
23/24 May
DORTMUND
2247-0319
FO F J Garvey and crew
25/26 May
DUSSELDORF
0007-0415
FO F J Garvey and crew
27/28 May
ESSEN
2247-0343
FS R King
Sgt K E L Farmelo
FS D E Curtin
Sgt D J Phelan
Sgt H Samme
Sgt R A Adams
Sgt E D McPherson
(Farmelo, Adams DFM and
McPherson DFM all killed in
action, 20 Jan 1944)
29/30 May
WUPPERTAL
2232-0320
FO M R Chick
FS A W Hicks (FE)
Sgt J W Slaughter (N)
Sgt B Turner (WOP)
PO C A S Drew (BA)
Sgt L P Howell (MU)
Sgt A Ellwood (RG)

11/12 Jun
MUNSTER
2335-0427
FO M R Chick and crew
FS S T Stacey in, Sgt Howell
out
12/13 Jun
BOCHUM
2327-0403
FO M R Chick and crew
Sgt Howell back
16/17 Jun
COLOGNE
2304-0311
FS M K Cummings
Sgt H W Luker
Sgt F W Wilcox
Sgt H W Cheshire
Sgt J Roughley
Sgt N Woodcock
Sgt R A Taylor
(all lost on 18 June in anoth-
er Lancaster)
19/20 Jun
MONTCHANIN
2243-0228
PO H Mappin
FS C E Wiggett
Sgt A J Boar
Sgt A A Crank
Sgt G A Livett
Sgt W Anderson
Sgt F W Turner
21/22 Jun
KREFELD
2349-0407
FO M R Chick and crew
22/23 Jun
MULHEIM
2309-0316
FL F J Garvey and crew
24/25 June
ELBERFELD
2302-0326
FL F J Garvey and crew
SL R J Manton (2P)
Sgt J A P Logan in, FS Thomas
out
28/29 Jun
COLOGNE
2304-0323
FL F J Garvey and crew
3/4 Jul
COLOGNE
2322-0407
FL F J Garvey and crew
8/9 Jul
COLOGNE
2255-0344
FL F J Garvey and crew
12/13 Jul
TURIN
2246-0816
FO W R Thompson
Sgt A W Belton
Sgt D E Potts
Sgt A Wilkes
Sgt P Henratty
Sgt R B Hicks
Sgt J H Tolman
24/25 Jul
HAMBURG
2216-0427
SL R J Manton
Sgt F S Chadwick
SL A G A Cochrane DFC
Sgt A E Evans

Sgt C Taylor
WO S T Stacey
DFC FO A J Ellis
(Manton, Cochrane, Evans,
Stacey and Ellis were all
killed in 1943)
25/26 Jul
ESSEN
2228-0307
FL F J Garvey
Brig Gen F L Anderson USAAF
SL A Price (N)
FS W L Webster (FE)
FS B H Turner
FS J A Cook
FS L L J Thomas
PO H A Ashton
27/28 Jul
HAMBURG
2235-0409
FL F J Garvey
Brig Gen F L Anderson USAAF
FS D E Potts in, SL Price out
29/30 Jul
HAMBURG
2214-0356
SL R J Manton and crew
12/13 Aug
MILAN
2130-0517
FL F J Garvey and crew
SL Georgeson in, Sgt Potts
out
14/15 Aug
MILAN
2127-0534
FL F J Garvey
FS S J Davis (2P)
SL N A Burt
FS W L Webster
FS B H Turner
FS J A Cook
FS L L J Thomas
FS H Sykes

No 467 (RAAF) Squadron
27/28 Sep
HANNOVER
1927-2050
PO A M Finch
Sgt V L Johnson (FE)
Sgt J W Nedwich (BA)
FO H C Ricketts (N)
FS G G Johnson (WOP)
FS R Smibert (MU)
FS R H Mark (RG)
29 Sep
BOCHUM
1800-2250
PO N M McClelland
FO J A Colpus (2P)
Sgt A W Martin (FE)
FO McCarthy (BA)
PO W Booth (N)
Sgt S Gray (WOP)
Sgt S G W Bethell (MU)
Sgt K L Worden (RG)
2/3 Oct
MUNICH
1815-0232
FL H B Locke
Sgt W G Holt
FS F F Townsend
Sgt H Hassell
Sgt L Butler
Sgt T Brooks
FS T Munro

3/4 Oct
KASSEL
1836-0047
FO J A Colpus
Sgt K Smith
Sgt S T Bridgewater
FO D J Stevens
Sgt P MacDonald
Sgt J L Brooks
Sgt F A Rutt
18/19 Oct
HANNOVER
1726-2241
PO N M McClelland and crew
PO H Griffin in, FO McCarthy
out
3/4 Nov
DUSSELDORF
1701-2122
PO N M McClelland and crew
Sgt C C Schomberg (2P)
10/11 Nov
MODANE
PO A Fisher
2056-0431
Sgt S L Smith
FS H Galloway
FS M H Rooney
FS T H Ronaldson
Sgt J H Rayns
FS F Beamish
18/19 Nov
BERLIN
PO N M McClelland and crew
1715-0136
FS D L Gibbs (2P)
22 Nov
BERLIN
PO N M McClelland and crew
1641-2312
23 Nov
BERLIN
1653-2324
PO N M McClelland and crew
26/27 Nov
BERLIN
1713-0059
FO J A Colpus and crew
Sgt L M Jackson in, Sgt
Brooks out

1944
15/16 Feb
BERLIN
1737-0025
PO J W McManus
Sgt J B McNab (FE)
Sgt I Stapleton (BA)
FS A R T Boys (N)
Sgt M Williams(WOP)
FS L C Vaughan RAAF
Sgt C K Fudge (RG)
19/20 Feb
LEIPZIG
2342-0706
PO J W McManus and crew
Sgt H Feltham in, Sgt McNab
out
20/21 Feb
STUTTGART
2349-0136
PO J W McManus and crew
FS A Summers in, Sgt Fudge
out
24/25 Feb
SCHWEINFURT
1832-0204

PO J W McManus and crew
McNab and Fudge back
25/26 Feb
AUGSBURG
1830-0215
PO J W McManus and crew
1/2 Mar
STUTTGART
2311-0719
PO J W McManus and crew
FO E G Strom in, FS Boys out
18/19 Mar
FRANKFURT
1857-0055
PO J W McManus and crew
22/23 Mar
FRANKFURT
1912-0024
PO J W McManus and crew
FS Boys back, FO Strom out
24/25 Mar
BERLIN
1852-2211
PO J W McManus and crew
25/26 Mar
AULNOYE
1920-0035
PO R E Llewelyn
Sgt L H Dixon
PO G W Venebles
Sgt W Prest
FS K Overy
Sgt F W Hammond
Sgt K W Ward
11/12 Apr
AACHEN
2025-0036
PO A B L Tottenham RAAF
Sgt R J Taylor
Sgt S Adam
FS J G Walsh
FS H A Cummins
Sgt T A Stevens
FS G G Podosky
18/19 Apr
PARIS/JUVISY
2111-0126
PO A B L Tottenham and crew
20/21 Apr
PARIS/LA CHAPPELLE
2333-0350
PO A B L Tottenham and crew
22/23 Apr
BRUNSWICK
2328-0455
PO A B L Tottenham and crew
24/25 Apr
MUNICH
2051-0630
PO A B L Tottenham and crew
26/27 Apr
SCHWEINFURT
2129-0627
PO A B L Tottenham and crew
28/29 Apr
ST MEDARD
2326-0655
PO A B L Tottenham and crew
3/4 May
MAILLY-LE-CAMP
2150-0323

PO T N Scholefield
Sgt R H C Burgess
FS F E Hughes
FO I Hamilton
FS R T Hillas
Sgt J D Wells
FS K W Stewart
6/7 May
SABLE-SUR-SARTHE
0035-0521
PO T N Scholefield and crew
10/11 May
LILLE (M/YDS)
2205-0133
PO T N Scholefield and crew
11/12 May
BOURG-LEOPOLD
2216-0152
PO T N Scholefield and crew
5/6 Jun
ST PIERRE DU MONT
0253-0705
FO I Fotheringham
Sgt P A Scratchley
Sgt G R S Miller
FO J M Beaton
FS F J Pottinger
Sgt C R Knapman
FO C A Phillips
(FO Fotheringham KIA 29 July)
6/7 Jun
ARGENTAN
2353-0352
FO I Fotheringham and crew
8/9 Jun
RENNES
2307-0518
FO I Fotheringham and crew
12/13 Jun
POITIERES
2218-0453
FS K V Millar
Sgt J H Barnes
FS A M Hughes
WO K A McKay
FS A M Meggs
FS A J Perkins
FS B Bromilow
14/15 Jun
AUNAY-SUR-ODON
2246-0321
FO I Fotheringham and crew
27/28 Jun
VITRY
2201-0532
FO I Fotheringham and crew
29 Jun
BEAUVOIR
1209-1534
FS M G Johnson
Sgt P Hounslow
FS R Dunn
FS N J Palfrey
FS J C Whitelaw
FS F S Cavenagh
FS E C Evans
4/5 Jul
ST LEU D'ESSERENT
2313-0336
FO W R Williams
Sgt A J Goodwin
Sgt D J MacDonald
Sgt L F C Weeks
FS J J Murray
FO J H Kitt
Sgt A H Cooper

7/8 Jul
ST LEU D'ESSERENT
2223-0312
PO M G Johnson and crew
14/15 Jul
PARIS/VILLENEUVE
2211-0455
PO M G Johnson and crew
18 Jul
CAEN
0357-0727
FS I R Cowan
Sgt F G W Mills
Sgt E E Warrender
Sgt B S Cox
FS C F Curtis
FS S Murphy
FS F K White
18/19 Jul
REVIGNY
2304-0418
PO M G Johnson and crew
17/18 Dec
MUNICH
1633-0142
SL E LeP Langlois DFC
FS J Scott
FO L W E Baines
FO A E Reid
FO E C Patten
FO C J Cameron
FL E C Ellis
(this crew failed to return, 4 March 1945)
18/19 Dec
GDYNIA
1720-0239
FO P K Shanahan
Sgt H C Compton
FS B T Stephens
FS L A Smith
FS A R Price
FS J H Schluter
FS R Schlenker
21/22 Dec
POLITZ
1644-0335
FO G H Stewart
Sgt F Baker
FS R Calov
FO R C Faulks
WO M J H West
FS D J Morland
FS F H Skuthorp
27 Dec
RHEYDT
1220-1713
FL M G Johnson and crew

1945
1/2 Jan
GRAVENHORST
MITTELLAND CANAL
1706-2344
FL W K Boxsell DFC
Sgt R J Bauchop
FS R L Pegler
FS V E Auborg
FS H H Leach
FS J R Stokes
Sgt W J Turnbull
4/5Jan
ROYAN
0112-0719
FO L W Baker RAAF
Sgt H V Price (FE)
FS P E Vertigan (BA)

FS A J Johnson (N)
FS R O Sayer RAAF (WOP)
FS A F Wallace (MU)
FS G J Collins (RG)
13/14 Jan
POLITZ
1634-0244
SL E LeP Langlois and crew
14/15 Jan
LEUNA-MERSEBURG
1629-0140
FO J J J Cross
Sgt K M Pope
FO R H Thomas
WO D F Edwards
FS W V Maurer
FS W K Perry
PO J E Brunskell
16/17 Jan
BRUX
1812-0355
FL F Lawrence DFC DFM
Sgt D R Baldry
WO P K Garvey
WO J S Hodgson
FS V J M McCarthy
Sgt E W Durrant
FO R F Chaplin
1/2 Feb
SIEGEN
1621-2232
FL F Lawrence and crew
2/3 Feb
KARLSRHUE
1950-0254
SL E LeP Langlois and crew
16/17 Mar
WURZBURG
1754-0134
FL K P Shanahan and crew
20/21Mar
BOHLEN
2333-0752
FO L W Baker and crew
Sgt N W Teggart and FO N B Ridings in
Sgt Price and FS Vertigan out
22 Mar
BREMEN
1113-1632
WC I H A Hay
FS W V Ward
FO N B Ridings
FO A R T Boys
Sgt G H Wing
Sgt C W Carter
Sgt P G Sanday
23/24 Mar
WESEL
1919-0051
FO L W Baker and crew
FO R H Thomas in, FO Ridings out
27 Mar
FARGE
1013-1456
FO L W Baker and crew
FS Vertigan back, FO Thomas out
4 Apr
NORDHAUSEN
0558-1307
WC I H A Hay
FO C H Nissen (2P)
FS L G W Barnes
FO L A Tabor
FS M J Miller

FO F G Bourke
FS A J Smith
FS A H Thomas
6 Apr
IJMUIDEN
0813-1131
SL W M Kynock DFC
Sgt D B Easton
FO R H Darwin
FO S Harwood
WO K R Morris
FS W T J George
WO R Watts
(mission aborted due to
allied troops taking occupa-

tion)
9 Apr
HAMBURG
1419-1937
WC I H A Hay
FS W V Ward
FO N B Ridings
FO A R T Boys
Sgt G H Wing
Sgt C W Carter
FO P G Sanday
16/17 Apr
PILSEN
2335-0817
FO R A Swift

Sgt G W Wrightson (FE)
FS J R Lewis (BA)
FS C M Wasson (N)
FS H S Stubbs (WOP)
FS T R King (MU)
FS K M Symonds (RG)
18/19 Apr
KOMOTAU
2336-0744
FO L W Baker and crew
FO N B Ridings in, FS
Vertigan out
23 Apr
FLENSBURG
FO L W Baker and crew

(aborted, ordered not to
bomb by Deputy Leader)
24 Apr
'EXODUS'/BRUSSELS
WC I H A Hay and crew
4 May
'EXODUS'/JUVIN-COURT
WC I H A Hay and crew
6 May
'EXODUS'/
JUVINCOURT
FL M G Bache and crew
12 May
'EXODUS'/BRUSSELS
FO L W Baker and crew

W4964
JOHNNY WALKER

A Mk I Lancaster, built at the Metropolitan Vickers plant, Trafford Park, Manchester, with four Merlin XX engines, W4964 left the A. V. Roe factory on 12 April 1943 and was assigned to No 9 Squadron at Bardney, Lincolnshire a few days later. Given the squadron code of WS and the individual letter J-Johnny, the latter led to its famous nose-art painting of the Johnny Walker whisky symbol with this firm's equally famous motto 'Still Going Strong' written beneath it.

Its first operational sortie was to Stettin. Ops came slowly at first, the eighth, to Dusseldorf on 11 June, causing Sergeant T. H. Gill and his crew some anxious moments over the target. Just after the bombs had gone down and Gill was keeping straight and level to take the bombing photo, the machine received a flak hit. The hydraulic pipe line was fractured causing the bomb doors to remain open. Flying home in this fashion, the emergency bottle was used for landing to put down the undercarriage and flaps. The next night the 4,000lb 'cookie' hung up over Bochum but it was eventually jettisoned over Munster on the way back.

J-Johnny had a couple of minor problems during its first dozen sorties, an unserviceable (u/s) rear turret caused an abort on 23 May and an intercom failure on 25 July followed by an electrical failure to the mid-upper turret, caused another. On 22 June W4964 went to ROS at Avro's for a major service, returning to No 9 Squadron on 17 July, and the aircraft then became Cat AC on 18 August – cause unknown – the day after the famous Pennemunde raid, returning to duty a month later. Otherwise, Johnny slowly began piling on the missions, including 12 ops to the German capital during the Battle of Berlin over the winter of 1943-44, although the trip on 2 January had to be aborted because the air speed indicator (ASI) iced-up and the starboard inner engine overheated.

Among Johnny's early captains and crews, the more regular was Sergeant C. P. Newton (later Flight Lieutenant Charles Newton DFC) who completed 16 trips in this aircraft during their tour, and Sergeant P. E. Plowright – later commissioned – who did 18 trips in Johnny and won the DFC. On 19 May 1944, Philip Plowright took Pilot Officer J. D. Melrose as second pilot, James Melrose later taking over Johnny. He went on to fly Johnny on 22 more ops – including the 100th. By that time Melrose was a Flight Lieutenant with the DFC.

By D-Day, Johnny had completed some 70 sorties and flew on the night of D-Day itself, marked as sortie number 71 on the nose, the bomb symbol having a 'D' on it. Johnny was Cat AC again on 30 July, going back to ROS but was back to the squadron on 4 August. By this time the total ops flown was nearing 100 but then the squadron was given a most important task, that of attacking the German battleship *Tirpitz* in Kaa Fjord, Norway. The Lancaster force from Nos 9 and 617 Squadrons would carry 12,000lb Tallboy bombs and the appropriately-named Johnny Walker bomb, or mine, a sort of oscillating mine designed for attacking capital ships in shallow water.

In order to make the attack, the two squadrons flew to Yagodnik, Russia, via Lossiemouth in Scotland, heading out on 11 September. From Yagodnik, via Archangel, the Lancaster force made the attack on 15 September – including Melrose in J-Johnny – one Tallboy, considered to be the one dropped by J-Johnny, hitting the battleship, while the mines and other Tallboy near-misses caused further damage. However, for J-Johnny, it also meant that 100 ops had now been recorded.

Johnny's last six operations were flown during September and early October 1944 and then the veteran was rested from war duties. W4964 was eventually struck off charge (SOC) on 2 November 1949.

Johnny had several interesting markings on its nose quite apart from the very distinctive figure of Johnny Walker. As members of its early crews received decorations so the relevant ribbon was painted in a vertical line. Two DFMs then two DFCs with the recipient's initials beneath them. The aircraft also carried three wound stripes for damage received, a chevron representing a year's duty on active service, the ribbon of the 1939-45 Star was also painted on, then a swastika which presumably was meant to represent a German aircraft, although no record of any claims seem to have survived. A searchlight represented the time an under-gunner (which No 9 Squadron had for a brief period) shot out a searchlight beam flying back at low level, then two more award ribbons, both DFMs. There is also a star – undoubtedly a red star denoting the trip to Russia.

Also at some stage a kangaroo was painted by the navigator's window just aft of the cockpit canopy (port side), denoting an Australian – in fact Flying Officer Jimmy W. Moore, the Aussie navigator in Melrose's crew. James Melrose also says there was another kangaroo by the wireless operator's window for his Aussie WOP Flying Officer R. G. Woolf. Among the ribbon awards were those for Sergeant J. H. Turner, Newton's navigator, Newton himself and Plowright.

To give the reader some indication of the attrition rate in Bomber Command during 1943-45 when these Lancasters were operating, James Melrose noted that of the nine crews who joined No 9 Squadron at the time he arrived, his was the only one left two months later.

1943	PO T Mellard	Sgt S G Bluntern	2345-0441
No 9 Squadron	Sgt G T M Gaines	FS H T Brown	Sgt G H Saxon
20/21 Apr	Sgt H G Watson	Sgt G Bartley	Sgt D C Ferris
' STETTIN	Sgt W R Barker	Sgt S Hughes	Sgt W C McDonald
2147-0524	**4/5 May**	Sgt L G Warner	Sgt R M Morris
WO W E Wood	DORTMUND	Sgt D B McMillan	Sgt J Reddish
Sgt C E Clayton	2202-0410	**12/13 May**	Sgt J C Owen
Sgt Chipperfield	Sgt J D Duncan	DUISBURG	Sgt J Buntin

13/14 May
PILSEN
2134-0540
WO W E Wood and crew
Sgt E L Crump in, Sgt
Chipperfield out
23/24 May
DORTMUND
2237-0205
Sgt G E Hall
Sgt L Field
Sgt W D Evans
Sgt E Colbert
Sgt O J Overington
Sgt K Chalk
Sgt H G Williams
(no attack, rear turret u/s; jet-
tisoned 'cookie' over North
Sea)
25/26 May
DUSSELDORF
2311-0412
Sgt T H Gill
Sgt M McPherson
Sgt R V Gough
Sgt B P Revine
Sgt W A Morton
Sgt K McDonagh
Sgt R McKee
27/28 May
ESSEN
2201-0305
Sgt T H Gill and crew
11/12 Jun
DUSSELDORF
2330-0448
Sgt T H Gill and crew
12/13 Jun
BOCHUM
2240-0345
Sgt T H Gill and crew
14/15 Jun
OBERHAUSEN
2247-0336
Sgt J A Aldersley
Sgt P Hall
Sgt P Webster
Sgt H Popplestone
Sgt G J Sinclair
Sgt H F Poynter
Sgt D G Fremblay
24/25 Jul
HAMBURG
2313-0415
Sgt C P Newton
Sgt J H Turner (N)
Sgt P Hall
Sgt E J Duck
Sgt J Ryan
Sgt W J Wilkinson
Sgt R McFerran
25/26 Jul
ESSEN
2159-0044
FS G A Graham
Sgt W Statham
PO D McDonald
Sgt R M Innes
Sgt A P Williams
Sgt H Altus
Sgt K Mellor
(aborted due to intercom fail-
ure, and M/U electrical fail-
ure; jettisoned 'cookie' safe)
27/28 Jul
HAMBURG
2303-0357

FS G Graham and crew
29/30 Jul
HAMBURG
2242-0400
FS C Ward
Sgt J Sutton
Sgt E Keene
Sgt G L James
Sgt G F K Bedwell
Sgt N F Nixon
Sgt W L Doran
2/3 Aug
HAMBURG
0015-0455
PO C P Newton and crew
10/11 Aug
NURNBERG
2216-0610
PO C P Newton and crew
12/13 Aug
MILAN
2154-0620
Sgt R A Knight
Sgt T Bradford
Sgt G A Munro
Sgt J W Noble
Sgt D G Connor
Sgt R E Jones
Sgt R G Nelson
15/16 Aug
MILAN
2036-0433
Sgt G T Hall and crew
17/18 Aug
PEENEMUNDE
2124-0451
PO C P Newton and crew
1/2 Oct
HAGEN
1841-0001
PO C P Newton and crew
2/3 Oct
MUNICH
1847-0252
PO C P Newton and crew
4/5 Oct
FRANKFURT
1842-0052
PO C P Newton and crew
7/8 Oct
STUTTGART
2059-0305
PO C P Newton and crew
8/9 Oct
HANNOVER
2303-0356
FS C Ward and crew
18 Oct
HANNOVER
1718-2236
FS C Ward and crew
20/21 Oct
LEIPZIG
1726-0038
FS C Ward and crew
22/23 Oct
KASSEL
1806-0029
FS M J Syme
Sgt Whiting
Sgt E Hubbert
PO J C Doughty
Sgt Cattley
Sgt Sorge
Sgt J Heron

3 Nov
DUSSELDORF
1654-2120
PO C P Newton and crew
10/11 Nov
MODANE
2053-0508
PO W M Reid
Sgt S W Richards
FO R D H Parker
Sgt D G Moir
Sgt B Harthill
Sgt C J Wilheim
Sgt G Brown
18/19 Nov
BERLIN
1742-0150
PO W M Reid and crew
22 Nov
BERLIN
1648-2317
FO C P Newton and crew
Sgt L T Fairclough in, Sgt
Duck out
23 Nov
BERLIN
1656-2350
FO C P Newton and crew
FS Allen in, Sgt Fairclough out
2 Dec
BERLIN
1644-2329
FO C P Newton and crew
PO H S Sandy (2P)
FL G Bell in, FS Allen out
3 Dec
LEIPZIG
0031-0803
Sgt P E Plowright
Sgt W C Lewis
Sgt N H B Lucas
FS R P Allen RCAF
Sgt H Hannah
Sgt F Corr
Sgt N F Wells
16 Dec
BERLIN
1639-2355
FO C P Newton and crew
Sgt E J Duck back, FL Bell out
23 Dec
BERLIN
0024-0731
FO C P Newton and crew
29/30 Dec
BERLIN
1657-0019
Sgt D P Proud
Sgt F Harman
Sgt D Carlick
Sgt L T Fairclough
Sgt W H Shirley
Sgt S L Jones
Sgt R L Biers

1944
1/2 Jan
BERLIN
0013-0820
Sgt D P Proud and crew
2/3 Jan
BERLIN
2349-0227
PO D H Pearce
Sgt C W Howe
FO J E Logan
FO W E Pearson

Sgt W R Doran
Sgt S L Jones
FS E A Thomas
(no attack, ASI iced-up and
starboard inner engine over-
heated, sortie aborted)
5/6 Jan
STETTIN
0019-0835
FO A E Manning
Sgt N F Burkitt
FO J W Hearn
FS P Warywodo
FL A G Newbond
Sgt Zamit
FS R C Hayter
14 Jan
BRUNSWICK
1640-2213
FO C P Newton and crew
PO H C Clark (2P)
20 Jan
BERLIN
1639-2338
Sgt D P Proud and crew
21/22 Jan
MAGDEBURG
2027-0311
Sgt D P Proud and crew
27/28 Jan
BERLIN
1732-0158
FO C P Newton and crew
1/2 Mar
STUTTGART
2331-0740
Sgt P E Plowright and crew
15/16 Mar
STUTTGART
1914-0258
Sgt P E Plowright and crew
Sgt W S Richardson in, Sgt
Lucas out
18/19 Mar
FRANKFURT
1913-0124
FS W R Horne
Sgt T W Powell
FS J J Shirley
Sgt J T Johnson
Sgt J H McReery
Sgt R A Morton
Sgt J S Parker
22/23 Mar
FRANKFURT
1902-0033
Sgt P E Plowright and crew
Sgt Lucas back
25/26 Mar
AULNOYE
1921-0035
FL J F Ineson
Sgt L C Margetts
FS H F Mackenzie RCAF
FO T L M Porteous
RNZAF Sgt R H Warren
Sgt H S Chappell
Sgt J Wilkinson
26/27 Mar
ESSEN
2004-0101
Sgt P E Plowright and crew
30/31 Mar
NURNBERG
2224-0619
Sgt P E Plowright and crew
10/11 Apr

TOURS
2255-0525
WC E L Porter
Sgt C E Bowyer
PO J Waterhouse
FO J McMaster
FS B Owen
FS J Michael
FS C R Bolt RCAF
(reached target but could not
identify aiming point so did
not bomb)
11/12 Apr
AACHEN
2033-0043
FL J F Ineson and crew
18/19 Apr
JUVISY
2051-0135
PO P E Plowright and crew
Lucas now PO, Allen now
WO
20/21 Apr
LA CHAPELLE
2316-0347
PO P E Plowright and crew
24/25 Apr
MUNICH
2059-0710
PO P E Plowright and crew
28/29 Apr
ST MEDARD-EN-JALLES
2255-0640
PO P E Plowright and crew
(squadron ordered to abort
over target due to smoke and
haze)
29/30 Apr
ST MEDARD-EN-JALLES
2228-0554
PO P E Plowright and crew
1/2 May
TOULOUSE
2141-0518
PO P E Plowright and crew
3/4 May
MAILLY-LE-CAMP
2203-0337
PO P E Plowright and crew
8/9 May
BREST/LANVEOC
2137-0231
PO P E Plowright and crew
10/11 May
LILLE
2206-0135
PO P E Plowright and crew
11/12 May
BOURG-LEOPOLD
2229-0156
PO P E Plowright and crew
(ordered to abort near target
area due to haze)
19/20 May
TOURS
2221-0339
PO P E Plowright and crew
PO J D Melrose (2P)
21/22May
DUISBERG
2251-0309
PO P E Plowright and crew
22/23 May
BRUNSWICK
2242-0520

PO J D Melrose
Sgt E C Selfe (FE)
FO J W Moore RAAF(N)
FO S A Morris (BA)
FO R G Woolf RAAF
Sgt E Hoyle (MU)
Sgt E E Stley (RG)
24/25 May
EINDHOVEN
2313-0208
PO H C Clark and crew
(ordered to abort due to
poor visibility)
27/28 May
NANTES
2258-0355
PO J D Melrose and crew
3/4 Jun
CHERBOURG
2315-0236
PO R S Gradwell
Sgt T Lynch
FO P E Arnold
FO R B Atkinson
RCAF Sgt J T Price
Sgt W F Best RCAF
Sgt L Sutton
5/6 Jun
ST PIERRE-DU-MONT
0311-0716
FO P D Blackham
Sgt J D Murrie
FO J Wenger RCAF
FO J D Elpick RCAF
FO G A White RCAF
Sgt V G A Stokes
Sgt J McHickey RCAF
6/7 Jun
ARGENTAN
2334-0325
FO J D Melrose and crew
8/9 Jun
RENNES
2306-0517
FO J D Melrose and crew
10/11 Jun
ORLEANS
2205-0335
FO J D Melrose and crew
14/15 Jun
AUNAY-SUR-ODON
2243-0320
FO J D Melrose and crew
21/22 Jun
GELSENKIRCHEN
2312-0350
FO J D Melrose and crew
23/24 Jun
LIMOGES
2241-0518
FO J D Melrose and crew
24/25 Jun
PROUVILLE/V1 SITE
2246-0228
FO P D Blackham and crew
27/28 Jun
VITRY LE FRANCOIS
2158-0519
FO J D Melrose and crew
29 Jun
BEAUVOIR/V1 SITE
1215-1520
FO J D Melrose and crew

4/5 Jul
CREIL
2325-0325
FO J D Melrose and crew
7/8 Jul
ST LEU D'ESSERENT
2240-0315
FO J D Melrose and crew
12/13 Jul
CULMONT-CHAL-ANDREY
2157-0635
FO J D Melrose and crew
18 Jul
CAEN
0407-0735
FO A M Morrison
Sgt A Aitkinhead
Sgt F Reid
FS L L Westmore
WO F Black
Sgt P Strachen
Sgt F Hooper
20/21 Jul
COURTRAI
2318-0244
FS W D Tweedle
Sgt C G Heath
FS E Shields
Sgt J W Singer RCAF
Sgt A Carson
Sgt J A Foot
Sgt K Mallinson
23/24 Jul
KIEL
2300-0456
FO J D Melrose and crew
25 jul
ST-CYR
1754-2151
FO J D Melrose and crew
26/27 Jul
GIVORS
2128-0613
FO J D Melrose and crew
28/29 Jul
STUTTGART
2201-0607
FO C E Scott
Sgt J E Simkins
Sgt L A Harding
Sgt L W Langley
Sgt E M Hayward
Sgt F A Saunders
Sgt L J Hambly
30 Jul
CAHAGNES
0632-1147
FO D McIntish
Sgt R V Cosser
Sgt N A Hawkins
Sgt P J Ramwell
Sgt P E Tetlow
Sgt J A Wood
Sgt G Owen
(not attacked, ordered to
abort due to cloud over tar-
get)
13 Aug
BREST
0833-1302
FL J D Melrose and crew
14 Aug
BREST
0832-1300
FL J D Melrose and crew

15 Aug
GILZE RIJEN A/F
0958-1321
FO B Taylor
Sgt D J Doherty
Sgt A L Cunningham
Sgt A M Holmes RCAF
FS K Burns
Sgt G C Freeman RCAF
Sgt G M Young RCAF
16 Aug
LA PALLICE
1623-2243
FL J D Melrose and crew
18 Aug
LA PALLICE
1124-1742
FL G C Camsell RCAF
Sgt W Andrews
Sgt P R Aslin
FO R H Thomas
Sgt D Beevers
Sgt W J Hebert RCAF
FS A E Boon RCAF
24 Aug
IJMUIDEN
1239-1531
FL G C Camsell and crew
27 Aug
BREST
1418-1844
FO K S Arndell RAAF
Sgt P H Jones
FS P E Campbell RNZAF
FO H W Porter
Sgt R Meads
Sgt J Brown
Sgt L J Richards
15 Sep
'TIRPITZ'
0950-1705
FL J D Melrose and crew
23/24 Sep
MUNSTER
F1907-0101 L J D Melrose
and crew
26/27Sep
KARLSRUHE
0120-0737
FO K S Arndell and crew
27/28 Sep
KAISERSLAUTERN
2150-0422
FO K S Arndell and crew
5 Oct
WILHELMSHAVEN
0750-1307
FO E C Redfern
Sgt J W Williams
Sgt R W Cooper
FO O P Hull
Sgt L G Roberts
Sgt W Brand
Sgt D Winch
6 Oct
BREMEN
1747-2230
FO A E Jeffs
Sgt C V Higgins
Sgt K C Mousley
Sgt H A Fisher RCAF
Sgt C M McMillan
Sgt W Thomas
Sgt G J Symonds

DV245
THE SAINT

Built at Metropolitan Vickers, Manchester, this machine was a Mk III equipped with four Merlin 28 engines. It went to No 32 MU at the end of August 1943 and was then assigned to No 101 Squadron at Ludford Magna, Lincolnshire, on 19 September.

Coded SR, with individual letter S-Sugar (also known as 'The Saint'), it made its first raid on the night of 7/8 October, to Stuttgart, with Flying Officer R. R. Leeder. While a variety of crews flew Sugar during the final months of 1943, Roy Leeder was the aircraft's more usual skipper, completing 16 ops (of their tour of 30) between October 1943 and 24 February 1944, when Leeder, now a flight lieutenant, completed his tour and went to No 1 Lancaster Finishing School (LFS). It was in fact his crew that suggested naming the aircraft 'The Saint' after *The Saint* books written by Leslie Charteris and the radio programme of the same name. Eric Bickley, the rear-gunner, supplied a sketch for the groundcrew artist featuring the yellow Saint astride a red bomb. At the end of their tour Leeder and James Turner received DFCs, Geoff Smith, Roy Brown, Percy Dought and Eric Bickley, DFMs.

Roy Leeder recalls the occasion when DV245 was used to fly for an experiment at Farnborough on 1 January 1944, and took up a lady boffin:

'In fact we flew with two boffins, Mrs Pearce and Mr McClellan. The purpose of the flight was to investigate the practicability of passing nitrogen - an inert gas - over the petrol in the fuel tanks in order to preclude the likelihood of fire in the event of the tanks being penetrated by flak or incendiary bullets.

'The question of valves freezing up was also involved and presumably that is why on the flight we endeavoured to get as high as possible to see if this would happen. [They reached 29,000ft!] In the event we heard nothing more of this experiment, so the idea must have been impracticable.

'All our shakiest do's were caused by appalling weather on return to England. In Sugar we frequently experienced difficulty from the closing down of the weather on our return from an operation. The fact that we survived can, I think, largely be attributed to the fact that Sugar must have been a remarkably stable aircraft. I can recall an occasion when I was trying in vain, visually to identify our airfield through low cloud, when over the intercom I heard Jim Turner, our navigator, screaming, "Watch your airspeed, skipper!" I must have inadvertently been pulling back on the stick, for the airspeed had sunk to well below stalling speed. I pushed desperately forward on the stick and our airspeed slowly built up. In theory we should have spun into the ground.'

Eric Bickley recalls they staggered up to 29,000ft at Farnborough, but were unable to reach the hoped-for 30,000. He also remembers that they should have been at Farnborough for three days but as they accidentally hit some scaffolding with a wing-tip whilst taxying, they were there for a week. Eric later began another tour with No 207 Squadron in 1945.

Once more a variety of crews flew DV245, before Flying Officer Harold Davies DFC emerged as the Lanc's more regular pilot – with 13 ops on her. 'Dave' Davies' bomb-aimer was Jack Kemp, who was to receive the DFM:

'The Nurnberg trip of 30 March 1944 was probably one of our most gruelling operations. We SAW more aircraft shot down than on any other sortie; at one point they were going down on either side of us, each no more than 400yds away, in less than a minute, but Sugar brought us back safely.

'Returning from Brunswick on 22 May somewhere east of the Ruhr a cone of searchlights snapped on dead ahead. Dave turned instinctively to starboard so he could see it, with the intention of sneaking past it, when a second cone sprang up requiring further deviation to starboard. This brought howls of protest from the navigator, "Ted" Barlow, who wanted us to make a port turn of 30 degrees. In all there were at least five cones in a line and we were heading about 330-345 degrees.

'Ted's protest went on until he was advised(!) to "have a look". He did and he was very quiet after that! To my knowledge he never looked out again on any raid.

'We decided to maintain the northerly heading and I was able to get a good fix on the coast. From this and the airplot, the wind was found to be 180 degrees out from the forecast. I believe a correction had been broadcast but we missed it as did many others. I heard that several aircraft ran out of fuel. The enemy searchlights brought us home that night.

'Two nights later we were going to Aachen. It was a warm night and facing downwind on the perimeter track, the engines were over-heating. On take-off three engines returned to normal but the starboard-outer radiator gauge went up to the stop. At 500ft, Bill Lees and Dave decided to shut it down, although a couple of attempts to restart it were made over the North Sea, but the vibration caused us more anxiety than the knowledge that we were on three engines, so we completed the trip like that.'

Sugar flew on the night of D-Day on a Special Duties patrol but had to abort when its engines began to overheat. This sort of thing was rare for Sugar, for the aircraft's aborts due to mechanical problems were few.

No 101 Squadron, of course, was a little out of the ordinary because by the autumn of 1943 its Lancasters were being equipped with ABC (Airborne Cigar) apparatus that could search out and then jam enemy R/T frequencies. Often an eighth crew member was carried to operate ABC, generally a specially-trained German speaking operator. These Lancasters usually carried a normal bomb-load, only slightly reduced due to the weight of the extra man and his apparatus, but the ABC-equipped Lancasters being distinguishable from others by their two large dorsal masts atop the fuselage.

DV245 operated during the Battle of Berlin over the winter of 1943-44, taking off nine times for the 'Big City' but had to abort twice through mechanical problems, on 2 and 16 December. By August 1944 it had notched up over 50 trips when Sergeant Sydney Bowater took over, going on to fly 22 ops in it. He went on to win the DFC and survived the war but died in a Shackleton crash when flying with No 205 Squadron in the South China Sea in 1958, having won the AFC in 1952 flying a Sunderland during an expedition to Greenland. His bomb-aimer was Gerry Murphy, who remembers:

'After our third op on 101, we were told the The Saint was to be our aircraft. I had mixed feelings because it had completed over 50 ops and it appeared to be a lucky aircraft, but it could be that its luck was due to run out.

'Like other crews we became very attached to our Lancaster. We did not like flying in another aircraft when she was in for service and we did not like other crews flying her when we were on leave. I think that she had only one vice: a reluctance to lose height as the Skipper soon discovered. She did not want to come down!

'We had no shattering experiences while flying The Saint but like most Lancasters she was holed by flak from time to time. On one occasion "Bow" Bowater nearly bought it when a piece of shrapnel came up through the floor and took away the oxygen mask clip on the side of his helmet. Another fraction of an inch and he would have gone. On another night we brought her back well holed and our ground crew were disgusted. They almost regarded her as their personal property whom they had lent to us!

'There were amusing incidents. One night on a bomb-run, the target was not quite in the bomb sight and I automatically called, "Dummy run!". There was no response for a moment and then I heard the voice of the rear gunner over the intercom: "Who is the dim-witted b****** who just called Dummy Run?". This was followed by ribald comments and improper suggestions by other members of the crew.

'A couple of nights later during an op the Skipper called me on the intercom and did not receive a reply. My oxygen tube had got fouled. The Skipper realised what might have happened, asked the engineer to check and then the comments started: "Leave him and we'll have no more dummy runs!". "Let the b****** die and we'll all live longer."

'Thirty-five years later, when we met again for the first time since 1944, we were parking our cars and being last to arrive I had some difficulty finding a space in a car park. Watched by the crew, I drove around and on my third circuit, Ted Reeves shouted, "I see you are still doing dummy runs!".'

In the autumn, Sugar began to vie with another Lancaster on the squadron, DV302 H-Harry, as to who would reach the century mark first. In the end, although Harry had started out on 100 raids first, several were aborted, so Sugar got there first with Harry still on 98.

Sugar completed its 100th operational sortie on the night of 5/6 January 1945, to Hannover, flown by a Canadian pilot from Regina, Saskatchewan, Flying Officer R. P. Paterson. In a typical wartime propaganda press release at the time of the 100th op, Paterson was reported as saying: 'I always say "give me Sugar any time and I'll fly her". Despite its age it can do as well as any aircraft I know and I think it climbs better than most. I don't think it's been hit by heavy flak yet. I'd be content to finish my tour in it.'

In spite of his remarks, the sortie to Hannover was only Paterson's second trip in this Lanc (the first being in early December!) and he only flew it once more – in March. DV245's usual skipper towards the end of its days was Flight Lieutenant K. Hanney and his crew, who had taken the aircraft out on 15 trips between December 1944 and April 1945. By the time Sugar reached its century it had had several engine changes and its operational flying time stood at over 720 hours.

Over Pforzheim on 23 February 1945, a twin-engined jet aircraft approached - probably a Me262. The bomb-aimer in the front turret fired three bursts at it and it caught fire, crashed and exploded.

Fortunately for them, neither Hanney nor Paterson was flying Sugar on the 23 March 1945 when 128 Lancasters of Nos 1 and 5 Groups attacked railway bridges at Bremen and Bad Oeynhausen, two Lancs being lost on the Bremen raid - one being Sugar. It was a daylight sortie and the Lanc was attacked and shot down by another Me262 jet. Of the seven-man crew, three were killed, including the pilot, Flying Officer Ralph Robert Little RCAF. By one of those strange coincidences, the rank and initials of Sugar's first operational skipper were exactly the same as its last.

1943	WO D M Windle (SO)	FS H I Howard	FS G W R Wright
No 101 Squadron	**8/9 Oct**	Sgt H James	T/Sgt E Jones USAAF
7/8 Oct	HANNOVER	FS W G Osmotherley	Sgt Tomachipolsky
STUTTGART	2240-0415	FS L H Fox	(crew missing 1/2 January
2025-0335	FO R R Leeder and crew	**20/21 Oct**	1944, all but Tippett, Wright
FO R R Leeder	**18/19 Oct**	LEIPZIG	and Tomachipolsky were
Sgt G F Smith (FE)	HANNOVER	1730-0055	with FL Robertson DFC on
PO J A Turner (N)	1710-2225	FL I Robertson	that sortie and all were
Sgt R S Brown (WOP)	SL J F Dilworth	Sgt T Calvert (FE)	killed)
Sgt B G Lyall (BA)	Sgt F Brookes	PO S I Kennedy (N)	**22/23 Oct**
Sgt P J Drought (MU)	FL F L South	PO A R Tippett	KASSEL
Sgt E W Bickley (RG)	WO W M Mitchell	Sgt B W Zeal (BA)	1755-0015

FO R R Leeder and crew
3/4 Nov
DUSSELDORF
1705-2130
FO R R Leeder and crew
18/19 Nov
BERLIN
1705-0155
FO R R Leeder and crew
22/23 Nov
BERLIN
1650-2325
FO R R Leeder and crew
26/27 Nov
STUTTGART
1705-0005
PO N A Marsh
Sgt F C G DeBrook
Sgt C G Kaye
Sgt D W Ince
PO K R Middleton
Sgt D R Glendinning
FS F H Quick
Sgt O Fischl
16/17 Dec
BERLIN
1610-1945
FO R R Leeder and crew
(aborted due to starboard
outer engine overheating; jet-
tisoned bombs)
20/21 Dec
FRANKFURT
1729-2235
FO R R Leeder and crew
23/24 Dec
BERLIN
0025-0740
FO R R Leeder and crew

1944
14 Jan
BRUNSWICK
1645-2215
FO R R Leeder and crew
20 Jan
BERLIN
1630-2325
FO R R Leeder and crew
21/22 Jan
MAGDEBURG
2000-0250
FO R R Leeder and crew
27/28 Jan
BERLIN
1730-0150
FS E T Holland RAAF
Sgt T Haycock
FS H Scott RAAF
FS A P Farquharson
FS I R Smith RAAF
FS F M McCarthy RAAF
Sgt V G Smith
FO H L Croisette
30 Jan
BERLIN
1715-2335
FO R R Leeder and crew
15/16 Feb
BERLIN
1717-0005
PO N A Marsh and crew
FS W S Ricketts RAAF in, FS
Quick out;
FL F C Bertlesham RCAF in,
Sgt Fischl out

19/20 Feb
LEIPZIG
2352-0718
FO R R Leeder and crew
Sgt R M McLeod in, Sgt Lyall
out;
Sgt J Davidson in, WO Windle
out
20/21 Feb
STUTTGART
2336-0658
FL R R Leeder and crew
WO Windle back, Sgt
Davidson out
24/25 Feb
SCHWEINFURT
1818-0213
FL R R Leeder and crew

(last but one op for FL
Leeder; tour-expired end of
February)
15/16 Mar
STUTTGART
1917-0150
WO T J Drew
Sgt S J Rodway
PO I M Bremner
Sgt C G Dudley
Sgt H N Merrion
Sgt J M Davies
Sgt F G Walter
(aborted when port inner
failed; unable to maintain
height)
18/19 Mar
FRANKFURT
1910-0105
FS J King
Sgt D H Perrett
FO W D Menger RCAF
Sgt A W Worts
FL H J Moore RCAF
Sgt T Bathgate
FS R L Williams RAAF
Sgt W J Childs
22/23 Mar
FRANKFURT
1830-0005
FO H Davies
Sgt W E Lees (FE)
FS E Barlow RNZAF (N) FS R
Pritchard RAAF
Sgt J H Kemp (BA)
Sgt M G Smith (SO)
Sgt R E Stace RAAF Sgt T
Jones (RG)
(MU)
24/25 Mar
BERLIN
1840-0220
FO H Davies and crew
26/27 Mar
ESSEN
1930-0130
FO H Davies and crew
Sgt A H Grainge in, Sgt Smith
out
30/31 Mar
NURNBERG
2210-0130
FO H Davies and crew
Sgt F W Balge RCAF in, Sgt
Grainge out
9/10 Apr
VILLENEUVE ST GEORGES
2100-0210

FL J A Keard
Sgt R Webster
FO A M Shannon RCAF
Sgt R J Crawford
Sgt J R Spowart
Sgt A Clarence
Sgt J E Worsford
10/11 Apr
TOURS
2320-0435
FO H Davies and crew (7)
Sgt Smith out; no 8th man
18/19 Apr
ROUEN
2200-0245
FO H Davies and crew (7)
20/21 Apr
COLOGNE
2325-0405
FO H Davies and crew (8)
Sgt F W Balge RCAF in
22/23 Apr
DUSSELDORF
2245-0350
FO H Davies and crew
24/25 Apr
KARLSRUHE
2142-0500
FO H Davies and crew
27/28 Apr
FRIEDRICHSHAVEN
2145-0610
FL J A Keard and crew
PO D C Frazer as 8th man
30/1 May
MAINTENON
2140-0235
PO R R Waughman
Sgt J Ormerod
PO A Cowan
Sgt I Arndell
FS N Westby
Sgt T Dewsbury
FS H S Nunn RCAF
9/10
MARDYCK
PO J N Brown
Sgt J W Offord
FO W R Cuthbertson
Sgt T Lyth
FS J Pritchard RAAF
Sgt D Urquhart RCAF
Sgt A T Couch RCAF
Sgt C V King RAAF
11/12 May
HASSELT
2150-0130
FO H Davies and crew
22/23 May
BRUNSWICK
2220-0410
FO H Davies and crew
24/25 May
AACHEN
2250-0425
(flew sortie on three
engines)
2/3 Jun
BERNEVAL LE GRAND
2340-0305
FO H Davies and crew
5/6 Jun
SD PATROL
2250-0400
PO J N Brown and crew
(eventually aborted due to

engines over-heating)
7/8 Jun
FORET DE CERISY
2345-0410
PO J N Brown and crew
11/12 Jun
EVREUX
0110-0530
FL N S Wedderburn
FS R Schofield
FO R Sidwell
FO W Patrick
FO E Hunter RCAF
Sgt H W Armishaw RFAC PO
R P Booth
14/15 Jun
LE HAVRE
2055-0115
PO J Harvey
Sgt J P Irvine
Sgt W R Osadchy RCAF
Sgt J L Sime
Sgt I M Hanon RCAF
Sgt A Wilson
Sgt F R Leveridge
Sgt W T Cooper
(all bombs failed to release;
eventually jettisoned HEs)
15/16 Jun
BOULOGNE
2120-0025
PO J Harvey and crew
24 Jun
LES HAYONS/V1 SITE
1555-1920
FS P J Hyland
Sgt J Hedges
Sgt C E Smith
Sgt J J Moore
Sgt T Crane
Sgt E R Brown
Sgt A W Tuuri RCAF
Sgt W Engelhardt
27/28 Jun
VITRY LE FRANCOIS
2149-0538
PO J Harvey and crew
31/1 Aug
FORET DE NIEPPE/ V1 SITE
2154-0118
FO L Bursell RNZAF
Sgt P J Clifford
Sgt F L S Sharp
Sgt E Reeves
Sgt W Woodridge
FS C W Austin RCAF
FS C L Robinson RCAF
PO R J Hardacre RAAF
3 Aug
TROSSY-ST-MAXIM
1145-1635
FO G M Atyso RCAF
Sgt C T Keeling
Sgt J W Lovatt
Sgt J F Anderson RCAF
FO B L Patterson RCAF
FS C F Pearce RCAF
H Balchin RCAF
PO F G D Smith
4 Aug
PAUILLAC
1335-2140
FO G M Atyso and crew
5 Aug
BLAYE
1430-2225
FO G M Atyso and crew

Sgt P D Kaye in, PO Smith
out
7/8 Aug
FONTENAY LE MARMION
2116-0132
FO G M Atyso and crew
26/27 Aug
KIEL
2035-0155
Sgt S Bowater
Sgt A W Oliver (FE)
Sgt F Campbell (N)
Sgt E Reeves (WOP)
Sgt G J H Murphy (BA)
Sgt J W Walsh (SO)
Sgt H Dickie (MU)
Sgt K R Dickinson(RG)
29/30 Aug
STETTIN
2126-0622
Sgt S Bowater and crew
31 Aug
ST RIQUIER
1347-1729
PO R Totham
Sgt G C Larkman
Sgt H W Emerson
FS H Johnson
Sgt R M Cavill
Sgt H Wynne
Sgt R M McPherson
Sgt W J Childs
3 Sep
GILZE RIJEN A/F
1548-1903
PO D H G Ireland RCAF
Sgt H J Black
PO J C Munro RCAF
FS F Coulson RCAF
Sgt C E Deatherage
Sgt E A J Davies RCF
Sgt J K Ladley RAAF
5 Sep
LE HAVRE
1616-1948
PO D H G Ireland and crew
6 Sep
LE HAVRE
1741-2114
PO D H G Ireland and crew
(mission aborted over target
by Master Bomber)
8 Sep
LE HAVRE
0640-1044
PO D H G Ireland and crew
10 Sep
LE HAVRE
1716-2045
Sgt S Bowater and crew
15/16 Sep
KIEL
2245-0346
Sgt S Bowater and crew
16/17 Sep
STEENWIJK
2217-0135
Sgt S Bowater and crew
17 Sep
WEST CAPELLE
1703-1913 Sgt S Bowater
and crew
20 Sep
CALAIS
1538-1913
Sgt S Bowater and crew

25 Sep
CALAIS
0709-0953
Sgt S Bowater and crew
26 Sep
CAP GRIS NEZ
1118-1428
Sgt S Bowater and crew
3 Oct
WEST CAPELLE
1339-1636
Sgt S Bowater and crew
6 Oct
BREMEN
1739-2206
Sgt S Bowater and crew
7 Oct
EMMERICH
1212-1632
Sgt S Bowater and crew
9 Oct
BOCHUM
1730-2248
PO F D McGonigle
Sgt J R McDowell
FS J E Knight RAAF
FS L Collins RAAF
FS W P Hart RAAF
Sgt R J Beckett
Sgt D Conroy
PO J K Armour RCAF
14 Oct
DUISBERG
0639-1108
Sgt S Bowater and crew
FL W K Parke RCAF (2P)
14/15 Oct
DUISBERG
0027-0557
Sgt S Bowater and crew
19/20 Oct
STUTTGART
2136-0413
Sgt S Bowater and crew
25 Oct
ESSEN
1305-1732
PO F D McGonigle and crew
28 Oct
COLOGNE
1335-1845
PO J W Hunting
Sgt J F Hailstones
FS W A Granger RAAF
FS G S Gayner RAAF
Sgt J Richardson
Sgt A J Clark
Sgt E L Pierson
Sgt J Rees
29 Oct
DOMBERG
1159-1451
Sgt S Bowater and crew
30 Oct
COLOGNE
1746-2339
Sgt S Bowater and crew
31 Oct
COLOGNE
1747-2257
Sgt S Bowater and crew
2 Nov
DUSSELDORF
1604-2141
Sgt S Bowater and crew
6 Nov
GELSENKIRCHEN

1124-1632
Sgt S Bowater and crew
9 Nov
WANNE EICKEL
0845-1323
FO R D Gray
Sgt T J Stynes
FS R C Brown RCAF
Sgt J H P Jones
PO W I Strachen RCAF
FS N G Hunt RCAF
FS P J Price RCAF
FS Glick RAAF
11 Nov
HOESCH-BENZIN
1545-2135
Sgt S Bowater and crew
16 Nov
DUREN
1240-1728
Sgt S Bowater and crew
18 Nov
WANNE-EICKEL
1558-2117
SL T J Warner
Sgt W Hartwill
Sgt R W Palmer
PO D W Weston
FS J H Symonds
Sgt E C Roberts
Sgt J H Jackson
Sgt H Felix
21 Nov
ASCHAFFENBURG
1533-2222
PO D H G Ireland and crew
Sgts J W Hodder and E J
Hartman in;
Sgts Davies and Ladley out
27 Nov
FREIBURG
1606-2248
FO J W Hunting and crew (7)
Sgt Rees out
2/3 Dec
HAGEN
1750-0017
FO R P Paterson RCAF
Sgt A J Wallis
FO M Ornstein
Sgt W Yeomans
Sgt M Dillon
Sgt A Greenhough
FO W E Thoroldson
FL C T Candy
RCAF
4 Dec
KARLSRUHE
1650-2305
FO K Hanney
Sgt A R Rogers
FS J L Daly RAAF
Sgt H Lyon
FS R Smith
Sgt E Lipscomb
Sgt D Ingram
5 Dec
SOEST
1811-2359
FO J W Hunting and crew
12 Dec
ESSEN
1638-2234
FO K Hanney and crew
15 Dec
LUDWIGSHAVEN
1439-2110

FO K Hanney and crew
17 Dec
ULM
1525-2257
FO K Hanney and crew
plus Sgt D Cooper (8)
28 Dec
BONN
1529-2138
FO W W Watt
Sgt R E Winstone
Sgt S R Allen
Sgt G A Stephens
FO A W Stuart
Sgt J A Slater
Sgt D Mortimer
(all killed in action 23
February 1945)
31 Dec
OSTERFELD
1503-2135
FO K Hanney and crew

1945
2 Jan
NURNBERG
1520-2337
FL R H H Wilder
Sgt J Dutton
FO J A Blackhall
Sgt K Gough
FO R Elliott
Sgt R J Brown
Sgt L Drill
Sgt A C Neve
5/6 Jan
HANNOVER
1848-0046
FO R P Paterson and crew
Sgts D Nelson and R Blitz in;
Sgt Wallis and FL Candy out
6 Jan
HANAU
1559-2206
FO H G H Meadows
Sgt D Thornhill
FO J T Burke
FS K Fitton
FS K R Rudge
Sgt R Fisher
Sgt J W Feast
16/17 Jan
ZEITZ
1746-0133
FO K E Roberts RAAF
Sgt L Schofield
Sgt S E Adams
WO W Hawke RAAF
Sgt A L Lister
Sgt S Arnold
Sgt H Campbell
Sgt O S Cleyn RCAF
22 Jan
DUISBERG
1713-2217
FO K Hanney and crew
1 Feb
LUDWIGSHAVEN
1600-2249
FO H I Davis
Sgt A H Woodier
Sgt S C Campbell
Sgt P Hammond
Sgt M H Pollard
Sgt J McCafferty
Sgt W Paul
Sgt W W Lemke RCAF

3 Feb
BOTTROP
1635-1824
FO K Hanney and crew
(operation aborted)
7/8 Feb
KLEVE
1911-0028
FO K Hanney and crew
13/14 Feb
DRESDEN
2143-0705
FO K Hanney and crew
PO J Kerr (2P) and Sgt F
Smith (8)
14/15 Feb
CHEMNITZ
2019-0510
FO K Hanney and crew (with-
out PO Kerr)
20/21 Feb
DORTMUND
2151-0445
FO K Hanney and crew
21/22 Feb
DUISBURG
1957-0154
FL R A Andrews
Sgt A Hammond
RCAF
FO E C Lobsinger RCAF Sgt W
Myers

FO E A Hamilton RCAF
FS J Mess RCAF
FS S A LaLonde RCAF
Sgt H van Geffen
23/24 Feb
PFORZHEIM
1607-0001
PO G Withenshaw
Sgt A H Halliday
FL L B H Lapointe
Sgt A Parsons
RCAF
FO J R Drewery RCAF
FS W H Robinson RCAF
FS J E Stead
PO E Graumann RCAF
2 Mar
COLOGNE
0709-1240
PO G Withenshaw and crew
PO Graumann out
5/6
CHEMNITZ
1653-0219
FL K Hanney and crew
Mar
FS R A Bird in, Sgt Rogers out
7/8 Mar
DESSAU
1713-0303
FO L W Rodger RCAF
Sgt W A Beach

FO A J Knebel RCAF
PO L A Pool
FO N A Jevne RCAF
FS J V Mills
Sgt R Krull
8/9Mar
KASSEL
1721-0110
FS J A Kell
Sgt J Ness RCAF
WO W Rusby
Sgt E Baines
Sgt C Adams
PO E Marshall
Sgt W West RCAF
11 Mar
ESSEN
1152-1655
FL K Hanney and crew
12 Mar
DORTMUND
1312-1905
FL K Hanney and crew
13/14 Mar
DAHLBURSCH
1745-2329
FL K Hanney and crew
16/17 Mar
NURNBERG
1741-0149
FO R P Paterson and crew

20/21 Mar
HEIDE
0207-0702
SL K Flint
Sgt J E Crawford
FO J P Bannister
FS I S Sangster
FS A Smith
FS C J Clark
FS E Iles
FO J C Wilson RCAF
21/22 Mar
BRUCH STRASSE
0058-0655
PO G Withenshaw and crew
Sgt L N Lemke in (8)
23 Mar
BREMEN
0658
FO R R Little RCAF
Sgt H J Clifton
FO J C Lee RCAF
Sgt H Woodard
FO W M Brooks RCAF
Sgt T S Nelson
Sgt T Churchill
(Failed to return - Ralph
Little, Jim Lee and Thomas
Churchill were buried in
Oldenburg Cemetery.)

DV302
HARRY

No 101 Squadron's second centenarian and for a time the rival of DV245, was this Metro Vickers-built Lancaster Mk I, completed by A. V. Roe in September 1943 with four Merlin 22 engines. Sent to No 32 MU at St Athan on 1 October, it was then assigned to No 101 Squadron at Ludford Magna, Lincolnshire, on 7 November. Given the squadron code letters SR, it became H-Harry, although sometimes referred to as H-Howe. Arriving during the Battle of Berlin it is no surprise to see that during the aircraft's first 25 operations, no fewer than 16 were against the Big City. In fact its first six ops were to Berlin and, of its first 18, 15 were to the German capital.

Its first trip was on 18 November, skippered by a New Zealander, Flying Officer D. H. Todd, later Flight Lieutenant DFC. Douglas Todd and three of his crew had been with No 98 Squadron as Vic Viggers RNZAF recalls:

'Our first operational sortie was with No 98 Squadron flying Mitchells, and attacking gun emplacements on the Dutch and French coasts. As volunteer crews for heavy bombers were being called for, we applied. After going through No 1667 Conversion Unit we went to No 101 Squadron and completed our tour of 30 ops – 11 of them to Berlin.

'I do remember Toddy had his aircraft preferences and referred to Harry as a very air worthy aircraft backed-up, of course, by a competent and caring groundcrew. Sadly Doug Todd was killed in a hit-and-run accident in Palmerston North, New Zealand, about 20 years ago.

'Our first special operator, Spafford, decided to change to another crew after our first four trips and was shot down and killed the very next night.'

Doug Todd took DV302 on nine of his first 16 sorties and at the end of their tour, Viggers, Ken Bardell and Harry Whittle all received DFCs, Stan Powell the DFM. DV302's more usual crew became that of Flight Sergeant E. T. Holland RAAF who did 20 trips in the aircraft. Flying Officer J. Kinman was the next regular skipper for 10 trips, although his first op in DV302 as second pilot on his experience trip was aborted soon after take-off on 12 June 1944 when the port outer engine failed.

Not surprisingly, the reader will find that several crews that flew 'The Saint' (DV245) also flew Harry, such as Flying Officers H. Davies and R. E. Ireland

RCAF. Like The Saint, Harry was operated as an ABC-equipped Lancaster, often flying with an eighth aircrew member, the German-speaking ABC operator whose job it was to locate German R/T wavelengths then confuse night fighter crews and their ground controllers by either jamming them or giving fake orders.

Harry became Cat AC on 5 July 1944 and went to ROS for overhaul but was back on 101's strength 10 days later. Pilot Officer Arthur E. 'Bill' Netting took over DV302 in August, flying no less than 29 sorties in it and received the DFC. Ernie Cantwell (who received the DFM) was Netting's navigator, and has some recollections of this Lanc which they called Howe:

'Shortly after joining 101 in July 1944 we were invited to take on H-Howe, a veteran of nearly 60 ops. We subsequently looked it over and unanimously decided we would take it in preference to a new one. If sentiment came into our decision it was a wise one, for H-Howe safely took us through 30 ops out of a tour of 32. Our WOP, Cyril Camplin, went sick and missed two trips so we did two extra in order that he would not complete his 30 with a strange crew. That cost him a few pints!

'Our first op was to Stettin, lasting some nine hours during which time we were coned in searchlights over the target. However, the "old boy" responded faithfully to the manoeuvres the skipper asked of him to escape the glare and hostility focused on us. We returned to base feeling that we had well and truly been initiated into what operational flying was all about.

'Another incident still fairly vivid was a daylight to Essen when we were alerted over the target by the mid-upper, that a stick of bombs was falling on us from an aircraft above. One of them struck a glancing blow to the port-inner and removed the propeller nose cone. Fortunately little further damage was inflicted and the faithful Merlin continued to function. We were later informed that it had most likely been a tail-fused 1,000-pounder.

'Our base at Ludford Magna was equipped with "Fido" and returning on another occasion in thick fog, it was used to aid our landing. It was quite an experience landing between tracks of burning petrol and I recall watching the skipper struggling to control against the turbulence.

'With regard to the fighter painted on the nose of Howe, we were told that the previous crew were credited with shooting down the first enemy aircraft with the .5 nose gun turrets fitted on him. Apparently the crew were entertained by the manufacturers for their gun's first success.'

Howe, or Harry, went on operating until 30 April 1945 when it went to No 46 MU at Lossiemouth. The aircraft achieved 98 ops by the time Sugar had done her 100th, Harry's own 100th coming on 7/8 January 1945, when Pilot Officer J. A. Kurtzer RAAF took it to Munich. In all it is reputed to have flown 121 operational sorties, although the list might indicate at least one more.

Howe had flown on the night of D-Day and on a raid on 31 July, when it was hit by flak on the way to the target wounding the bomb-aimer, Flight Sergeant A. L. Plimmer. Despite a painful wound to his right arm, he continued to direct his pilot - Kinman - to complete an accurate bombing attack. For this episode Albert Plimmer received the DFM. On another raid on 5 October, Howe's crew were frustrated when the bomb doors failed to open over the target. However, after that sortie the aircraft only aborted once, when its Gee became u/s on 2 February 1945.

After its 100th sortie there appears to be few regular crews flying the aircraft which may well have been a sign of its ageing, but it still achieved some long trips, notably to Chemnitz, Dessau, Nurnberg and an eight-hour sortie to Potsdam on 14/15 April 1945.

With 121 bombs painted on its nose, plus a long service medal ribbon applied by his faithful groundcrew (and what looks like a Ju88 falling in flames, although there is no surviving reference to any successful combat), DV302 was finally Struck off Charge on 15 January 1947.

One of the aircraft's Merlin engines survives. A former Lancaster pilot named Mackenzie Hamilton owned a 1930 Rolls-Royce Merlin Competition Roadster, but by 1946 its Kestrel engine was beyond repair. Knowing that Lancasters were being broken up at Lossiemouth, Hamilton went along and managed to obtain DV302's port inner, which he managed to squeeze into the car's chassis. Hamilton was killed in a flying accident in 1965. In 1990 its new owner bought the car for £48,000, plus commission and VAT!

1943	FO B Wilkinson (SO)	**23/24 Dec**	FS O'Dwyer back, FS Teitz out
No 101 Squadron	(this crew, except FO	BERLIN	FS P S W Napier RAAF in, FO
18/19 Nov	Wilkinson, killed in action	0035-0815	Raine out
BERLIN	14/15 January 1944)	FS D Langford	**20/21 Jan**
1710-0140	**26/27 Nov**	Sgt H Swift	BERLIN
FO D H Todd RNZAF	BERLIN	Sgt J E Price	1640-0001
Sgt S Powell (FE)	1700-0110	Sgt F Urch	FS L Kidd
FO W Frazer (N)	FO D H Todd and crew	FS H W Davy	FS J Hall (FE)
FO V C Viggers RNZAF	PO W M Shubie and FO P J	Sgt G A Wilby	FS F T Sanderson RAAF
Sgt K H Bardell (BA)	W Raine in	Sgt H R Riley	Sgt T M Cavender RAAF
Sgt H R Whittle (MU)	FS O'Dwyer and FO Spafford	Sgt D Hoffman	FS K F Stanton RAAF
FS W M O'Dwyer	out	**29 Dec**	Sgt J A Watt RAAF(BA)
FO G L Spafford (SO)	**2/3 Dec**	BERLIN	Sgt R Wilson (RG)
RNZAF(RG)	BERLIN	1710-2355	PO O Fischl (SO)
22 Nov	1657-0027	FO D H Todd and crew	**21/22 Jan**
BERLIN	FS K B Corkhill	FS O'Dwyer back, FL Hill out	MAGDEBURG
1700-2340	Sgt E A F Cole		2010-0310
FS E J Trotter	Sgt K G Thompson	**1944**	FS A J Sandford
Sgt J Rawcliffe	Sgt H Street	**1/2 Jan**	Sgt A H Smallman
FO D Holder	Sgt R M Gundy	BERLIN	Sgt E Barron
Sgt J T Broad	Sgt L P Swales	0015-0815	Sgt T D Simpson
Sgt V Pullen	Sgt F G Welsh	FO D H Todd and crew	Sgt R T Ottewell
Sgt A Kirton	Sgt G E H Schultz	Sgt P J Drought in, FS	Sgt K Bartholomew
Sgt K Archibald	**16/17 Dec**	O'Dwyer out	Sgt E H Alcock
23 Nov	BERLIN	**2/3Jan**	Sgt E H Manners
BERLIN	1625-0015	BERLIN	**27/28 Jan**
1700-2329	FO D H Todd and crew	2345-0650	BERLIN
FL T W Rowland	FS O'Dwyer back, PO Shubie	FO D H Todd and crew	1740-0205
Sgt R Allison (FE)	out	FS R Teitz RAAF in, Sgt	FS K D Corkhill and crew
FO D G Higgs (N)	**20 Dec**	Drought out	Sgt H Manser and Sgt P N D
Sgt P H Lamprey (WOP)	FRANKFURT	**14 Jan**	Skingley in
FS W H Yuill (BA)	1720-2325	BRUNSWICK	Sgt Street and Sgt Schultz out
Sgt R Bateman (MU)	FO D H Todd and crew	1645-2235	**28/29 Jan**
Sgt H G Clements (RG)	FL W F J Hill in, FS O'Dwyer out	FO D H Todd and crew	BERLIN

51

0020-0835
FO D H Todd and crew
Sgt G T J Heath in, FS
O'Dwyer out
30/31Jan
BERLIN
1720-0005
FS K D Corkhill and crew
15/16 Feb
BERLIN
1730-0000
FS E T Holland RAAF
Sgt T Haycock
FS H Scott RAAF
FS A P Farquharson RAAF
FS I R Smith RAAF
Sgt A M McCartney
Sgt V G Smith RAAF
20/21 Feb
STUTTGART
2345-0730
FS E T Holland and crew plus
Sgt H Docherty (SO)
24/25 Feb
SCHWEINFURT
2030-0500
FS E T Holland and crew
25/26Feb
AUGSBURG
2120-0440
FS E T Holland and crew
15/16Mar
STUTTGART
1910-0330
PO E T Holland and crew
FS R J Hardacre in, Sgt
Docherty out
18/19 Mar
FRANKFURT
1900-0045
PO E T Holland and crew
PO L P Whymark in, Sgt
McCartney out
22/23 Mar
FRANKFURT
1825-0010
PO E T Holland and crew
24/25 Mar
BERLIN
1900-0150
PO E T Holland and crew
Sgt M G Smith in, FS
Hardacre out (SO)
26/27 Mar
ESSEN
1940-0025
PO E T Holland and crew
Sgt R W Sharp (2P)
30/31Mar
NURNBERG
2215-0635
PO E T Holland and crew
FS Hardacre back, Sgt Smith
out
10/11 Apr
AULNOYE
2330-0455
PO E T Holland and crew
FS Hardacre out
11/12 Apr
AACHEN
2020-0020
PO E T Holland and crew
FS Hardacre back
18/19 Apr
ROUEN
2145-0155

FS E A Askew
Sgt T B Johnstone
Sgt C W Rolfe
Sgt S N Hewitt
Sgt W S Newman
Sgt E J Delaney RCAF
Sgt S Morford
20/21Apr
COLOGNE
2320-0355
PO E T Holland and crew
22/23Apr
BRUNSWICK
2305-0500
PO D C Rippon
Sgt R A Smith
Sgt R W Snell
Sgt B Whitehead
FO F J Lynain
Sgt R D Stack
Sgt W J R Hunter
PO H H King RCAF
24/25 Apr
KARLSRUHE
2120-0400
PO E T Holland and crew
26/27 Apr
SCHWEINFURT
2130-0630
PO E T Holland and crew
27/28 Apr
FRIEDRICHSHAVEN
2150-0630
PO K Fillingham
Sgt C D Goodliffe
FO S J Licquorish RCAF
Sgt P R Medway
FO K D Connell RCAF
Sgt J Soulsby
Sgt J Law
FS A M Marks
3/4 May
MAILLY-LE-CAMP
2200-0400
PO T A Welsby
Sgt J A Parker
FO F A Pierce
Sgt T R Hurst
Sgt M T Boreham RCAF
Sgt R Cox
Sgt G A Wallace
FS R Laurie RAAF
21/22 May
DUISBURG
2215-0250
PO R R Waughman
Sgt J Ormerod
PO A Cowan
Sgt I Arndell
FS N Westby
Sgt T Dewsbury
FS H S Nunn RCAF
Sgt E H Manners
22/23 May
DORTMUND
2225-0320
PO R R Waughman and crew
24/25 May
AACHEN
2350-0430
PO E T Holland and crew
Sgt W O Ross in, PO
Whymark out
27/28 May
AACHEN
0010-0425
PO E T Holland and crew

2/3 Jun
TRAPPES
2340-0335
PO E T Holland and crew
5/6Jun
SD PATROL
2220-0535
FL R N Knights
Sgt W Perry
FS B Pinner
FS M E Bromley
FS F W Morgan
Sgt J ?
FS H S Mann RCAF
FO H L Croisette
8/9 Jun
FORET DE CERISY
2325-0340
PO E T Holland and crew (7)
12/13 Jun
GELSENKIRCHEN
2245-0255
PO L P Bateman RCAF
Sgt A Fazachisley
WO W E Buie RCAF
FS E Crook
FO F C Brooks RCAF
WO J J Byrne
Sgt J V Browne
Sgt H Docherty
14/15 Jun
LE HAVRE
2048-0007
PO D C Rippon and crew
24 Jun
LES HAYONS/V1 SITE
1605-1930
FO J Kinman
Sgt H Mawson
FO M Gisby
Sgt E E Evans
Sgt A L Plimmer
Sgt T Wright
Sgt S C Trevis
Sgt J P Auer
25 Jun
LIGESCOURT/V1 SITE
0730-1115
FO J Kinman and crew
27/28 Jun
VAIRES
0045-0500
FO J Kinman and crew
29 Jun
SIRACOURT/V1 SITE
1210-1555
FO J Kinman and crew
30/1Jul
VIERZON
2224-0323
FO J Kinman and crew
4/5 Jul
VILLENEUVE ST
GEORGE
2256-0413
FO J Kinman and crew
23/24 Jul
KIEL
2240-0345
FO J Kinman and crew
24/25Jul
STUTTGART
2240-0350
FO J Kinman and crew
(mission aborted due to rear
turret becoming u/s)

28/29 Jul
FORET DE NIEPPE
0225-0540
FO J Kinman and crew
31/1 Aug
FORET DE NIEPPE
2155-0105
FO J Kinman and crew
FL D M MacDonald RCAF in,
Sgt Evans out
(FS A L Plimmer wounded -
awarded DFM)
29/30 Aug
STETTIN
2105-0610
PO A E Netting
Sgt T C Brown (FE)
Sgt E H A Cantwell (N)
Sgt C H Complin (WOP)
Sgt P R Gunter (BA)
Sgt M A Ordway (MU)
Sgt V R Burrill RCAF
Sgt J E C Lodge (SO)
(RG)
3 Sep
GILZE RIJEN A/F
1549-2011
PO A E Netting and crew
6 Sep
LE HAVRE
1749-2059
PO A E Netting and crew
(mission abandoned on
direction of controller)
8 Sep
LE HAVRE
0647-1048
PO R Totham
Sgt G C Larkman
Sgt H W Emerson
FS H Johnson
FS R M Caville
Sgt H Wynne
Sgt R M McPherson
Sgt Hawkins
(mission again abandoned
on instruction due to weath-
er)
10 Sep
LE HAVRE
1701-2037
PO A E Netting and crew
11/12 Sep
DARMSTADT
2108-0241
PO A E Netting and crew
16/17Sep
HOPSTEN
2130-0124
PO A E Netting and crew
20 Sep
CALAIS
1544-1851
PO A E Netting and crew
25 Sep
CALAIS
0651-1044
PO A E Netting and crew
(mission abandoned due to
weather)
26 Sep
CALAIS
0651-1044
PO A E Netting and crew
26 Sep
CAP GRIS NEZ
1059-1353

PO A E Netting and crew
5/6 Oct
SAARBRUCKEN
1850-0116
PO W A McClanaghan
Sgt C G Vicary RCAF
PO J R Balcombe
WO L F Kennedy
Sgt J D Lamb
Sgt E R Boyd
Sgt F R Fletcher
PO E Grauman
(mission abandoned over target when bomb doors failed to open)
6 Oct
BREMEN
1746-2215
FO L G James
Sgt W G Orr
FS L R ColemanRCAF
FS G H Taylor
FS R S Davies
FS Gordon
FS G Williams RCAF
Sgt L R Hall
7 Oct
EMMERICH
1216-1643
FO L F James
Sgt A Yorrington
FO R D Irving RCAF
Sgt L L Wright
PO H G Bullock RCAF
Sgt E W Dean RCAF
Sgt A R Walker
Sgt D Burnett
14 Oct
DUISBERG
0648-1139
FO R E Ireland RCAF
Sgt H J Black
FS R E Hine RCAF
PO J C Munro RCAF
FS F Coulson RCAF
FS G E Deatherage RCAF
FS K A Davies RCAF
FS J K Ladley RAAF
14/15 Oct
DUISBERG
2236-0341
PO A E Netting and crew
Sgt B C Reeves in, Sgt Complin out
15 Oct
WILHELMSHAVEN
1731-2145
PO A E Netting and crew
19/20 Oct
STUTTGART
2132-0415
PO A E Netting and crew
Sgt Reeves out, Sgt Complin back
23 Oct
ESSEN
1619-2151
PO A E Netting and crew
25 Oct
ESSEN
1207-1745
PO A E Netting and crew
28 Oct
COLOGNE
1327-1835
PO A E Netting and crew

30 Oct
COLOGNE
1740-2355
PO A E Netting and crew
31 Oct
COLOGNE
1805-2303
PO A E Netting and crew
2 Nov
DUSSELDORF
1601-2144
SL I McLeod-Selkirk
Sgt F Milton
FO R C Pitman
Sgt A Pringle
FO G W Hess RCAF
Sgt F W Hughes
Sgt R Hurley
Sgt H Van Geffen
4 Nov
BOCHUM
1713-2202
PO A E Netting and crew
6 Nov
GELSENKIRCHEN
1122-1626
PO A E Netting and crew
9 Nov
WANNE EICKEL
0746-1301
PO A E Netting and crew
11 Nov
HOESCH/DORTMUND
1601-2143
PO A E Netting and crew
16 Nov
DUREN
1242-1736
PO A E Netting and crew
21 Nov
ASCHAFFENBURG
1523-2216
PO A E Netting and crew
27 Nov
FREIBURG
1621-2308
PO R A Lowe RCAF
Sgt E J Rogers
FS B J Brophy RCAF
Sgt W A Perry
FO R G Masters RCAF
Sgt C J Mountjoy
Sgt A J Clutchy RCAF
29 Nov
DORTMUND
1209-1729
PO V W Ashes RAAF
Sgt G Kennedy
FS A Watt
FS W J Gurner RAAF
Sgt A W Robinson
Sgt J P Fellows
Sgt J Pringle
30 Nov
DUISBURG
1654-2204
FO R A Harrison
Sgt J Breare
FS R J Swain
FS K E McDonnell
Sgt G Hillman
Sgt D J Mack
Sgt S Stephens
Sgt F Smith
2 Dec
HAGEN
1736-2348

FO A E Netting and crew
4 Dec
KARLSRUHE
1651-2318
FO A E Netting and crew
6/7 Dec
LEUNA/MERSEBURG
1645-0035
FO A E Netting and crew
PO H I Davies (2P)
12 Dec
ESSEN
1610-2156
FO A E Netting and crew
15 Dec
LUDWIGSHAVEN
1425-2056
FO A E Netting and crew
17 Dec
ULM
1518-2326
PO K J Brookin
Sgt H J Hadingham
Sgt F Edwards
Sgt J Rogerson
FS V A Maki
Sgt J Brown
Sgt S Borthwick
28 Dec
BONN
1531-2059
WC M H deL Everest
Sgt T Murray
FO A Jeffcoat
Sgt R Whiteford
FO R Shepherd
Sgt W H Horner
Sgt J W Hodder
29 Dec
SCHOLVEN
1508-2110
FO C G Rogers
Sgt J Gorman
FS R Newnham RAAF
PO R Rowland
Sgt E A Mason
Sgt W J Calvert
Sgt R T Small
PO J McDermott

1945
2 Jan
NURNBERG
1504-2309
FO W A McClanaghan and crew
FO R Inceberg RCAF and FO S Brookes in
Sgt Lamb and PO Grauman out
5/6 Jan
HANNOVER
1851-0040
FO W A McClanaghan and crew (7)
FO Brookes out
6 Jan
HANAU
1557-2210
FO W A McClanaghan and crew
7/8 Jan
MUNICH
1849-0337
PO J A Kurtzer RAAF
Sgt J Harris
FS G H Honeysett

FS J W Fletcher RAAF
RAAF
FO A L Woodhart
FS J C Dyke RAAF
FS J S Alexander RAAF
Sgt J O S Fochs RCAF
16/17 Jan
BRUX
1744-0300
FO W A McClanaghan and crew (8)
Sgt H Felik in
28/29 Jan
STUTTGART
1940-0309
FO W A McClanaghan and crew
Sgt H Wells in, Sgt Felik out
1 Feb
LUDWIGSHAVEN
1546-2233
FO W A McClanaghan and crew (7)
Sgt Wells out
2 Feb
WIESBADEN
1546-2233
FO W A McClanaghan and crew (8)
Sgt Van Geffen in
(abandoned at 2127 hours - Gee u/s)
7/8 Feb
KLEVE
1905-0031
PO J D Robinson
Sgt S A Lewis
FS C G Scott RCAF
WO N W Berger RCAF
FO J R Weinfield RCAF
FS S A Spence RCAF
FS J C Spencer RCAF
14/15 Feb
CHEMNITZ
2015-0520
FO H J West
PO K D J Ward
FL W O Mitchell
Sgt R A Kemshall
Sgt H D Budd
Sgt G A Duncan
Sgt R F Milroy
FO H Walker RCAF
20/21 Feb
DORTMUND
2039-0415
FO H I Davies
Sgt A H Woodier (FE)
FS F Campbell
Sgt F Hammond
FS M H Pollard
Sgt J McCafferty
Sgt W Paul
Sgt L W Lemke RCAF
21/22 Feb
DUISBURG
1940-0125
FO H J West and crew
FO H G Sheath RCAF in, FL Mitchell out
23 Feb
PFORZHEIM
1550-2339
FO H J West and crew
1 Mar
MANNHEIM
1140-1822

FO E W Mackay RCAF
Sgt N H Allpress
FO G L Hillier RCAF
FS J Kevac RCAF
FO R G Millward RCAF
Sgt R S Hadden
Sgt B Girwing RCAF
2 Mar
COLOGNE
0647-1219
FO R L Black RCAF
Sgt J E McDonald
FS W V McIntyre RCAF
Sgt E S Stanley
FS C E Wade RCAF
FS G Nutkins RCAF
FS C P Rolfe RCAF
5/6 Mar
CHEMNITZ
1633-0228
PO P G L Collett RAAF
FS R Matin
FS W H Horner RAAF

FS A Conden
FS M Hann RAAF
Sgt M Smedley RAAF
PO A W Thompson
FO C D Ruppell RAAF
7/8 Mar
DESSAU
1657-0255
FO W K Weightman
Sgt R D Bond
PO J D Browne
Sgt R Burtaft
PO E Davies
Sgt A P Ainley
Sgt C Amory
8/9Mar
KASSEL
1716-0048
FO W K Weightman and crew
13 Mar
DAHLBURSCH
1740-2330
FO H I Davies and crew

15/16 Mar
MISBURG
1723-0106
PO P G L Collett and crew
FO J C Wilson RCAF in, FO
Ruppell out
16/17 Mar
NURNBERG
1730-0210
FL J R Cooke RCAF
Sgt J G Wheeler
FS F Graham
FS A Atkinson
PO A R Nevin RCAF
Sgt R F Poulson
Sgt R K O'Brien
FS J W Engler RCAF
18/19 Mar
HANAU
0059-0759
PO P G L Collett and crew
20/21 Mar
HEIDE

0211-0774
FO E W Mckay and crew (8)
FO A L Schaefer RCAF in (SD)
31 Mar
HAMBURG
0615-1146
FL H J West and crew
Sgt C Lambie and FO H G
Sheath in
PO Ward and FL Mitchell out
9/10 Apr
KIEL
1938-0131
FL H J West and crew
14/15 Apr
POTSDAM
1824-0305
FL H J West and crew
18 Apr
HELIGOLAND
1039-1452
FL H J West and crew

ED588
GEORGE

A Mk III Lancaster with four Merlin XX engines, ED588 was assigned to No 97 Squadron at Coningsby, Lincolnshire, on 8 February 1943 but within two days it had gone Cat AC and was not returned to 97 until 6 March. By the time it was deemed operational the squadron had moved to Woodhall Spa.

The squadron records show its first operational sortie was flown on the night of 26 March 1943 with Sergeant K. Brown and his crew taking it to Duisburg. Although coded with 97's OF, it is shown as having individual letter 'E' in the records but by the next night it was being shown as 'H'. In all it completed eight operations with No 97 Squadron during March and April, six by Brown including a trip to Kiel on 4/6 April during which another Lancaster fired a burst at it north of Heligoland, but missed.

On 10/11 April a sortie to Frankfurt was abandoned after about 1 1/2 hours due to engine trouble. After a trip to Italy a couple of nights later, ED588 was re-assigned to No 50 Squadron at RAF Skellingthorpe, Lincolnshire, where its new squadron codes 'VN' were applied, together with the individual letter G-George.

In the guise of G-George, ED588 made its first sortie with No 50 Squadron on the last night of April with Sergeant D. A. Duncan, who became its more-or-less regular pilot until August, completing perhaps 16 sorties in it, with some others being aborted due to technical problems. Here some of these were noted as 'Not Completed', so one can assume that these ops were not counted or allowed. During this period George was hit and damaged by flak on at least two occasions.

Flying Officer F. B. M. Wilson flew George several times during the winter, including three ops to Berlin, with the aircraft going to the Big City on 15 occasions during Bomber Command's Battle of Berlin. On one trip – 2/3 December – the mid-upper gunner's oxygen failed and he fell unconscious but recovered later. In early 1944 Flight Sergeant E. Berry RAAF took George regularly to targets in Germany and France – 18 being consecutive during February-April. Berry was commissioned half-way through this period. In May and June he added another six trips, thereby completing his tour, most of which had been flown in G-George. Ernest Berry received the DFC. Then Flying Officer Howell Enoch took over the aircraft, completing 23 sorties of his tour in this machine.

Fortunately for Enoch and his crew, they went on leave just before the end of August 1944, and the very next mission George failed to return from Königsberg on the night of the 29th/30th. For Flying Officer Tony Carver and his crew,

Königsberg proved fatal and all but one of the crew were killed, most if not all those lost being buried in Palsjockrkogard Cemetery in Sweden. Enoch returned to complete his tour in October and received the DFC.

During George's period on No 50 Squadron, it became Cat AC twice, 1 September 1943 having returned damaged from Berlin, and 12 April 1944 after a raid on Aachen, having an overhaul at No 58 MU, ROS, but was back on ops again on the 18th of that month.

G-George appears devoid of any nose-artwork other than a steadily increasing number of bomb markings in rows of 10. After 10 rows, the 100th bomb was added after a trip to St Leu on 4 July 1944, the 11th and then a 12th row was added beneath the cockpit. One of the last photographs of George was taken with 125 bombs showing and it was supposed to have been lost on its 128th op. George gave very little trouble, rarely aborting through engine or other malfunctions; it just plodded on until finally it went on one trip too many and failed to get home. It is now known that the raid on Königsberg met with heavy night fighter opposition and in all 15 Lancasters were lost, G-George being one of them. Had it not failed to return, and with several months of the war still to go, ED588 might well have become the Lanc with the most sorties flown.

1943
No 97 Squadron
26/27 Mar
DUISBURG
1901-2320
Sgt K Brown
Sgt H Hogg
Sgt F H Alexander
Sgt D H Meade
Sgt J Curry
Sgt J T Sullivan
Sgt L C Boyton
27/28 Mar
BERLIN
2026-0304
Sgt K Brown and crew
28/29 Mar
ST NAZAIRE
2000-0136
Sgt G W Armstrong
Sgt E Bellis (FE)
Sgt J J Mansfield(N)
Sgt David (BA)
Sgt R J Williams (WOP)
Sgt S Blackhurst (MU)
Sgt A Laing (RG)
29/30 Mar
BERLIN
2148-0433
Sgt K Brown and crew
3/4 Apr
ESSEN
1914-2345
Sgt K Brown and crew
Sgt Freedman in, Sgt Blackhurst out
4/5 Apr
KIEL
2103-0232
Sgt K Brown and crew Sgt T J Roche and FS Broomfield in, Sgts David and Freedman out

8/9 Apr
DUISBURG
2059-0225
Sgt K Brown and crew
9/10 Apr
DUISBURG
2017-0054
FS W McLeod
Sgt F J Horsham
FS Gillespie
Sgt B L May
Sgt J Curry
FO E A Adams
Sgt J Baker
10/11 Apr
FRANKFURT
0023-0206
Sgt K Brown and crew (abandoned due to engine trouble, jettisoned bombs)
13/14 Apr
SPEZIA
2043-0708
Sgt K Brown and crew
Sgt Baker (2P)
Lt D S Watt in, Sgt Roche out

No 50 Squadron
30/1 May
ESSEN
0037-0459
Sgt D A Duncan
Sgt E Poulter
Sgt W J Evans
Sgt R L Hayter
Sgt I C Dooley
Sgt J Fulton
Sgt K D White
4/5 May
DORTMUND
2233-0359
Sgt D A Duncan and crew

13/14 May
PILSEN
2201-0458
Sgt D A Duncan and crew (hit by heavy flak over target, holed 14 times in fuselage and tail)
23/24 May
DORTMUND
2304-0343
Sgt D A Duncan and crew
25/26 May
DUSSELDORF
2344-0355
Sgt D A Duncan and crew Sgt Curtis in, Sgt White out
27/28 May
ESSEN
2217-0033
Sgt D A Duncan and crew Sgt White back (aborted due to rear-turret failure, sortie not completed)
29/30 May
WUPPERTAL
2257-0325
FO A N Hollis
Sgt D S Adshead
FO R Palmer
Sgt T G Cheshire
Sgt R G Yates
Sgt R A Kemp
Sgt W Walker
11/12 Jun
DUSSELDORF
2342-0423
Sgt D A Duncan and crew
12/13 Jun
BOCHUM
2214-0333
Sgt D A Duncan and crew
24/25 Jun

WUPPERTAL
2311-0350
PO R L Hendry
Sgt ?
FO K Toner
Sgt P A Chapman
Sgt A McDonnell
Sgt Cousins
Sgt Dalrymple
25/26 Jun
GELSENKIRCHEN
2305-0335
Sgt D A Duncan and crew
28/29 Jun
COLOGNE
2335-0416
Sgt D A Duncan and crew
Sgt Howie (2N)
8/9 Jul
COLOGNE
2244-0229
Sgt D A Duncan and crew FO D H Simpson (2N) (returned early, R/T failure, sortie not completed)
24/25 Jul
HAMBURG
2241-0326
Sgt D A Duncan and crew
Sgt Page (2N)
25/26 Jul
ESSEN
2200-0234
PO K Ruskell
Sgt T S Bradley
FO D H Simpson
FS E E Howell
Sgt J H Blatt
Sgt G A Lewis
PO M Dicks
27/28 Jul
HAMBURG

2236-0335
PO K Ruskell and crew
29/30 Jul
HAMBURG
2240-0329
FS J W Thompson
Sgt E W Lowe
Sgt S Chapman
Sgt C R Corbett
Sgt A E Nicholson
Sgt J Conlon
Sgt R A Wyllie
2/3 Aug
HAMBURG
0013-0534
FS J W Thompson and crew
9/10 Aug
MANNHEIM
2310-0154
Sgt D A Duncan and crew
Sgt E G Lloyd (2P)
(aborted due to oxygen failure, bombs jettisoned)
10/11 Aug
NURNBERG
2153-0543
Sgt D A Duncan and crew
FS L A J McLeod (2P)
17/18 Aug
PEENEMUNDE
2151-0422
Sgt D A Duncan and crew
22/23 Aug
LEVERSKUSEN
2120-0149
Sgt D A Duncan and crew
FO G M Brown (2P)
27/28 Aug
NURNBERG
2106-0454
Sgt D A Duncan and crew
30/31 Aug
MUNCHEN GLADBACH
0019-0402
Sgt D A Duncan and crew
31/1 Sep
BERLIN
2030-0358
FS J L Heckendorf
Sgt J Henderson
PO K T Dale
Sgt A D Hope
Sgt R C Turner
Sgt G K Luton
Sgt D Hall
(tailplane, fuselage and starboard wing holed by flak)
23/24 Sep
MANNHEIM
1905-0200
FO G M Brown
Sgt G Smith
Sgt T Watson
Sgt B F Tutt
Sgt D Little
Sgt P W Green
Sgt R W Sindon
27/28 Sep
HANNOVER
1933-0134
FO G M Brown and crew
28/29 Sep
BOCHUM
1838-2343
FO G M Brown and crew
1 Oct
HAGEN

1835-2353
FO G M Brown and crew
Sgt L T Dewhurst in, Sgt Green out
2/3 Oct
MUNICH
1853-0310
FO G M Brown and crew
Sgt Green back
3/4 Oct
KASSEL
1832-0052
FS J W Thompson and crew
4/5 Oct
FRANKFURT
1832-0052
FO G M Brown and crew
7/8 Oct
STUTTGART
2040-0310
FO G M Brown and crew
3 Nov
DUSSELDORF
1720-2144
FO F B M Wilson
Sgt H W Felton
PO J B H Billam
Sgt J Short
FS P F J Harrington
FS J W Newman
FS I W Anderson
(attacked by two FW190s over target after dropping bombs)
10/11 Nov
MODANE
2049-0502
FO F B M Wilson and crew
FSs S O Smith and S Proctor in;
Sgt Short and FS Anderson out
18/19 Nov
BERLIN
1720-0129
FO F B M Wilson and crew
Sgts E J Gunn and Bateman in;
FSs Smith and Proctor out
22 Nov
BERLIN
1653-2326
FO F B M Wilson and crew
Sgts L C Green and C Baker in;
Sgts Gunn and Bateman out
23 Nov
BERLIN
1718-2355
FO F B M Wilson and crew
Sgts E J Gash and J P Flynn in;
Sgts Green and Baker out
2/3 Dec
BERLIN
1639-0022
Sgt R C Thornton
Sgt J Clark
Sgt D R Neal
Sgt J Webb
Sgt L A Taylor
PO C A Maddin
Sgt C Seddon
16/17 Dec
BERLIN
1645-0030
FS R A Leader

Sgt E D Rosenberg
FO A R Candy
FS S J R Lewis
Sgt D R Tupman
PO L Stevens
Sgt T F Coulson
20/21 Dec
FRANKFURT
1726-2305
FO M J Beetham
Sgt E D Moore
FO W J Swinyard
Sgt K Payne
Sgt J C A Rodgers
Sgt A L Bartlett
Sgt F G Ball
23/24 Dec
BERLIN
2344-0818
PO D A Jennings
Sgt I T Stephens
Sgt T R D Carroll
FS G W Hughes
Sgt H C P_____?
Sgt E W Turton
Sgt C King
29/30 Dec
BERLIN
1717-0016
FS C Erritt
Sgt G Jones
FS N P Delayon
FS F A Taylor
Sgt H W J Lineham
Sgt P A Green
FS T M Williamson

1944
1/2 Jan
BERLIN
2350-0810
FS C Erritt and crew
2/3 Jan
BERLIN
2347-0726
FS C Erritt and crew
5/6 Jan
STETTIN
0006-0919
FS C Erritt and crew
14 Jan
BRUNSWICK
1716-2230
FS C Erritt and crew
(this crew went to No 83 Squadron PFF in January)
20/21 Jan
BERLIN
1634-0010
Sgt L D Wort
Sgt H R Sinton
Sgt W Jennings
Sgt G H Fry
Sgt D Gapp
Sgt A W Russell
Sgt D R Oliver
21/22 Jan
MAGDEBURG
2037-0304
SL T W Chadwick
FS W J Beesley
WO J Watt
PO SA R Verrier
FS G F Graham
FO H A Hughes
FS A M MacDonald

27/28 Jan
BERLIN
PO L Creed
Sgt S G Attwood
Sgt D Gressop
Sgt C A Hern
Sgt H R S Elder
Sgt L C Hagben
Sgt I G Evans
28/29 Jan
BERLIN
0050-0816
PO L Creed and crew
30 Jan
BERLIN
1725-2349
PO L Creed and crew
15/16 Feb
BERLIN
1739-0104
PO W V Amphlatt
Sgt K Tonks
FO T G Evans
FS W T Newton
Sgt F W Kendrick
FO A Arnold
FS H Jackson
19/20 Feb
LEIPZIG
2355-0741
FS G A Waugh
Sgt G Prince
Sgt D A Chaston
Sgt R J Dunn
Sgt D L Seblin
Sgt D C Lynch
Sgt R F Thibedeau
(this crew failed to return 31 March; all but Sgt Dennis Chaston survived)
20/21 Feb
STUTTGART
0001-0731
FS G A Waugh and crew
24/25 Feb
SCHWEINFURT
2030-0510
FS E Berry RAAF
Sgt B R W Holmes
FL L Howarth
Sgt P M N Honeywood
Sgt R Hamilton
FS W Bull
Sgt R Phillips
25/26 Feb
AUGSBURG
1833-0246
FS E Berry and crew
1/2 Mar
STUTTGART
2334-0752
FS E Berry and crew
15/16 Mar
STUTTGART
1934-0316
FS E Berry and crew
18/19 Mar
FRANKFURT
1916-0104
FS E Berry and crew
22/23 Mar
FRANKFURT
1908-0049
FS E Berry and crew
24/25 Mar
BERLIN
1914-0308

FS E Berry and crew
26/27 Mar
ESSEN
2007-0109
PO E Berry and crew
30/31 Mar
NURNBERG
2232-0543
PO E Berry and crew
5/6 Apr
TOURS
2033-0350
PO E Berry and crew
10/11 Apr
TOURS
2319-0501
PO E Berry and crew
11/12 Apr
AACHEN
2038-0037
PO E Berry and crew
18/19 Apr
JUVISY
2106-0156
PO E Berry and crew
20/21 Apr
PARIS/LA CAPELLE
2312-0353
PO E Berry and crew
22/23 Apr
BRUNSWICK
2325-0523
PO E Berry and crew
24/25 Apr
MUNICH
2052-0736
PO E Berry and crew
26/27 Apr
SCHWEINFURT
2142-0639
PO E Berry and crew
28/29 Apr
ST MEDARD-EN-JALLES
2314-0620
PO E Berry and crew
29/30 Apr
ST MEDARD-EN-JALLES
2243-0602
PO A Handley
FO T E Archard
Sgt R S Garrod
Sgt D Bissett
Sgt C T Brown
Sgt C Whitelock
Sgt G E Gilpin
6/7 May
LOUAILLE
0050-0521
PO W V Amphlatt
Sgt H Snowling
FO T G Evans
FO D A Bacon
FS G E Jagger
Sgt F W Kendrick
Sgt D R Oliver
(PO William Amphlatt failed
to return on 22 May and was
killed in action)
10/11 May
LILLE
2212-0130
PO E Berry and crew
11/12 May
BOURG-LEOPOLD
2238-0215

PO E Berry and crew
19/20 May
TOURS
2209-0504
PO E Berry and crew
21/22 May
DUISBURG
2219-0349
PO E Berry and crew
12/13 Jun
BRUNSWICK
2225-0436
PO L W Pethick
Sgt J Potter
FO J D Bell
Sgt W V Wallace
Sgt L Taylor
Sgt R Marlow
Sgt T R Mackenley
14/15 Jun
AUNAY
2218-0241
PO E Berry and crew
19/20 Jun
WATTEN/V1 SITE
2249-0108
FO H W T Enoch
Sgt G Pritchard
FO A George
Sgt J Hugh
Sgt E Garstang
Sgt H R Southcott
Sgt E Goymer
(recalled over target due to
cloud)
21/22 Jun
GELSENKIRCHEN
2314-0350
FO H W T Enoch and crew
23/24 Jun
LIMOGES
2240-0535
FO H W T Enoch and crew
24/25 Jun
PROUVILLE/V1 SITE
2248-0231
FO H W T Enoch and crew
27/28 Jun
VITRY
2140-0543
FO H W T Enoch and crew
Sgt W Hyde (2P)
29 Jun
BEAUVOIR/ V1 SITE
1233-1538
FS T G Curphy RCAF
Sgt I W Lewis
Sgt C F (DA?) Allen
FO G E Sholte RCAF
Sgt H Lambert
FO M L G Lovering
Sgt J I Fisher RCAF
(FS Thomas Curphy and crew
failed to return 29 July, all
but the rear-gunner being
killed)
4/5 Jul
ST LEU D'ESSERENT
2337-0322
FO H W T Enoch and crew
7/8 Jul
ST LEU D'ESSERENT
2244-0309
FO H W T Enoch and crew

12/13 Jul
CULMONT
2205-0558
FO H W T Enoch and crew
15/16 Jul
NEVERS
2207-0532
FO H W T Enoch and crew
18 Jul
CAEN
1423-1746
FO H W T Enoch and crew
18/19 Jul
REVIGNY
2300-0428
FO H W T Enoch and crew
20/21 Jul
COURTRAI
2314-0223
FO H W T Enoch and crew
24/25 Jul
DONGES
1220-0410
PO D L Haynes RAAF
Sgt E H Blackwell
Sgt G C Remsberry
FS P H C Lucas
FS G D Currey RAAF
Sgt E P Gavin
Sgt W I Warrington RAAF
26/27 Jul
GIVORS
2125-0639
PO D L Haynes and crew
28/29 Jul
STUTTGART
2204-0607
PO D L Haynes and crew
30 Jul
CAHAGNES
0559-1031
PO D L Haynes and crew
31 Jul
JOIGNY
1739-2301
PO D L Haynes and crew
1 Aug
SIRAUCOURT/V1 SITE
1516-1826
PO D L Haynes and crew
2 Aug
BOIS DE CASSON/V1 SITE
1425-1826
PO D L Haynes and crew
FS J L Shortal in, Sgt Gavin
out
3 Aug
TROSSY-ST-MAXIM/V1 SITE
1145-1630
PO D L Haynes and crew
(Douglas Haynes, with four
of his usual crew, failed to
return 12/13 August; Haynes,
Philip Lucas, George Currey
and Bill Warrington were all
killed and buried in Bad-Oha
Cemetery, Belgium)
5 Aug
ST LEU DE'ESSERENT
1050-1540
FO H W T Enoch and crew
7/8 Aug
SECQUEVILLE/V1 SITE
2132-0154
FO H W T Enoch and crew

9/10 Aug
CHATELLERAULT
2056-0305
FO H W T Enoch and crew
11 Aug
BORDEAUX
1209-1939
FO H W T Enoch and crew
12/13 Aug
RUSSELSHEIM
2137-0312
FO H W T Enoch and crew
FS P D A Lorimer (2P)
(FS Lorimer and his own
crew failed to return on 13
August daylight mission to
Bordeaux)
13 Aug
BORDEAUX
1635-2347
FO H W T Enoch and crew
14 Aug
BREST
1745-2242
PO R J Odgers RNZAF
Sgt K Cox
Sgt L Clarke
Sgt S D Bartlett
Sgt P Donnelly
FS G T Alcock
Sgt E Neville
(PO Richard Odgers and crew
failed to return on 12
September; he, Patrick
Donnelly and Edward Neville
were buried in Bad Tolz
Cemetery, Durnbach)
16/17 Aug
STETTIN
2055-0531
FO H W T Enoch and crew
18 Aug
FORET DE L'ISLE-ADAM
1130-1615
FO H W T Enoch and crew
19 Aug
LA PALLICE
0513-1145
FO H W T Enoch and crew
25/26 Aug
DARMSTADT
2028-0514
FO H W T Enoch and crew
26/27 Aug
KONIGSBERG
2039-0719
FO H W T Enoch and crew
29/30 Aug
KONIGSBERG
2058-
FO A H Carver.
Sgt R R Clifford
Sgt F G Plowman
Sgt D A MacDonald
Sgt E Match
Sgt W R Campbell
Sgt R W Bysouth
(FO Anthony Carver, Ronald
Clifford, Frederick Plowman,
Donald MacDonald, William
Campbell and Raymond
Bysouth all killed; Sgt Match
appears to have survived)

58

ED611
UNCLE JOE

Built by A. V. Roe as a Mk III with four Merlin 28 engines, ED611 arrived at No 5 MU on 14 February 1943 and was assigned to No 44 (Rhodesia) Squadron at Waddington, Lincolnshire, on 5 April 1943, the squadron moving to nearby Dunholme Lodge the following month. Given the squadron code letters KM, its individual letter was U-Uncle. It is unclear if the aircraft was known merely as U-Uncle, or 'Uncle Joe' at this early stage. We know it became Uncle Joe later but whether that was due to its change of squadron (to 463), when its code letter became JO, has not been confirmed.

Joseph Stalin, the Russian leader (often referred to as Uncle Joe by the media), was very much in the news during this period, and at some stage U-Uncle became known as Uncle Joe, Stalin's portrait being painted on the nose, superimposed on the red star of Russia. Once ED611 began operations, its groundcrew began painting not tiny bomb symbols beneath its cockpit, but tiny stars – yellow for night sorties, red for day sorties.

From photographs it appears that after 65 'stars' had been painted on the side of Uncle Joe, in five and a half rows of 10 and five, more five-star rows were begun further forward, just aft of the front turret. After nine rows of these had been completed - making 45, which brought the tally to 100, the front five-star rows were painted out and the previous rows were completed in tens. What is not certain is where the additional 15 stars were marked (the aircraft is understood to have completed 115 ops), as the 10 rows of 10 completely fill the area from cockpit edge to bomb doors.

When ED611 flew its 100th sortie, someone chalked on the nose '100 UP TONIGHT' under Stalin's picture and it was shortly after this that the five and a half rows of 10 were extended to show 10 rows of 10, with, presumably the five-star rows painted out. The other famous picture shows these 100 stars while '104 NOT OUT' was then written under the likeness of Stalin.

Its first sortie was a long haul to Spezia in northern Italy on 13/14 April 1943, with Flying Officer D. M. Moodie in command. Moodie and crew went to No 97 Squadron in July and flew another Lanc that was to achieve its century, EE176, on seven of her trips.

Uncle became Cat AC on 21 August 1943 after being damaged by an Me109 attack on the Peenemunde raid, 18/19 August. All its starboard side was shot-up, the starboard outer engine knocked out, the rear-turret hydraulics failed, the intercom made u/s and the starboard wheel-tyre burst. Later the aircraft came under attack from an Me210 but Uncle's pilot, Pilot Officer Deryck Aldridge, having dived steeply, got the damaged Lanc under control, climbed again and then, with Sammy Holmes the rear-gunner coming up to say his turret was almost useless, they headed for home, flat out at 230kts on three engines, low over Denmark. It was their 26th op and Uncle's 24th or 25th. Aldridge received the DFC for this effort, his navigator, Flying Officer Desmond B. F. Heslop, collecting one too.

Uncle was out of action until 18 September, returning to No 44 (Rhodesia) Squadron where it went on to complete 43 ops, but it also had several encounters with night fighters during its tours of duty. One was on 4/5 October going to Frankfurt when the rear-gunner, Sergeant D. H. Watts, spotted a Ju88 at 300 yds. He and the mid-upper, Pilot Officer A. Rimmer, both opened fire as the pilot, Flying Officer C. D. Wiggin, put the Lanc into a diving turn to port. Fortunately for the Lanc crew, the night fighter had been burning its navigation lights which quickly went out as the gunners opened up. The '88 did not return fire and it was last seen diving away to starboard.   Another combat occurred on the night of 21/22 January 1944 over Magdeburg. The rear-gunner, Sergeant D. Chalmers, spotted a Me109 at 700yds and ordered the pilot to take evasive action. As Pilot Officer R. M. Higgs did so, the '109 pilot attacked and both gunners - Flight Sergeant H. S. Tiller was in the mid-upper turret - returned fire. Strikes were seen on the '109 which was then observed to go underneath with black smoke coming from its engine and was not seen again. Two minutes later another '109 was spotted attacking another Lancaster off to port and the rear-gunner opened fire. The other bomber dived and the '109 broke off and headed away and down. David Chalmers and Henry Tiller both received the DFM, but Chalmers was later killed in action.

After taking part in the Battle of Berlin - 10 trips - Uncle was assigned to No 463 (RAAF) Squadron at Waddington on 4 February 1944, its new codes being JO. Some of the early ops recorded for Uncle were shown under the serial number ME611, but fortunately there was no such serial number for a Lancaster, so for ME611 one can read ED611 with impunity.

Uncle began its second career on 19/20 February with an op to Leipzig and the next month flew on the final Berlin raid. Missing flying on the pre-D-Day night, Uncle was taken out by an extraordinary crew on the night of 6/7 June. It would appear a senior officer wanted to be a part of this great day, for the crew comprised a wing commander, a squadron leader, two flight lieutenants, two junior officers and a flight sergeant. From No 5 Group Headquarters was Wing Commander George Geoffrey Petty DSO, DFC, who had been an RAF apprentice in 1931, commissioned in 1940, awarded the DFC in 1941 and the DSO in 1943. One of his earlier sorties had been a daylight to Milan in October 1942.

During this D-Day period Uncle's usual crew was that captained by Pilot Officer N. W. Sanders RAAF who had arrived at the squadron on 12 April 1944. On the night of 8/9 June they had a tyre burst on take-off and great difficulty was experienced in getting off the ground but Sanders made it. Over the target they met heavy ground opposition, causing them to make three dummy bomb-runs before letting the bombs go on the fourth run - on which they were considerably shot-up. Noel Sanders even managed a good landing on one wheel back at base, and apparently it should have been their night off! Sanders received the DFC on completion of his tour.

Another crew lost both outer engines over Brest on 14 August, but Flight Sergeant A. G. Stutter made an excellent return and landing. However, Uncle was now nearing 100 ops, the magic figure being reached (one assumes) on a raid in mid-September. A count through the Form 541s could indicate the 100th op being flown on 12/13 September, but when the aircraft went for a major refit at the end of that month the groundcrew chalked '104 Ops' on its nose. That would indicate another with a further nine flown after its return, making 113, but in total 115 have been credited to Uncle. There is nothing in the squadron records to confirm the date of the 100th op, the only evidence is the photograph of the aircraft with '104 not out' on its nose with the words 'Good-bye Old Faithful' chalked on the port bomb-door. As this legend must have been written when it went off for refit at the end of September, that total seems in error.

Having became Cat B at the end of September, Uncle went off for a major refit at Avro's, coming back to No 463 (RAAF) Squadron on 15 December 1944. If the total of 115 ops is correct, only nine more can be found after its return, so once again, either the 100th is in question or some of its final sorties are either not recorded or recorded incorrectly. By counting the ops below, the 100th does appear to be that flown to Stuttgart on 12/13 September 1944, captained by another of this Lanc's more regular skippers, Pilot Officer T. A. Perry.

Uncle's last sortie came on 8/9 February 1945, to Pölitz with Flying Officer M. S. Wickes RAAF in command. Prior to bombing they were attacked by two Ju88s which resulted in No 1 port fuel tank catching fire and the hydraulics being severely damaged. The fighters were eventually evaded and being just a few minutes from the target, they carried on and bombed.

Immediately afterwards, Milton Wickes had to feather the port inner engine but the fuel tank was still burning and it stood out like a beacon attracting another night fighter. Wickes put the Lanc into a corkscrew manoeuvre and while the fire dimmed it did not go out, so Wickes ordered everyone to prepare to bale out. Making one last attempt he dived Uncle steeply and the fire was finally extinguished.

They were now heading home on three engines and with the bomb doors hanging down. The artificial horizon was u/s, 10 degrees of flap was also down and the intercom to both gunners knocked-out. However, the rear-gunner had succeeded in shooting down the Ju88 in the second attack, with the aid of the mid-upper. The navigator brought them out over Denmark and because of their fuel shortage they only just made Carnaby, with a very damaged aeroplane that was anything but air worthy. Milton Wickes later received the DFC for this sortie.

It was a sad but gallant finale to Uncle's operational tour of duty. The machine now went off to be repaired and, still Cat B on 9 March, went to RIW, then to AW/CN on 22 June, ending up at West Freugh as a hack aircraft with the Bombing Trials Unit. It was finally Struck off Charge on 22 June 1947.

1943
No 44 (Rhodesia) Squadron
13/14 Apr
SPEZIA
2041-0627
FO D M Moodie
Sgt L F Melbourne (FE)
Sgt J J Bundle (N)
Sgt H W N Clausen
Sgt T F Stamp (WOP) (BA)
Sgt L A Drummond
Sgt F A Hughes (RG) (MU)
14/15 Apr
STUTTGART
2217-0050
Sgt L J Ellis
Sgt R L LePage
PO W A Rollings
Sgt J Brown
Sgt A C Ellis
Sgt R Williams
Sgt S S McClellan
(aborted, bombs jettisoned over North Sea; hydraulic leak to rear turret)
16/17 Apr
PILSEN
2117-0546
Sgt J O Pennington
Sgt D Morrison
Sgt J R Hewitt
Sgt L Hawkes
Sgt D Betts
Sgt W A Harrall
Sgt G B Homewood
18/19 Apr
SPEZIA
2107-0601
FL R D Robinson
Sgt E E C Hayward
FO A A StC Miller
Sgt L A T Parsons
PO D Hartung
Sgt G M E Weller
Sgt D S Mindel
20/21 Apr
STETTIN
2130-0528
Sgt G N Stephenson
Sgt J A Robinson
FS E G More
Sgt A R Smith
Sgt D Betts
Sgt S R Hopkin
Sgt W J Beggs
26/27 Apr
DUISBURG
0048-0453
FS C Shnier
Sgt A N Gibbons
Sgt N Laidler
Sgt H T Wigley

Sgt P C Evans
Sgt D E Croft
Sgt B G Knoesen
30/1 May
ESSEN
0033-0501
FO W D Rail
Sgt H K Underwood
Sgt A T C Bromwich
Sgt W C Digby
Sgt R C Boardman
Sgt R S A Walker
Sgt G Batty
4/5 May
DORTMUND
2144-0315
Sgt D R Aldridge
Sgt T Phillips
FO D B F Heslop
Sgt J A Dellow
Sgt R W West
Sgt D J N Palmer
Sgt T S Holmes
12/13 May
DUISBURG
0015-0420
Sgt D R Aldridge and crew
Sgt D Wolensky in, Sgt Palmer out
13/14 May
PILSEN
2139-0554
Sgt D R Aldridge and crew
Sgt Palmer back
23/24 May
DORTMUND
2250-0400
Sgt D R Aldridge and crew
25/26 May
DUSSELDORF
2346-0403
Sgt D R Aldridge and crew
29/30 May
WUPPERTAL
2257-0402
Sgt D R Aldridge and crew
11/12 Jun
DUSSELDORF
2322-0407
Sgt D R Aldridge and crew
20/21 Jun
FRIEDRICHSHAVEN
2158-0741
FO D M Moodie and crew
8/9 Jul
COLOGNE
2249-0441
PO D R Aldridge and crew
9/10 Jul
GELSENKIRCHEN
2226-0041
PO D R Aldridge and crew

(bombs jettisoned at 0009hrs over North Sea; starboard outer and port inner engines u/s, port generator, RSJ unit and MU turret, all u/s)
12/13 Jul
TURIN
2233-0745
Sgt D A Rollin
Sgt J C Blackmore
Sgt T B Malia
Sgt E J Tocher
Sgt J Barker
Sgt B Chew
Sgt R Standing
24/25 Jul
HAMBURG
2234-0324
PO D R Aldridge and crew
25/26 Jul
ESSEN
2214-0291
Sgt R M Campbell
Sgt J G Watkins
PO L G Poperwell
Sgt A H Thompson
Sgt J Graham
Sgt H C Macannick
Sgt W Phillips
29/30 Jul
HAMBURG
2214-0309
PO D R Aldridge and crew
2/3 Aug
HAMBURG
2324-0435
PO D R Aldridge and crew
12/13 Aug
MILAN
2150-0635
FS H G Norton
Sgt J H Stevens
Sgt S D Stait
Sgt F Thompson
Sgt E E Greenfield
Sgt R G Martin
Sgt W A Whalley
15/16 Aug
MILAN
2029-0453
FS H G Norton and crew
17/18 Aug
PEENEMUNDE
2132-0458
PO D R Aldridge and crew
Sgt D Wolensky in, Sgt Palmer out
(badly shot-up by Me109 and Me210. PO Aldridge tour-expired October and posted to No 84 OTU)

4/5 Oct
FRANKFURT
1810-0039
FO C D Wiggin
Sgt A Jones
FO R H Maury
FO C G Rogers
Sgt A Dicken
PO A Rimmer
Sgt D H Watts
18/19 Oct
HANNOVER
1709-2228
PO D A Rollin and crew
20/21 Oct
LEIPZIG
1709-0043
PO D A Rollin and crew
3/4 Nov
DUSSELDORF
1707-2121
PO R L Ash
Sgt G F Ives
PO C G Whitehead
FS H Cushom
Sgt B H White
Sgt J Murphy
18/19 Nov
BERLIN
1658-0130
PO R M Higgs
Sgt J E Cowan
FS T W Black
Sgt J W Would
Sgt V G Williams
FS H S Tiller RAAF
Sgt D Chalmers
22/23 Nov
BERLIN
1649-2355
PO R M Higgs and crew
23/24 Nov
BERLIN
1743-2006
FO W F Newell and crew
(aborted mission when unable to climb above 9,000ft by East Coast)
26/27 Nov
BERLIN
1731-0115
PO R M Higgs and crew
3/4 Dec
LEIPZIG
0032-0740
PO R M Higgs and crew
16 Dec
BERLIN
1628-2348
PO R M Higgs and crew
Sgt D D Orme in, Sgt Cowan out

20/21 Dec
FRANKFURT
1700-2251
PO N F Lyford
Sgt A Semple
Sgt J R Tijou
FS G Owen
Sgt G A Ford
Sgt H Marrs
Sgt A Wainwright
23/24 Dec
BERLIN
0017-0748
PO N F Lyford and crew
29 Dec
BERLIN
1645-2331
PO R M Higgs and crew
Sgt Cowan back

1944
14 Jan
BRUNSWICK
1636-2004
PO R M Higgs and crew
(aborted due to port outer
engine u/s and starboard
outer overheating; 'cookie'
jettisoned over North Sea at
1815hrs)
20 Jan
BERLIN
1634-2327
PO R M Higgs and crew
Black, Would and Williams
now POs
21/22 Jan
MAGDEBURG
1941-0244
PO R M Higgs and crew
27/28 Jan
BERLIN
1731-0155
FS E Barton
Sgt J C Thompson
FO G F Garland
Sgt F H Barnes
Sgt T M Willett
Sgt R W Joy
Sgt L J Hummell
28/29 Jan
SYLT
0026-0530
FS E Barton and crew
(rear turret became u/s,
supercharger gear u/s; severe
icing with turrets and on
windscreen. Main target was
Berlin and it is assumed
Uncle bombed Sylt as a sec-
ondary target)
30/31 Jan
BERLIN
1703-0015
PO B M Hayes
Sgt J M Ella
Sgt C Dean
FO E Dunn
Sgt W K Walker
Sgt K V Radcliffe
Sgt W G Perrie
No 463 (RAAF) Squadron
19/20 Feb
LEIPZIG
2357-0731
PO V H Trimble
Sgt A J Marles

FL A Williams
Sgt B T Aldworth
Sgt W J Nixon
Sgt R McNaughton
FS J P Lawrence
24/25 Feb
SCHWEINFURT
2049-0445
PO V H Trimble and crew
FO A W Herriott in, FL
Williams out
25/26 Feb
AUGSBURG
1840-0208
PO V H Trimble and crew
1/2 Mar
STUTTGART
2307-0636
PO V H Trimble and crew
15/16 Mar
STUTTGART
1932-0250
PO V H Trimble and crew
18/19 Mar
FRANKFURT
1906-0121
PO V H Trimble and crew
22/23 Mar
FRANKFURT
1858-0054
PO B W Giddings
Sgt A Pritchard (FE)
FS W Webb (N)
FS C Clements (BA)
FS R Bethel (WOP)
FO J McGill (MU)
FS W Seale (RG)
24/25 Mar
BERLIN
1927-0235
FO B A Buckham
Sgt W Sinclair
FO R W Board
FS L F Manning
FS E L Holden
Sgt F Burton
FS Moorhead
10/11 Apr
TOURS
2315-0413
PO K H Robertson
Sgt R J Patrick
Sgt D Ball
FO F N Chandler
FS S E W Smith
PO C G Parker
FS N J Bowman
18/19 Apr
JUVISY
2053-0211
FS R W Page
Sgt S R Crate
WO W W Fair
FO J Braithwaite
FS E R Brown
Sgt R Guide
FO C H Noakes
26/27 Apr
SCHWEINFURT
2148-0609
WC R Kingsford-Smith DFC
Sgt R Fairburn
FO N H Kobeike
FS B W Webb
FO M J McLeod
FO J E R Rees
FO R M Croft

28/29 Apr
ST MENARD-EN-JALLES
2313-0620
PO N W Sanders RAAF
Sgt D A Brett RAF
FS M B Greacen
FO E Rosenfeld
FS E O Davis
FO G H Swindells
Sgt G Elliott RAF
(aborted due to haze over
target and brought bombs
back)
1/2 May
TOULOUSE
2152-0536
PO N W Sanders and crew
3/4 May
MAILLY-LE-CAMP
2153-0316
PO N W Sanders and crew
31/1 Jun
SAUMUR
2351-0502
PO N W Sanders and crew
3/4 Jun
FERME D'URVILLE
2312-0252
PO A B Tottenham
Sgt R J Patrick
FS D Ball
FO F N Chandler
FS S E W Smith
PO C G Parker
PO N J Bowman
6/7 Jun
ARGENTAN
2329-0353
WC G G Petty DSO DFC
SL Evans
FL Archer
FO D Falgate
FL K M D Lyons
PO Aitken
FS Flood
8/9 Jun
RENNES
2310-0509
PO N W Sanders and crew
(hit and damaged by flak
over target, having burst a
tyre on take-off)
15/16 Jun
CHATELLERAULT
2133-0409
PO N W Sanders and crew
21/22 Jun
GELSENKIRCHEN
2315-0351
PO N W Sanders and crew
23/24 Jun
LIMOGES
2234-0524
PO N W Sanders and crew
PO M F Sweeney (2P)
24/25 Jun
PROUVILLE/V1 SITE
2234-0221
PO N W Sweeney and crew
(held in searchlights for 5
minutes over target)
27/28 Jun
VITRY
2202-0528
PO M J Roe
Sgt W N Hill
Sgt A E Fox

Sgt T F Byrne
FS N S Palmer
FS T R Ryan
Sgt K C Butcher
29 Jun
BEAUVOIR/V1 SITE
1218-1532
PO M J Roe and crew
4/5 Jul
ST LEU D'ESSERENT
2322-0327
PO M J Roe and crew
7/8 Jul
ST LEU D'ESSERENT
2215-0304
PO N W Sanders and crew
12/13 Jul
CULMONT/ CHALINDREY
2152-0555
PO N W Sanders and crew
15/16 Jul
NEVERS
2218-0528
PO N W Sanders and crew
18 Jul
CAEN
0359-0734
FS T A Perry
Sgt F A England
FS G S Carlton
FS W T Clayton
FS H E Williams
FS C T Forrester
Sgt L Oxley
18/19 Jul
REVIGNY
F2306-0416
O D C Gundry
Sgt P O F Wadsworth
FS E J Sincock
RAFF
S E P Fallon
FS V J Sheldt
Sgt N M Davidson
FS E Burke
19 Jul
THIVERNY/V1 SITE
1916-2332
PO N W Sanders and crew
20/21 Jul
COURTRAI
2331-0242
PO N W Sanders and crew
24/25 Jul
DONGES
2235-0402
FO I J Dack
Sgt H E Lee
FS J F Maple
FS J R McWilliam
Sgt A Easton
FS R Coward
Sgt C F Kirby
26/27 Jul
GIVORS
2135-0614
PO N W Sanders and crew
28/29 Jul
STUTTGART
2206-0553
FO M F Sweeney
Sgt H R Carpenter
FS H H Mirtha
FO G T Gill
FS G A Russ
FS W N Robinson
FS R W Palmer

31 Jul
RILLY-LA-MONTAGNE
1719-2228
PO N W Sanders and crew
1 Aug
MONT CANDON/ V1 SITE
1636-2133
PO N W Sanders and crew
(mission abandoned due to
fog over target)
3 Aug
TROSSY ST MAXIM
1202-1631
PO N W Sanders and crew
5 Aug
ST LEU D'ESSERENT
1040-1542
FS A G Stutter
Sgt H Walsh
FS P L Wilkinson
FS P O'Loughlin
FS D J Browning
FS M F Woodgate
FS H R Holmes
6 Aug
BOIS DE CASSON
0932-1331
PO N W Sanders and crew
7/8 Aug
SECQUEVILLE
2120-0200
PO T A Perry and crew
9/10 Aug
CHATELLERAULT
2043-0322
PO N W Sanders and crew
11/12 Aug
GIVORS
2029-0451
PO N W Sanders and crew
12/13 Aug
RUSSELSHEIM
2115-0320
PO T A Perry and crew
14 Aug
BREST
1757-2229
FS A G Stutter and crew
(lost both outer engines but
made excellent landing)

15 Aug
GILZE RIJEN A/F
0939-1335
PO T A Perry and crew
16/17 Aug
STETTIN
2138-0527
FO M F Sweeney and crew
18 Aug
L'ISLE ADAM
1148-1632
PO T A Perry and crew
25/26 Aug
DARMSTADT
2037-0501
PO N W Sanders and crew
26/27 Aug
KONIGSBERG
2020-2219
PO K E Tanner
Sgt K W Yates
FS R W Dent
FS W A Hundy
WO K W A Brock
FS D M Wills
FL A A Fraser
(captain became ill so mis-
sion aborted)
29/30 Aug
KONIGSBERG
2035-0700
PO N W Sanders and crew
(Sanders completed tour and
sent to No 5 LFS)
31 Aug
ROLLENCOURT
1618-2006
PO T A Perry and crew
11 Sep
LE HAVRE
0522-0925
PO T A Perry and crew
11/12 Sep
DARMSTADT
2058-0238
PO T A Perry and crew
12/13 Sep
STUTTGART
1901-0156
PO T A Perry and crew

17 Sep
BOULOGNE
0811-1129
PO K W Reilly
Sgt R W Jones
FS A R MacKenzie
FO B C Howard
FS R N Hall
Sgt A D Pryer
FS R N Battye
18 Sep
BREMERHAVEN
1814-2328
PO K W Reilly and crew
19/20 Sep
RHEYDT
1901-0001
PO K W Reilly and crew
23/24 Sep
DORTMUND-EMS CANAL
1908-0023
PO T A Perry and crew
26/27 Sep
KARLSRUHE
F0059-0704 O L C Peart
Sgt P Kerns
FS J Palmer
FS A E Parkes
FS K N Brandwood
FS G C Bowler
FS P J Rogers
27/28 Sep
KAISERSLAUTERN
2200-0442
FO L C Peart and crew
27 Dec
RHEYDT
1216-1636
FO M S Wickes
Sgt L R Botting (N)
FO R S Brownlee
WO F K Brett
FS R S Jenkins
Sgt F H Boddy
Sgt L D Cottroll
31 Dec
HOUFFALIZE
0224-0732
FL B S Martin and crew
(two runs made over target, a

road bottleneck in a valley,
but target not identified so
attack not carried out)

1945
1 Jan
DORTMUND-EMS CANAL
0735-1400
FO M S Wickes and crew
4/5 Jan
ROYAN
0047-0737
FO M S Wickes and crew
7/8 Jan
MUNICH
1656-0207
FO M S Wickes and crew
13/14 Jan
POLITZ
FO1639-0335 G F Lincoln
Sgt J Clift
FS D Pickford
FS W E Hooper
FS D M Stubing
FS Hutchinson
FS W C Faust
14/15 Jan
MERSEBERG-LEUNA
1632-0215
FO M S Wickes and crew
1/2 Feb
SIEGEN
1617-2237
PO E Foster
Sgt C Beighton
Sgt D J Starling
Sgt H K Hurst
FS F Pentiman
Sgt H Gooder
Sgt G E Howitt
8/9 Feb
POLITZ
1655-0300
FO M S Wickes and crew
(Aircraft badly damaged by
night fighter attacks; one
Ju88 shot down)

ED860
NUTS

Another A. V. Roe, Manchester-built Mk III, equipped with four Merlin 28 engines, ED860 was produced early in 1943 and assigned to No 156 Squadron at Warboys, Huntingdonshire, on 14 April. Once in the hands of its new groundcrew, the Lanc was given the squadron codes of GT. What is not known for certain is what the individual aircraft letter was at this stage. However, No 156 Squadron was part of the Pathfinder Force and ED860 served with this unit until August, taking part in some 23 operations.

According to records, it was assigned to No 61 Squadron at Syerston, Nottinghamshire, on 20 August, although No 156 Squadron records the aircraft's last op on 22 August and another on the 23rd, which may be in error. It seems more likely that its last op with 156 was to Milan on 15 August and its first with No 61 Squadron was on 23 August, as a 156 combat report dated 27 August shows ED860, while No 61 Squadron clearly shows ED860 on ops to the same target with them. This is yet another example of poor record keeping, with a clerk merely putting down either a continuation of what had gone before, or a new aircraft with the same individual letter as the old machine. Another theory is that ED860 could have been aircraft 'Y' as this same combat report for 27 August showing ED860 also records the aircraft letter as 'Y'.

However, once on No 61 Squadron ED860 certainly had the individual letter N-Nuts on its fuselage, and it seems possible that it was 'N' on 156, too, because of the clerical errors in the Form 541.

ED860 started its career on No 61 Squadron with a trip to the Big City as the Battle of Berlin was just starting. Once this Battle had got into its stride N-Nuts went to the Big City on a further 19 occasions, although one was aborted due to engine problems. Few crews flew this Lanc for long, although Flying Officer H. Scott seemed to fly it more regularly than anyone else, until Pilot Officer E. A. Stone took over in early 1944, operating in the aircraft until May.

Bernard Fitch and crew flew it on seven trips of their tour; his navigator was Sid Jennings who remembers:

'Our first op was a mining sortie off Texel. Despite intense AA fire from flak ships we dropped our mines from 750 to 1,000ft but because of our height, the flak came at us horizontally!

'On one of our Berlin trips I was acting as Windfinder, and we had been given a NNE wind at 70mph, whereas I soon discovered the true winds were

120mph at 040 degrees. I sent this back to Bomber Command HQ but they just couldn't believe it and only adjusted the wind as 75mph at 030 degrees. Thus most bombers came back via the Ruhr, and as a result a lot of aircraft were brought down.

'On another occasion our compass was found to be 60 degrees off so I had to get a shot from the astrodome on Polaris and use the astro compass from then on, but we made it to the target and back.'

Bernard Fitch, who like his navigator would receive the DFC, remembers the windy Berlin trip, as due to the winds they had to do a dummy run over Berlin and on both runs the headwind was so strong they were practically standing still during the approach. The rear turret was frozen, so when a night fighter came at them the gunner could only order a corkscrew; they lost 4,000ft. He also remembers the 18/19 November trip to Berlin for another reason:

'We were due to go on leave on Friday 19 November, and I was due to get married in Bedford the next day. The squadron was comfortably established at Syerston, and best uniforms were at the cleaners in nearby Newark. Out of the blue we were moved to Skellingthorpe on the 16th.

'Things were not back to normal when ops on the 18th turned out to be the opening of the Battle of Berlin, which didn't augur well for a stand-down (and leave) next day.

'It was a hell of a rush to get off in time in ED860. Something had to go wrong and we lost our upper escape hatch on take-off, so for most of us the trip was colder than usual. But we made it although it took us 7 1/2 hours.

'Next morning we retired to bed not knowing whether it was leave or ops. Thus we – and I especially – were very relieved on being called as arranged at 2pm, to learn of a stand down! It took much longer than our flight time to Berlin to travel to Bedford, but at least I got to the church on time, albeit in my working uniform. At the time of writing, we have just reached our Golden Wedding anniversary.'

ED860 missed D-Day due to major servicing but on the night of 18/19 July 1944 the Lanc and its crew of Flying Officer D. G. Bates came under night fighter attacks over Revigny. After one inconclusive action, the mid-upper gunner, Pilot Officer J. Fletcher, saw an aircraft coming in astern, gave the pilot the order to corkscrew as he opened fire, the rear-gunner, Sergeant D. Hancock also opening fire. Bullets were seen to strike the enemy aircraft which burst into flames and dived to the ground where both gunners saw it explode, witnessed also by the pilot and bomb-aimer. As the pilot came round to try for another bomb-run, they came under fire from another Lancaster's gunners but suffered no damage. By this time the target markers were no longer visible and so the attack had to be abandoned.

In August 1944, Flying Officer N. E. Hoad and his crew, who had commenced their tour in July, began to fly ED860, making six trips in all. Norman Hoad recalls:

'The "honour" of flying ED860 was something of a dubious privilege in that in some respects it was well past its sell-by-date. However, as a new boy on the squadron it was not for me to pick and choose and I had to make the best of it. Both outer engines had a nasty habit of over-heating for which there was no remedy but to throttle back. Apart from having to fly the machine with engines out of sync, the most serious consequence was to trail slowly behind the rest of the stream.

'On the submarine pen operation on 11 August [Bordeaux] we were so far behind we attacked virtually on our own. Not wishing to make a lonely way home across enemy-held territory, I took the decision to drop down to sea level and make the return flight the long way round at wave top height. Some way out in the Bay of Biscay we encountered a Dornier 14 flying-boat going the other way but both of us chose to ignore the other!

'The last time I flew "N" was, of course, on 29 August in the raid on Königsberg. Here again her foibles were a pain. The Lancaster's auto-pilot was pneumatically operated and on this occasion the aircraft was thrown into a violent pitch every time an attempt was made to engage. This is not the best way to treat a heavily loaded aircraft and the result was that it had to be flown manually for the 10 hours 45 minutes the flight lasted.

'Approaching the target area the fighters were extraordinarily active and though I did not see them myself, I was constantly told by my gunners to corkscrew. The 5 Group corkscrew should, in theory, enable a pilot to maintain a roughly straight track but a continuous series of 60-degree banked turns in either direction, coupled with dives and climbs through 1,000ft, can and did lead to some deviation in our route. So much so in this case that I found myself totally out of position to make the bombing-run on the prescribed heading so I decided to go round again rather than waste my bombs.

'It all took a certain amount of time and as at Bordeaux, I found myself over the target area more or less alone. At some stage in all this the aircraft sustained various hits, the most serious of which was in the starboard wing-root between the fuselage and the starboard main fuel tank. To this day I do not know whether this was merely flak damage or, as I think more likely, cannon fire from an upward-firing night fighter. At all events I was given a new paddle-bladed Lancaster for my next operations!'

N-Nuts became Cat AC after this raid, and did not come back onto the serviceable list until 18 October. In the event it went on only one more operation and then it was made Cat E on the 28th, to be ingloriously Struck off Charge and scrapped on 4

November 1944, without seeing the victory it had helped to achieve. ED860 had flown nearly 1,032 hours.

Norman Hoad noted in his log-book that the six ops he flew in N–Nuts had been numbered 118, 119, 120, 121, 124 and 129, so that its final mission had to be its 130th. Therefore the 100th sortie must have been flown in late June 1944, during the period Flying Officer B. S. Turner and his crew were flying the Lanc. Basil Turner took N–Nuts on some 15 ops during his tour and won the DFC.

ED860 had no special nose-artwork, just two columns of 10-line bomb symbols for the 100 trips, then another column of tens starting under the first one. Curiously there were two swastika markings painted on, denoting two successes against enemy night fighters, although only the victory on 18/19 July is known for certain. As explained earlier, the combat on 27 August 1943 – when the Lancaster crew 'attacked' a He111 on their way to Nurnberg – had to be another aeroplane. N–Nuts did have encounters with Ju88s over Stuttgart on 7/8 October 1943, but no claims appear to have been made.

1943
No 156 Squadron
20/21 Apr
STETTIN
2135-0455
PO J M Horan
FS R J Atkin FS E P Fast
Sgt J C Chapman
FO D C A Sanders
Sgt G G Forbes
Sgt J R Curtis
26/27 Apr
DUISURG
0035-0435
PO J M Horan and crew
30/1 May
ESSEN
0042-0459
FL D T Muir
PO H G Innes
Sgt G W Bramley
Sgt H J Folland
Sgt R H Wedd
Sgt M Haslegrove
Sgt W B MacKinley
4/5 May
DORTMUND
2253-0454
FO B F Smith
PO W J Smith
FO R E Goodwin
Sgt B Marshall
FO J A Philps
Sgt L J Jones
FO S Hayes
13/14 May
PILSEN
2210-0532
Sgt L W Overton
Sgt D P Clements
Sgt J A M Arcari
Sgt T Cable
Sgt D M Davies
Sgt F Sunderland
Sgt A Barnett
23/24 May
DORTMUND

2255-0324
FS D L Wallace
Sgt T H Harvey
Sgt W H Moore
Sgt R J Jackson
Sgt R J Twinn
FS D Ross
Sgt H A Lister
25/26 May
DUSSELDORF
2350-0424
FS D L Wallace and crew
29/30 May
WUPPERTAL
2243-0340
Sgt L W Overton and crew
11/12 Jun
MUNSTER
2346-0419
FL R E Young
PO T Burger (N)
Sgt E G Hodges (WOP)
Sgt J K Carlton (FE)
Sgt T H Evans (BA)
Sgt J S Goodman (MU)
Sgt J W Boynton (RG)
12/13 Jun
BOCHUM
2331-0352
FL R E Young and crew
16/17 Jun
COLOGNE
2241-0210
FL R E Young and crew
(aborted due to severe icing making front turret and R/T u/s)
24/25 Jun
ELBERFELD
2314-0341
FL R E Young and crew
28/29 Jun
COLOGNE
2306-0340
Sgt C W Wilkins
Sgt H Holman
Sgt E S Crabbe

Sgt G E Milton
Sgt A Cloud
Sgt R Lobb
Sgt E J Cahill
3/4 Jul
COLOGNE
2247-0328
FL A L McGrath
FO R P Wright
FO J Facey
Sgt D L Wilkie
FO C R Johnson
Sgt J Prochera
Sgt G D Aitken
8/9 Jul
COLOGNE
2255-0353
FL R E Young and crew
24/25 Jul
HAMBURG
2221-0436
FL R E Young and crew
27/28 Jul
HAMBURG
2201-0359
FL R E Young and crew
29/30 Jul
HAMBURG
2203-0345
FL R E Young and crew
2/3 Aug
HAMBURG
2336-0531
FL R E Young and crew
FS G A Lindsey in, Sgt Boynton out
7/8 Aug
TURIN
2147-0609
WO G Denwood
PO F R Kennedy (N)
Sgt E J Cutter
Sgt A L Barlow
FS J C Ross
WO J E Barnham
FS J A O Lovis
(PO Kennedy was to die of

natural causes in Ely Hospital in September; he also received the DFC)
9/10 Aug
MANNHEIM
2339-0443
WO G Denwood and crew
10/11 Aug
NURNBERG
2135-0503
WO G Denwood and crew
15/16 Aug
MILAN
2135-0503
WO G Denwood and crew
15/16 Aug
MILAN
2035-0401
PO P A Coldham
FO A P Stevens
Sgt A A P Bland
Sgt T E Rees
FS G R Robinson
FS N Warwick
FS G H Pascoe
No 61 Squadron
23/24 Aug
BERLIN
2030-0401
Sgt M C Lowe
FS J G McAlpine (2P)
Sgt C R Moffitt (FE)
FO R J Clarke (N)
Sgt M W McPhail (BA)
Sgt D A Turner (WOP)
Sgt A W Dearden (MU)
Sgt H A White (RG)
27/28 Aug
NURNBERG
2102-0429
FS J G McAlpine
Sgt E Vine
FS H W Harris
Sgt V A Martin
PO S Heald
Sgt H S Oldfield
Sgt B H Varey

31/1 Sep
BERLIN
2033-0355
Sgt E Willsher
Sgt T J Hurdiss
Sgt J E Gripton
Sgt R C Everest
Sgt B W Bell
Sgt R E Salter
Sgt J Fenswick
2/3 Sep
MINING-TEXEL
2017-2315
FO B C Fitch
Sgt J W Taylor (FE)
FO S A Jennings (N)
PO A Lyons (BA)
Sgt C Kershaw (WOP)
FS H W Pronger RAAF
Sgt L W Cromarty(RG) (MU)
3/4 Sep
BERLIN
2014-0337
Sgt E Willsher and crew
5/6 Sep
MANNHEIM
1953-0220
Sgt E Willsher and crew
6/7 Sep
MUNICH
1958-0417
PO E Willsher and crew
22/23 Sep
HANNOVER
1845-0007
PO E Willsher and crew
23/24 Sep
MANNHEIM
1857-0137
PO E Willsher and crew
(19 year-old Dick Willsher
and his crew, having flown
15 trips with No 61
Squadron, went to No 617
Squadron in October.
Willsher won the DFC after
completing a total of 31 ops)
27/28 Sep
HANNOVER
1948-0119
FS J G McAlpine and crew
29/30 Sep
BOCHUM
1821-2308
FS J G McAlpine and crew
1/2 Oct
HAGEN
1829-0024
FO H N Scott
Sgt A E Harris
PO S Halliwell
Sgt S R Knight
Sgt W C McDonald
Sgt E F Dunn
Sgt C A Haig
2/3 Oct
MUNICH
1840-0258
FO H N Scott and crew
Sgt S P Nicholas in, Sgt Dunn
out
4/5 Oct
FRANKFURT
1833-0104
PO F J McLean
Sgt W D French
Sgt A Pestell

Sgt B R Greatree
Sgt W Brown
Sgt R T Charles
Sgt R D Murdock
7/8 Oct
STUTTGART
2026-0312
FS J G McAlpine and crew
(four encounters with Ju88
night fighters)
8/9 Oct
HANNOVER
2249-0413
FO H N Scott and crew
Sgts F J Davis, D B Chalk and
G Allan in;
PO Halliwell, Sgts Dunn and
Haig out
18 Oct
HANNOVER
1729-2301
FO H N Scott and crew
Sgt F J Peack in, Sgt Haig
back;
Sgts Davis and Chalk out
20/21 Oct
LEIPZIG
1730-0101
FO B C Fitch and crew
22/23 Oct
KASSEL
1810-0024
FO H N Scott and crew
3 Nov
DUSSELDORF
1718-2211
FO B C Fitch and crew
10/11 Nov
MODANE
2026-0516
FO H N Scott and crew
Sgt J C Homewood in, Sgt
Peack out
18/19 Nov
BERLIN
1750-0111
FO B C Fitch and crew
Sgts G R Jeffrey and H S
Rosher in;
PO Lyons and FS Pronger out
22/23 Nov
BERLIN
1700-2355
FO H N Scott and crew
FL C T Fuller in, Sgt
Homewood out
23/24 Nov
BERLIN
1724-0016
FO H N Scott and crew
26/27 Nov
BERLIN
1743-0040
FO H N Scott and crew
2 Dec
BERLIN
1652-2121
PO D Paul
FS R A Coulter
FS R A Griffin
Sgt D H Miller
Sgt R F Brazier
Sgt S Billington
Sgt P McGibney
(aborted, loss of oil pressure
and power, jettisoned the
'cookie')

16 Dec
BERLIN
1622-2339
FO H N Scott and crew
FS R A Griffin in, FL Fuller out
20 Dec
FRANKFURT
1717-2256
FO B C Fitch and crew
23/24 Dec
BERLIN
0005-0741
FO H N Scott and crew
FL Fuller back, FS Griffin out
29 Dec
BERLIN
1654-2338
FO H N Scott and crew

1944
1/2 Jan
BERLIN
2353-0730
FO H N Scott and crew
2/3 Jan
BERLIN
2326-0656
FO B C Fitch and crew
Sgt L C Whitehead (MU) in,
Sgt Rosher out
5/6 Jan
STETTIN
0001-0837
PO V McConnell
Sgt T Powell
FO A J Watts
Sgt H S Vickers
Sgt W Surgey
Sgt G H Bradshaw
Sgt W J Throsby
14 Jan
BRUNSWICK
1647-2207
FO B C Fitch and crew
FS L B Rolton in, Sgt Jeffrey out
20 Jan
BERLIN
1620-2316
FL H N Scott and crew
FL J A McDonald in, Sgt
Knight out
21/22 Jan
BERLIN
2003-0237
FL H N Scott and crew
27/28 Jan
BERLIN
1734-0210
PO E A Stone
Sgt A Dick
FS J F Mills
Sgt W J Sinclair
Sgt T Francis
Sgt G E Cunningham
Sgt A Kane
28/29 Jan
BERLIN
2345-0738
FL H N Scott and crew
FS G A Leslie in, FL McDonald
out
30 Jan
BERLIN
1738-2324
FL H N Scott and crew
15 Feb
BERLIN

1656-2348
FL H N Scott and crew
Sgt J H Eastwood (2P)
FL McDonald back, FS Leslie
out
19/20 Feb
LEIPZIG
2327-0634
FL H N Scott and crew
FS Leslie back, FL McDonald
out;
FL J Breakey (MU) in, Sgt
Allen out
20/21 Feb
STUTTGART
2329-0701
FL H N Scott and crew
FL McDonald back, FS Leslie
out;
Sgt W Walker in, FL Breakey
out
24/25 Feb
SCHWEINFURT
2045-0414
FL H N Scott and crew
FS Leslie back, FL McDonald
out; FS H W Pronger in, Sgt
Walker out
25/26 Feb
AUGSBURG
1822-0203
FL H N Scott and crew
(the 'cookie' hung-up, only
dropped incendiaries)
1/2 Mar
STUTTGART
2315-0737
PO E A Stone and crew
10/11 Mar
CHATEAUROUX
1959-0135
PO E A Stone and crew
15/16 Mar
STUTTGART
1905-0311
PO E A Stone and crew
18/19 Mar
FRANKFURT
1915-0042
PO E A Stone and crew
PO D Carbutt (2P)
22/23 Mar
FRANKFURT
1857-0042
PO E A Stone and crew
24/25 Mar
BERLIN
1841-0135
PO E A Stone and crew
30/31 Mar
NURNBERG
2214-0627
PO E A Stone and crew
5/6 Apr
TOULOUSE
2021-0356
PO E A Williams
Sgt AS B Woodvine
FS D Bresley
FS A D Anderson
Sgt J L Parker
FS E Parry
Sgt S A Gardner
10/11 Apr
TOURS
2213-0313
PO E A Stone and crew

11/12 Apr
AACHEN
2037-0014
PO E A Stone and crew
18/19 Apr
JUVISY
2105-0109
PO E A Stone and crew
20/21 Apr
PARIS
2308-0355
PO E A Stone and crew
22/23 Apr
BRUNSWICK
2328-0456
PO E A Stone and crew
24/25 Apr
MUNICH
2044-0707
PO E A Stone and crew
26/27 Apr
SCHWEINFURT
2123-0628
PO E A Stone and crew
Sgt H W Cooper (2P)
FL R B G Murphy RAAF (N) in,
FS Mills out
28/29 Apr
ST MEDARD-EN-JALLES
2304-0641
PO E A Stone and crew
PO D Street (2P)
FO R P Kayser in, FL Murphy
out
29/30 Apr
ST MEDARD-EN-JALLES
2228-0544
PO E A Stone and crew
FL R B G Murphy in, FO
Kayser out
3/4 May
MAILLY-LE-CAMP
2204-0311
PO E A Stone and crew
FO J H Dyer in, FL Murphy
out
6/7 May
LOUAILLES
0036-0515
PO E A Stone and crew
Sgt Mills back, FO Dyer out
8/9 May
BREST
2139-0221
PO E A Stone and crew
FL J Breakey in, FS Pronger
out
10/11 May
LILLE
2205-0135
PO D Street
Sgt C F Waghorn
FS D K Grant
FS T W Boothby
Sgt P F Hadden
Sgt R Gilbert
Sgt T W Brown
19/20 May
TOURS
2229-0430
PO E A Stone and crew
Sgt T E Hunt (FE) in, Sgt Dick
out
21/22 May
DUISBURG
2225-0306
PO E A Stone and crew

Sgt D Dunkley in, Sgt Hunt
out
22/23 May
BRUNSWICK
2239-0428
PO R J Auckland
Sgt J Slome PO J Moran
FS D G Patfield
Sgt E V Jackson
Sgt J Miller
Sgt G R Chinery
24/25 May
EINDHOVEN
2310-0205
PO R J Auckland and crew
(ordered not to bomb due to
poor visibility, jettisoned
bombs over N.Sea)
27/28 May
NANTES
2310-0437
FO B S Turner
Sgt R Brown
WO G W James RCAF
FO E Jones RAAF
Sgt H Edwards
Sgt G W McDonell RCAF
Sgt N M Pettis RCAF
Jun
2337-0519
Sgt F E Hardy
FO B A Little
WO G McLaughlin
Sgt K G Merrifield
Sgt J F Blane
Sgt W I Quigley
19/20 Jun
WATTEN
2255-0031
FO B S Turner and crew
(aborted, recalled by con-
troller, jettisoned bombs)
21/22 Jun
GELSENKIRCHEN
2312-0406
FO B S Turner and crew
23/24 Jun
LIMOGES
2243-0516
FO B S Turner and crew
24/25 Jun
PROUVILLE/V1 SITE
FO B S Turner and crew
FL I M Pettigrew in, Sgt
Brown out
27/28 Jun
VITRY
2138-0538
FO B S Turner and crew
Sgt Brown back, FL Pettigrew
out
29 Jun
BEAUVOIR/V1 SITE
1200-1525
FO B S Turner and crew
4/5 Jul
ST LEU D'ESSERENT
2335-0320
FO B S Turner and crew
14/15 Jul
VILLENEUVE
2204-0508
FO B S Turner and crew
15/16 Jul
NEVERS
2226-0532
FO B S Turner and crew

18 Jul
CAEN
0350-0714
FO B S Turner and crew
18/19 Jul
REVIGNY
2305-0435
FO D G Bates
Sgt G Farrow
FS A N Hughes
PO G O Cameron
Sgt F W Cotton
PO J Fletcher
Sgt D Hancock
(did not bomb owing to
night fighter combat)
20/21 Jul
COURTRAI
2307-0225
FS W R McPherson
Sgt W J Scott
FO P P Brosko RCAF
Sgt R G McMillan RCAF
Sgt G H Postins
Sgt J A McFie RCAF
Sgt D T Currie RCAF
24/25 Jul
DONGES
2224-0558
FO C N Hill RCAF
Sgt C L Goss
FO D Hayley RCAF
FO G S Machen RCAF
Sgt M R Vagnolini
Sgt G R Mildren
Sgt J C O'Neill
25 Jul
ST CYR
1740-2212
FS A H Harrison RCAF
Sgt E A Walker
Sgt W D Beacon RCAF
Sgt J E Heffernan RCAF
Sgt R L Taylor
Sgt J M Fraser RCAF
Sgt J M Stewart
(damaged by flak)
26/27 Jul
GIVORS
2124-0640
FS A H Harrison and crew
28/29 Jul
STUTTGART
2213-0556
FS A H Harrison and crew
30 Jul
CAHAGNES
0604-1015
FS A H Harrison and crew
(aborted, ordered not to
bomb; brought bombs back)
31 Jul
RILLY LA MONTAGNE
FL B S Turner and crew
1725-2220
1 Aug
SIRACOURT/V1 SITE
1500-1830
FL B S Turner and crew
(mission abandoned as
instructed by controller due
to fog)
2 Aug
BOIS DE CASSEN/V1 SITE
1445-1915
FL B S Turner and crew
FS C Baldwin, (3AG)

5 Aug
ST LEU D'ESSERENT
1045-1525
FL B S Turner and crew
6 Aug
BOIS DE CASSEN/V1
0940-1425
FO E R Church
Sgt D Dunkley
FO S A Fleming
WO W Lewis
Sgt T Moffatt
Sgt F Kohut
FS A J Anderson
(mission aborted over target
as instructed)
7/8 Aug
SECQUEVILLE
2126-0131
FO N E Hoad
Sgt C S Webb (FE)
FO K O W Ball (N)
FO W H Pullen (BA)
Sgt G P Boyd (WOP)
Sgt N England (MU)
Sgt G Anslow (RG)
9/10 Aug
CHATELLERAULT
2045-0320
FO N E Hoad and crew
11 Aug
BORDEAUX
1201-1939
FO N E Hoad and crew
Sgt C V Embury in, Sgt
Anslow out
12/13 Aug
RUSSELSHEIM
2130-0315
FO N E Hoad and crew
14 Aug
BREST
1752-2248
WO D R Souter RCAF
Sgt H J Ockerby
Sgt J F Duncan
Sgt F Carling
Sgt C Darn
Sgt B McCormack
Sgt F Maudesley
15 Aug
GILZE RIJEN A/F
1008-1341
WO D R Souter and crew
16/17 Aug
STETTIN
2114-0505
FO N E Hoad and crew
18 Aug
BORDEAUX
1129-1620
FO J S Cooksey
Sgt D E Rose
Sgt E P Meaker
Sgt T Stevenson
Sgt S J Scarrett
Sgt A R Marshall
Sgt N W Salter
19 Aug
LA PALLICE
0526-1154
FO J S Cooksey and crew
25/26 Aug
DARMSTADT
2056-0535
WO D R Souter and crew

26/27 Aug
KONIGSBERG
2013-0708
WO D R Souter and crew
29/30 Aug
KONIGSBERG

2035-0722
FO N E Hoad and crew
(damaged in target area during night fighter action)
23 Oct
FLUSHING

1443-1756
FO L A Pearce
Sgt J B Murray
FS R B Pettigrew RAAF
Sgt D A Barker
Sgt A E Perry

Sgt A Barker
Sgt R Gillanders

ED888
MIKE SQUARED

Built as a Mk III by A. V. Roe at Manchester, initially equipped with four Merlin 28 engines, ED888 came off the production line early in 1943 and by 20 April it had arrived on No 103 Squadron, 'B' Flight, based at Elsham Wolds, Lincolnshire.

Given the squadron codes of PM, it became M-Mother when the individual aircraft letter went on, but the aircraft had, at least initially, no nose-artwork other than a steadily increasing number of bomb symbols. Mother flew on 54 ops with No 103 Squadron, the aircraft being 'awarded' a DFC when it completed its 50th trip in November 1943.

The initial part of ED888's career was fairly routine, the first time something out of the ordinary occurred was on a trip to Nurnberg on 27/28 August. The navigator's oxygen plug was accidentally pulled out as the wireless-operator went by his compartment, resulting in the navigator passing out. When he eventually came-to, he had lost all track of their position so they were forced to abort, jettisoning their bombs in the Channel.

ED888 went to Berlin six times with No 103 Squadron, being hit by flak on one of them and having to abort another when the starboard outer engine packed-up. It also had encounters with night fighters. During a raid on Mannheim on 23/24 September, the mid-upper spotted one with a nose light, but it quickly went out and the fighter disappeared when he fired at it. A FW190 approached them on 20/21 October but again a burst from the mid-upper turret sent the fighter off into the darkness.

The aircraft's regular skipper during her first months with the squadron was Sergeant, later Pilot Officer, Denis W. Rudge, who flew it on 23 ops and received the DFC; the navigator, George Lancaster, was given the DFM. The mid-upper gunner was Charlie Baird:

'We were an all-NCO crew so ate, slept, flew and imbibed together, and we had a very good groundcrew whom we valued very highly. They were Bert Booth, fitter; Bob Draper, rigger; Tom Gean, aircraft hand. They were very protective of ED888 and didn't care to see a few holes in her. We were a very alert crew and were never complacent from start to finish of our tour. We had no idle chit-chat from take-off to return and when we spoke it was all business. I'm certain this led to our success in surviving.

'On several occasions when I saw an enemy aircraft, immediately Den went into action; he was a great pilot and only 19 years old. Our theory was that if the German pilot broke off his attack on realising we were awake, then he would go off and look for someone less alert. One night we took a Wing Commander Slater as second dickie and on landing he told our skipper we were too damned quiet and had no animation. Den told us of his remarks and only said to us, "bugger the animation", let's carry on as we are!

'Very few crews were finishing their 30 ops at this period, many coming off early, a lot going down on the 28th or even 30th. We came to our 30th on 23 August 1943 and we were briefed for the Big City - Berlin! What a hot-bed to finish on; the adrenalin was flowing freely. This is how that evening went.

'Main meal 7pm, final briefing 7.30, driven out to "Mother of them all", engines tested and general last minute checks, then shut down engines. It was while we were lying on the hard-standing with the groundcrew having a last smoke and idle chatter as other crews were testing their aircraft that one, about 500yds from us, had a negative earth and its bombs fell off on trying to start engines.

'As soon as we saw the first incendiaries catch alight we all dashed for the slit-trench which hadn't been used for months. I was first in with all the others on top of me. We were barely in when the aircraft blew up, a h*** of a blast of hot air and debris began dropping about us.

'Out we got, skip and engineer ran around checking for damage, none apparent, so all aboard. At this time I became aware of a dreadful odour, and there it was; someone had used the trench as a toilet and I had landed on it. So out with knife and cut the right knee out of my Sidcot flying suit, then giving it the heave-ho down the flare chute.

'By now – 8.25 – we were taking off and were wondering if this was all a bad omen for our last trip. We all had very mixed thoughts. Anyhow, M-Mother took us to Berlin and whilst many aircraft went down around us, we had a charmed run; 58 aircraft were lost on this raid.

'On our return we were all very sad at the thought of never flying ED888 again. All we could do was to give her a pat and while it may sound daft, there were lumps in some of our throats.

'Her first op was on 4 May 1943, Nick Ross being the pilot, a Scot like me, and a few weeks later, while we were on leave, another Scottish pilot, Bob Edie, took her on a raid, so she had a good Scots influence during her first 32 trips: a great aircraft. She should be in some museum today, rather than being broken up as she was.'

What turned out to be its last sortie with No 103 Squadron – at least in 1943 – was a raid on Berlin on 26/27 November, with Flying Officer G. S. 'Taffy' Morgan bringing the Lanc safely home. Gomer Morgan received the DFC in 1944, 15 of his ops being flown in ED888.

The aircraft was then assigned to No 576 Squadron which had been formed from 'C' Flight of No 103 Squadron on 25 November. Thirteen crews went from No 103 Squadron, including Morgan and crew, while another four came from 101, but the new squadron also operated from Elsham Wolds. ED888's fuselage letters were changed to UL-V, becoming V-Victor.

With a nominal gender change, the first trip with 576 was back to Berlin and ED888 did another eight trips to the Big City. When attacked by a Dornier Do217 on one of them, 28/29 January 1944, its gunners damaged the night fighter.

On the night 29/30 December 1943, when going to Berlin, ED888 took a Mr (Sergeant) Benjamin Frazier along, a US Army war correspondent. His account of the raid later appeared in the *Yank* magazine (see Appendix A).

During March 1944 ED888 was once more given the letter 'M' – this time becoming Mike. In fact all 576's aircraft were marked with a small '2' beside the individual letter, so it was not long before ED888 became known as 'Mike Squared', although it also became known as 'The Mother of them All' once its total ops began to rise and rise.

ED888's first recorded trip as M2 was to Essen on 26/27 March 1944, but on crossing the North Sea the starboard inner packed-up and while the engineer was trying to feather it, he feathered another engine by mistake. Consequently the aircraft lost a lot of height before he got the third engine going again, and once the pilot had regained the lost height he decided to continue on to the target. Their troubles were not over for just outside the target area they were attacked by a Ju88 and badly shot-up, but managed to loose the night fighter. Then they were attacked by another '88 but no more damage was sustained.

Recrossing the enemy coast, a second engine failed, but asking for help they were located by a Mosquito that escorted them back over the sea and to the RAF fighter base at West Malling in Kent where they landed safely.

Back on strength in April, the Lanc's usual skipper became Pilot Officer Jimmy Griffiths who flew Mike on 29 of its 30 ops between then and the end of July when he finished his tour, including Mike's 99th op.

Jimmy Griffiths, a Scot, soon discovered that ED888 was the oldest, most clapped-out aircraft of the squadron and asked his Flight Commander if he could get another aircraft. A few days later they received a new one only to experience a problem with it on a test-flight causing them to bale out. So then it was back to ED888. It became 'their' aircraft and in fact nobody else wanted to fly it; even when they were on leave, nobody operated in it!

Griffiths' rear-gunner, a Welshman named D. W. 'Taffy' Langmead, shot down two German night fighters, a Me410 from point-blank range on 24/25 June, then a Ju88 on their 30th and last trip while on the bomb-run over Revigny marshalling yards on 14/15 July. On this occasion they had just got a message to abort but the bomb-aimer, Charlie Bint, had the target in his sights and Jimmy Griffiths told him to bomb. Just then a Ju88 flew right across them and came in behind. Taffy and the '88 opened fire at the same time, but the 88's fire stopped abruptly and the fighter spun down into some cloud.

Pilot Officer James B. Bell was flying his first op with Griffiths this night, ED888's 99th, then took over Mike, doing 32 straight trips, including the 100th on 20 July. By October when he completed his tour he too received the DFC. Shortly after its 100th op, Mike had the ribbon of the DSO painted on by the Station Commander (Mike already had the DFC), and by now the aircraft also carried two swastikas for the night-kills. Above each was the name 'Taffy', for Taffy Langmead who had scored them.

On the last day of October, Mike returned to No 103 Squadron, when No 576 Squadron moved to Fiskerton, being coded PM-M2. It had now flown a total of 131 sorties, marked in two lots of five, 10-bomb rows. Mike was to complete a further nine trips with 103, the additional bomb symbols initially being marked with a row of 10, then below that eight, below that six, then four, two and finally one, which made the 131 while with No 576 Squadron. When completing the last nine with 103, the rows were 'filled-in' to make it three rows of 10, one of six, one of two and finally one: total 139. It is not certain how or if the 140th bomb was marked. When the Lanc had done 131 ops, its DFC ribbon had the rosette of a bar painted on it.

Mike became Cat AC on 26 January 1945 and went to Avro's for an overhaul and although it returned to 103 on 3 February it did not fly on further ops prior to becoming Cat B on the 20th. By early August Mike was at No 10 MU and finally Struck off Charge on 8 January 1947 and reduced to scrap.

1943
No 103 Squadron
4/5 May
DORTMUND
2202-0302
WO N R Ross
FS J A B Cooper
Sgt L McLellan
Sgt G Hickson
Sgt A J S Girling
Sgt T A Platt
Sgt M R Tuxford
12/13 May
DUISBURG
0010-0450
Sgt D W Rudge
Sgt G Lancaster (N)
Sgt T W Catton (BA)
Sgt H T Greenwood
Sgt S T Robinson (FE)
(WOP)
Sgt C R Baird (MU)
Sgt J D Kirkpatrick
13/14 May
BOCHUM
2336-0521
Sgt D W Rudge and crew
23/24 May
DORTMUND
2250-0358
Sgt D W Rudge and crew

25/26 May
DUSSELDORF
2339-0447
Sgt D W Rudge and crew
Sgt D Williamson in, Sgt Catton out
27/28 May
ESSEN
2302-0344
Sgt D W Rudge and crew
29/30 May
WUPPERTAL
2250-0406
Sgt D W Rudge and crew
Sgt R P Orme in, Sgt Wilkinson out
11/12 Jun
DUSSELDORF
2304-0333
Sgt D W Rudge and crew
12/13 Jun
BOCHUM
2235-0341
Sgt D W Rudge and crew
14/15 Jun
OBERHAUSEN
2222-0310
Sgt D W Rudge and crew
16/17 Jun
COLOGNE
2236-0337

Sgt R J Edie
Sgt E H J Suarez
PO J C Maxwell
Sgt W Fawley
Sgt E S Boorman
Sgt E C Benham
Sgt J May
21/22 Jun
KREFELD
2305-0317
SL T O Prickett DFC
FL W C Longstaff
Sgt L Pulfrey
Sgt S Foster
Sgt J Terrans
Sgt W J Miller
FS J L Betty
22/23 Jun
MULHEIM
2302-0323
Sgt D W Rudge and crew
24/25 Jun
WUPPERTAL
2239-0340
Sgt D W Rudge and crew
25/26 Jun
GELSENKIRCHEN
2244-0328
Sgt D W Rudge and crew
28/29 Jun
COLOGNE

2258-0350
WO E J Presland
FS L D Groome
Sgt G Aitken
Sgt S G Staplehurst
Sgt C D Hornby
Sgt E R Foster
FS E E Piper
3/4 Jul
COLOGNE
2237-0401
Sgt D W Rudge and crew
8/9 Jul
COLOGNE
2219-0400
Sgt D W Rudge and crew
9/10 Jul
GELSENKIRCHEN
2235-0425
Sgt D W Rudge and crew
12/13Jul
TURIN
2207-0338
Sgt D W Rudge and crew
24/25Jul
HAMBURG
2207-0338
Sgt D W Rudge and crew
25/26 Jul
ESSEN
2155-0222

Sgt D W Rudge and crew
27/28 Jul
HAMBURG
2211-0315
WO E J Presland and crew
29/30 Jul
HAMBURG
2246-0539
Sgt H Campbell
Sgt M Hartley
Sgt D J McGrath
Sgt T W Moore
Sgt J J Robshaw
Sgt C O'Neill
Sgt W H Chambers
2/3 Aug
HAMBURG
2355-0543
Sgt H Campbell and crew
7/8 Aug
TURIN
2053-0505
Sgt D W Rudge and crew
9/10 Aug
MANNHEIM
2309-0450
Sgt D W Rudge and crew
10/11 Aug
NURNBERG
2146-0459
Sgt D W Rudge and crew
WC Slater (2P)
12/13 Aug
MILAN
2108-0548
Sgt D W Rudge and crew
15/16 Aug
MILAN
2001-0423
Sgt D W Rudge and crew
22/23 Aug
LEVERKUSEN
2101-0306
Sgt D W Rudge and crew
FS M McMahon (2P)
23/24 Aug
BERLIN
2024-0306
Sgt D W Rudge and crew
(Rudge completed his tour in
September, went to No 1667
CU and received the DFC on
being commissioned)
27/28 Aug
NURNBERG
2132-0201
FS M McMahon
Sgt W P S Hersham
FO A G MacDonald
Sgt A R Fleming
Sgt H K Grant
Sgt G E Crawford
Sgt T Thompson
(aborted when navigator lost
their position after
passing-out)
31/1 Sep
BERLIN
2029-0428
FS H Campbell and crew
FO R E Ault and Sgt P W
Alderton in;
Sgts McGrath and Robson
out
(aircraft hit by flak)
3/4 Sep
BERLIN

2001-2200
FS H Campbell and crew
(aborted; starboard outer
failed)
5/6 Sep
MANNHEIM
2000-0250
FS H Campbell and crew
Sgt McGrath back, FO Ault
out
6/7 Sep
MUNICH
2006-0405
FS H Campbell and crew
22/23 Sep
HANNOVER
1918-0019
PO M P Floyd
Sgt R H Mansfield
Sgt E Benroy?
Sgt J W Brewster
Sgt C J Fuller
Sgt L Marsh
Sgt W Watters
23/24 Sep
MANNHEIM
1913-0145
Sgt T Gallacher
Sgt T Dixon
FS G Beike
Sgt H Crook
Sgt J A Bonsonworth
Sgt F C Read
Sgt F Child
27/28 Sep
HANNOVER
1944-0118
Sgt T Gallacher and crew
29 Sep
BOCHUM
1845-2308
L F J Hopps
Sgt R S Imeson
PO N Olsberg
Sgt F J Roberts
Sgt R Thomas
Sgt N James
Sgt R E Black
1/2 Oct
HAGEN
1856-0024
FS M J Graham
Sgt G P Rae
Sgt W J C Keigwinarea
Sgt W J Condick
Sgt P Harris
Sgt D Roberts
Sgt P J Daly
4 Oct
LUDWIGSHAVEN
1839-2203
Sgt B B Lydon
Sgt E Buxton
Sgt E Benroy
Sgt T C Forster
Sgt S J Edwards
Sgt J A Brewster
Sgt A P Collins
(returned early when star-
board outer failed)
7/8 Oct
STUTTGART
2041-0315
FS H Campbell and crew
8/9 Oct
HANNOVER
2315-0359

FS H Campbell and crew
18 Oct
HANNOVER
1737-2301
FO G S Morgan
Sgt J R Mearns (N)
FS N A Lambell RAAF
PO E M Graham (FE)
Sgt J R O'Hanlon (WOP)
Sgt S S Greenwood (MU)
Sgt C E Shilling (RG)
20/21 Oct
LEIPZIG
1742-0050
FO G S Morgan and crew
(attacked by FW190, no dam-
age)
22 Oct
KASSEL
1815-2109
FO G S Morgan and crew
(abandoned sortie owing to
navigator experiencing oxy-
gen failure at Dutch coast)
3 Nov
DUSSELDORF
1722-2146
FO G S Morgan and crew
10/11 Nov
MODANE
2035-0454
FO G S Morgan and crew
18/19 Nov
BERLIN
1740-0135
FO G S Morgan and crew
22 Nov
BERLIN
1702-2335
FO G S Morgan and crew
26/27 Nov
BERLIN
1719-0135
FO G S Morgan and crew
No 576 Squadron
3/4 Dec
LEIPZIG
0000-0720
WO C C Rollins
FS J R Henningham
Sgt E D Roff
FO H L Rees
Sgt J Rutter
Sgt R Hammond
Sgt L S Sumak
Sgt M A Frost
16 Dec
BERLIN
1625-2339
FO G S Morgan and crew
Sgt A C Blackie (2P)
Sgt D Roberts in, Sgt
Greenwood out
20 Dec
FRANKFURT
1703-2246
FO G S Morgan and crew
Sgt R E Rogers in, Sgt Roberts
out
23/24 Dec
BERLIN
0020-0735
FO G S Morgan and crew
Sgt B Frazier, US War
Correspondent
Sgt R Harris in, Sgt Rogers
out

29 Dec
BERLIN
1700-2330
FO G S Morgan and crew
Sgt A Newman in, Sgt Harris
out

1944
1/2 Jan
BERLIN
0020-0745
Sgt D G Mann
Sgt J Anderson
Sgt R Mosley
FO B N J Price
Sgt F D Robbins
Sgt P T Lalor
Sgt R McManus RCAF
2/3 Jan
BERLIN
2330-0800
FO G S Morgan and crew
FO J M Shearer RNZAF (2P)
FO K J Risi RCAF in, Sgt
Newman out
5/6 Jan
STETTIN
0000-0930
FL C A B Johnson
Sgt G B Valetine
FO H Gerus RCAF
Sgt N H Morris
Sgt R W Owen
Sgt J P Duns
14 Jan
BRUNSWICK
1635-2220
FO G S Morgan and crew
20 Jan
BERLIN
1610-2325
FO G S Morgan and crew
Sgt Greenwood back, FO Risi
out
21/22 Jan
MAGDEBURG
1940-0230
FO G S Morgan and crew
28/29 Jan
BERLIN
PO E H Childs
Sgt V E T White
Sgt R Johnstone RCAF
Sgt E Bardsley
Sgt H R Bowles
Sgt C M Brewster
Sgt C A Gifford
(combat with Do217 night
fighter)
(Edward Childs and his crew
failed to return from Berlin
the following night and it
appears that all except Sgt
White were killed)
30 Jan
BERLIN
1704-2345
Sgt R R Read
Sgt A Taylor
Sgt M A Sarak RCAF
FO G Hallows
FO W Murphy
Sgt G A Coon RCAF
Sgt Hodson
(Read was later commis-
sioned and won an immedi-
ate DSO in May 1944)

76

24/25 Feb
SCHWEINFURT
1820-0220
FL P E Underwood
Sgt R J A Boon
FO E C Espley
Sgt J A Hildreth
Sgt A E Evans
Sgt H R Lawrence
Sgt L Washer
25/26 Feb
AUGSBURG
1820-0220
FL P E Underwood and crew
18/19 Mar
FRANKFURT
1910-0125
FS V A Sheerboom
Sgt H R Piper
FS R J Tinsley
FO W Woodfine
Sgt C E Harris
Sgt L V C LaBelle
Sgt J G DelaMothe
26/27 Mar
ESSEN
2000-0115
FS C G Wearmouth
Sgt D R Willis
Sgt J W Carter
FO H T Wilson
Sgt A MacDonald
Sgt J Graham
Sgt S J Bott
(flew to target on three
engines and attacked by
night fighters)
24/25 Apr
KARLSRUHE
2150-0420
PO J S Griffiths
Sgt J D Hawkeswood (FE)
Sgt D C Bint (BA)
Sgt T Atherton (N)
Sgt W J McCarthy(WOP)
Sgt T Jago (MU)
Sgt D W Langmead (RG)
26/27 Apr
ESSEN
2307-0356
PO J S Griffiths and crew
27/28 Apr
FRIEDRICHSHAVEN
2144-0650
PO J S Griffiths and crew
3/4 May
MAILLY-LE-CAMP
2222-0422
PO J S Griffiths and crew
7/8 May
RENNES ST JACQUES
2145-0305
PO J S Griffiths and crew
Sgt D S Lowe in, Sgt Jago out
9/10 May
MARDYCK
2240-0140
PO J S Griffiths and crew
Sgt S S Greenwood in, Sgt
Lowe out
11/12 May
HASSELT
2206-0156
PO J S Griffiths and crew
WO R S Pyatt RCAF in, Sgt
Greenwood out

19/20 May
ORLEANS
2200-0255
PO J S Griffiths and crew
Sgt Jago back, WO Pyatt out
21/22 May
DUISBURG
2235-0325
PO J S Griffiths and crew
22/23 May
DORTMUND
2230-0330
PO J S Griffiths and crew
FL C F Jenkinson RCAF in, Sgt
Bint out
24/25 May
AACHEN
0020-0440
PO J S Griffiths and crew
Sgt Bint back, FL Jenkinson
out
27/28 May
AACHEN
2359-0445
PO J S Griffiths and crew
6/7 Jun
VIRE
2220-0345
PO J S Griffiths and crew
9/10 Jun
FLERS A/F
0025-0545
PO J S Griffiths and crew
12/13 Jun
GELSENKIRCHEN
2240-0315
PO J S Griffiths and crew
14/15 Jun
LE HAVRE
2025-0025
PO J S Griffiths and crew
16/17 Jun
STERKRADE
2305-0350
PO J S Griffiths and crew
17/18 Jun
AULNOYE
2345-0410
PO J S Griffiths and crew
22 Jun
MIMOYECQUES/V1 SITE
1350-1715
PO J S Griffiths and crew
23/24 Jun
SAINTES
2145-0525
PO J S Griffiths and crew
24/25 Jun
FLERS/V1 SITE
0125-0515
PO J S Griffiths and crew
(attacked by Me410 night
fighter which was shot
down)
27/28 Jun
CHATEAU BENAPPE/
V1 SITE
0115-0450
PO J S Griffiths and crew
29 Jun
DOMLEGER/V1 SITE
1130-1515
PO J S Griffiths and crew
30 Jun
OISEMONT/
NEUVILLE/V1 SITE
0540-0940

WO H D Murray
Sgt C C Addams
Sgt R Lee
Sgt D A Barnes
Sgt F Thackeray
Sgt P J Taylor
Sgt W O Kenyon
(this crew failed to return on
26 August, flying their 29th
op. Harold Murray and all but
Sgt Addams were killed)
2 Jul
DOMLEGER/V1 SITE
1155-1540
PO J S Griffiths and crew
4/5 Jul
ORLEANS
2159-0400
PO J S Griffiths and crew
7 Jul
CAEN
1920-2230
PO J S Griffiths and crew
12/13 Jul
REVIGNY
2115-0700
PO J S Griffiths and crew
14/15 Jul
REVIGNY
2100-0525
PO J S Griffiths and crew
PO J B Bell (2P)
(attacked by Ju88 which was
shot down)
20 Jul
WIZERNES
1925-2255
PO J B Bell
Sgt H C Gore (FE)
Sgt T E Seabrook (BA)
Sgt R Hughes (N)
Sgt R E Badger (WOP)
Sgt S G Parry (MU)
Sgt P Turton (RG)
23/24 Jul
KIEL
2245-0355
PO J B Bell and crew
30 Jul
CAHAGNES
0645-1035
PO J B Bell and crew
31 Jul
LE HAVRE
1810-2140
PO J B Bell and crew
(mission aborted due to
cloud over the target)
3 Aug
TROSSY-ST-MAXIM
1140-1615
PO J B Bell and crew
5 Aug
BLAYE
1430-2230
PO J B Bell and crew
7/8 Aug
FONTENAYE
2110-0115
PO J B Bell and crew
10 Aug
DUGNY
0910-1415
PO J B Bell and crew
11 Aug
DOUAI

1345-1745
PO J B Bell and crew
25 Aug
RUSSELSHEIM
2005-2355
PO J B Bell and crew
(aborted due to rear turret
failure)
26/27 Aug
KIEL
2000-0200
PO J B Bell and crew
29/30 Aug
STETTIN
2055-0540
PO J B Bell and crew
31 Aug
AGENVILLE
1300-1725
PO J B Bell and crew
3 Sep
EINDHOVEN A/F
1535-1920
PO J B Bell and crew
5 Sep
LE HAVRE
1625-1955
PO J B Bell and crew
6 Sep
LE HAVRE
1705-2035
PO J B Bell and crew
8 Sep
LE HAVRE
0620-0920
PO J B Bell and crew
10 Sep
LE HAVRE
1630-2020
PO J B Bell and crew
12/13 Sep
FRANKFURT
1820-0135
PO J B Bell and crew
16/17 Sep
LEEUWARDEN A/F
0025-0355
PO J B Bell and crew
17 Sep
FLUSHING
1623-1910
PO J B Bell and crew
20 Sep
CALAIS
1455-1820
PO J B Bell and crew
23 Sep
NEUSS
1825-2345
PO J B Bell and crew
24 Sep
CALAIS
1645-1200
PO J B Bell and crew
27 Sep
CALAIS
0850-1200
PO J B Bell and crew
3 Oct
WESTKAPELLE
PO J B Bell and crew
1300-1600
4/5 Oct
SAARBRUCKEN
1810-0045
PO J B Bell and crew

14 Oct
DUISBURG
2235-0340
PO J B Bell and crew
14/15 Oct
DUISBURG
2235-0340
PO J B Bell and crew
19/20 Oct
STUTTGART
2120-0400
PO J B Bell and crew
22 Oct
ESSEN
1605-2200
FL H Leyton-Brown
Sgt R A Hawkins
FS G Peterson
Sgt L Peters
Sgt E Johnson
Sgt J P McCullen
Sgt G Lester
25 Oct
ESSEN
1250-1755
FO R N Crowther
Sgt R F H Wilshire
Sgt W Thorpe
Sgt H L Basson
Sgt B T Wilmot

Sgt B Rink RCAF
FS J R Scarfe RCAF
No 103 Squadron
6 Nov
GELSENKIRCHEN
1146-1712
PO A J Mosley RCAF
Sgt P L Thompson
Sgt G Marriott
FO K C Dunn RCAF
Sgt R J Evans
Sgt L A Wostenholme
Sgt R E Ward
9 Nov
WANNE EICKEL
0741-1314
FO A V J Vernieuwe (Belg)
Sgt W N Wells (FE)
FO R H Seaton RCAF(BA)
FS G H A Othem (N)
FS W J M Baillie RAAF
Sgt G H Relf (MU)
Sgt T I Quinlan (RG)
16 Nov
DUREN
1304-1745
FL R A Butts RCAF
Sgt A D Berndt
FS B C McGregor RCAF
FO C E Kramer RCAF

Sgt D H Penny RCAF
FS W D Stuckey
FS J R Murphy
18 Nov
WANNE EICKEL
1551-2222
WO W J McArthur RCAF
Sgt H Canoy
FS M H Horne RCAF
Sgt M Greenstein RCAF
FO R J Loughead
Sgt D J McAuley RCAF
Sgt D F Campbell
21 Nov
ASCHAFFENBURG
1544-2212
PO G W Henry RAAF
Sgt K Foster
Sgt M M Bertee RAAF
FO H S Mitchell RAAF
FS K C McGinn RAAF
FS H J Porter RAAF
FS J W Grice RAAF
27 Nov
FREIBURG
1548-2303
FS J C Cooke RCAF
Sgt E W McGrath
FO G T Mortimore
FS J A Goff RCAF

WO F R Hill RCAF
FS J M C McCoubrey RCAF
FS M O Orr RCAF
29 Nov
DORTMUND
1152-1742
FL R A Butts and crew
4 Dec
KARLSRUHE
1641- 2304
FO A S Thompson
Sgt R C Pain
Sgt J M Peace
Sgt W H Tromp
Sgt A J Crampin
FS J C Rochester RCAF
FS D G Kyle
24 Dec
COLOGNE
1448-2100
FL S L Saxe RCAF
Sgt J J Bent
FS J A Wright
FO H Shatsky RCAF
WO M D Benadetto RCAF
Sgt A C Clark
FS R C Snell RCAF

ED905
FOX/AD EXTREMUM

A Manchester-built Mk III, ED905 came off of the production line in early 1943 and was assigned to No 103 Squadron at Elsham Wolds, Lincolnshire, the day after ED888 went there – 21 April 1943. Given the squadron code letters PM and individual letter X-X-ray it began operations in May.

Her first captain was a Belgian pilot – one of the relatively few who flew in Bomber Command – Flying Officer F. V. P. van Rolleghem, so it is no surprise to note that during the aircraft's period with No 103 Squadron, X-ray had crossed Belgian and British flags painted on its nose, the artist being LAC John Lamming. Two rows of bombs were painted beneath the cockpit, the third symbol being a mine on a parachute, denoting a 'Gardening' sortie.

. Having taken ED905 on its first raid, Van Rolleghem became the Lanc's regular pilot over those first months with No 103 Squadron. In fact of ED905's 26 trips with 103 he took it on 21 of them.

Florent Van Rolleghem was 30 years-old and had joined the Belgian Air Force pre-war. He escaped to England in 1941 to become a heavy-bomber pilot with the RAF and joined No 103 Squadron at the end of April 1943. He won the DFC and Belgian Croix de Guerre with the squadron and completed his first tour in September 1943, then went on to complete a second tour totalling 42 ops in all, winning the British DSO. He eventually rose to become a Lieutenant-General in the Belgian Air Force in 1970. He died in 1983.

ED905 certainly had a number of incidents during its period of service. On the seventh mission it was hit by heavy flak which damaged the port inner engine and oil system, while putting holes in the bomb-doors, undercarriage, wings and rudders. Six trips later – the 13th – ED905 was hit by flak again, holing several parts of its airframe.

When C Flight of No 103 Squadron became No 166 Squadron after the latter re-equipped with Lancasters in September 1943, ED905 was re-assigned to the new unit, changing its squadron codes to AS but retaining the letter 'X'. The aircraft had a quiet start but soon the adventures with enemy defences and other problems started once again.

Over Dusseldorf on 3/4 November the Lanc was attacked by a night fighter sustaining considerable damage and, although both gunners opened up, they could make no claims. Returning to duty after this damage had been repaired, its next trip was to Berlin on 29 December, with Pilot Officer J. Horsley in command, flying his own 22nd op. Forty miles inside the Dutch coast the starboard inner engine failed but they carried on. Upon reaching Berlin they found the bomb-doors would not open, so they had to make a second bomb-run but again the bombs failed to release, so finally they jettisoned them.

Over the Big City the port engine was then hit by flak, caught fire and had to be feathered, while the artificial horizon and directional gyro both went u/s. Horsley had to drop to below 5,000ft to fly home and although they were worried by flak and searchlights when they got back to the Dutch coast they got home – two hours after their ETA. Joe Horsley and his navigator Ken Cornwell both received immediate DFCs for their efforts. These two men along with the rest of the crew failed to return from another Berlin raid on 28 January 1944 (in ND382) flying on their 29th sortie, both being killed.

Oddly enough, the Berlin raid of 28 January was when ED905 was next on the duty roster after repairs, Flight Sergeant R. B. Fennell completing this trip in the aircraft, but he then failed to return on 30/31 March – the ill-fated Nurnberg raid when 96 Bomber Command aircraft were lost. Roy Fennell was killed too.

The next trip – yet another to Berlin – on 30 January, had to be abandoned as both turrets went u/s, and a raid on Stuttgart on 1 March was also aborted, due to losing the starboard outer engine. In between, on 24/25 February, Flight Sergeant R. V. F. Frandson and crew had to jettison the 'cookie' as they were unable to gain operational height, but carried on afterwards and dropped their incendiaries on the target. On 3/4 May, shortly before bombing the military camp and dump at Mailly-le-Camp, with Pilot Officer P. J. Wilson DFC in command, they were attacked by a Me110 night fighter. They had just been ordered to orbit the rendezvous point by the controller when the '110 was spotted by the rear-gunner, Sergeant C. G. M. Meadows. He ordered the captain to dive to port and both bomber and fighter opened fire simultaneously – the mid-upper, Pilot Officer B. W. G. Felgate joining in. The '110 was trying to follow the diving curve but after more gunfire it was seen to pour out black smoke, going down out of control and on fire to crash below.

Due for a major service ED905 went to No 54 MU on 26 May but on completion was assigned, on 10 June 1944, to No 550 Squadron where the aircraft became (BQ) F-Fox (shortened from the more usual F-Foxtrot), the previous 'F' (ME556) having failed to return from a sortie on D-Day. By this time the Lanc had undergone a new paint-job, the twin flags and original bomb log being painted out. The flags were replaced by a colourful crest and the motto 'Ad Extremum' beneath it. Unhappily the origins like many others have either not survived or are unknown to anyone except those who originated them during he war. However, photographs of the crest show two foxes' heads (obviously referring to the phonetic name), a female face and a foaming pot of ale. The latter two were undoubtedly often in the minds of young aircrew. There also appears to be a shield with the cross of St

Andrew upon it and if this is correct then there must have been a Scottish influence somewhere.

Fox's 'new' bomb symbols were now marked in rows of 15 under the cockpit – back-dated but omitting the parachute mine. The 45th op would indicate the Lanc's first daylight sortie as it is in a lighter colour, presumably the usual yellow, with red being reserved for the night sorties. Although by the time Fox had done 70 trips, its other lighter-painted bombs do not correspond with the daylight raids it went on. Above the rows of bombs was also stencilled a RAF catch-phrase of the period 'Press on Regardless'.

The 45th operation and the first daylight trip was an attack on a V1 site in the Pas de Calais, flown on 22 June. On its next raid Fox lost the port outer engine and the Gee apparatus but the crew carried on and bombed. On their return the port inner began to run rough but they made it back.

Fox's next light and dark coloured bombs, denoting day and night ops, do not appear to tie-up, but the day ops were against V1 rocket sites in France. Whatever was happening, the bombs were increasing regularly over the summer and Fox's more-or-less regular pilot was Pilot Officer D. A. Shaw.

In all ED905 is supposed to have flown 100 sorties – the 100th being noted in some records as being flown on 4 November 1944, with Flying Officer V. B. Ansell at the controls, the target Bochum. However, there is a photograph of Shaw sitting in Fox's cockpit with 99 bombs on the nose and another picture of the Lanc supposedly taking-off for her 100th trip – with David Shaw – which is also dated 4 November. Clearly both pilots didn't take Fox to Bochum and it is more likely that Shaw would have done so. It would seem that the 100th op was in fact flown on 2 November – to Dusseldorf – with Shaw in command, but taking a new crew that night. Shaw was at the end of his tour (he received the DFC soon afterwards) and no doubt would have liked, if not insisted, that he take ED905 on its 100th, and one way would have been to take a sprog crew on one of their first trips. The 4 November raid to Bochum is noted as being flown by aircraft 'F', but the serial number shows it to be NG250, which could have been a replacement 'F'.

Retired from operational flying, ED905 went to No 1 LFS on 4 November (note the date) then, on 10 November, to No 1656 CU. Here the aircraft spent its last days, bringing its flying hours to over 628, but on 20 August 1945 a pilot allowed the old Lanc to swing on landing, the undercarriage collapsed and a crash followed. Written-off as at that date ED905 passed into history.

1943	PO R K McLeod	**23/24 May**	Sgt A S Wheeler
No 103 Squadron	**13/14 May**	DORTMUND	Sgt R H J Rowe
12/13 May	BOCHUM	2225-0315	FS P J F Vivers
DUISBURG	2337-0504	FO F V P van Rolleghem and	Sgt W C Gillespie
2352-0510	FO F V P van Rolleghem and	crew	**11/12 Jun**
FO F V P van Rolleghem	crew	**25/26 May**	DUSSELDORF
Sgt G H Agar	**18/19 May**	DUSSELDORF	2300-0341
Sgt W Carling	MINING-BIARRITZ	2341-0418	FO F V P van Rolleghem and
Sgt T Proctor	2152-0553	FO D W Finlay	crew
Sgt P O Vickers	FO F V P van Rolleghem and	Sgt J H McFarlane	**12/13 Jun**
Sgt R White	crew	Sgt I D Fletcher	BOCHUM

2229-0353
FO F V P van Rolleghem and
crew
(hit and damaged by heavy
flak)
14/15 Jun
OBERHAUSEN
2207-0411
FO F V P van Rolleghem and
crew
16/17 Jun
COLOGNE
2210-0400
FO F V P van Rolleghem and
crew
21/22 Jun
KREFELD
2259-0310
FO F V P van Rolleghem and
crew
22/23 Jun
MULHEIM
2258-0314
FO F V P van Rolleghem and
crew
24/25 Jun
WUPPERTAL
2259-0419
FO A H Langille
PO E L G Grant
PO C B Reynolds
PO D Powers
Sgt R L Hollywood
Sgt G J Wallis
PO J H Addison
25/26 Jun
GELSENKIRCHEN
2300-0330
FL F V P van Rolleghem and
crew
(hit and damaged by heavy
flak)
7/8 Aug
TURIN
2052-0515
FL F V P van Rolleghem and
crew
9/10 Aug
MANNHEIM
2329-0558
Sgt J E James
Sgt W G Bell
Sgt D J Edwards
Sgt E M L Davis
Sgt J J Robshaw
Sgt A V Collins
Sgt W W O'Malley
10/11 Aug
NURNBERG
2223-0607
Sgt J E James and crew
12/13 Aug
MILAN
2126-2356
PO R Atkinson
Sgt G F Brice
Sgt R J H Littleton
Sgt A L Norman
Sgt H K Garewal
Sgt C Campbell
FS R L Taggart
4/15 Aug
MILAN
2053-0619
PO R Atkinson and crew
17/18 Aug
PEENEMUNDE

2113-0354
FL F V P van Rolleghem and
crew
22/23 Aug
LEVERKUSEN
2104-0358
FL F V P van Rolleghem and
crew
3/24 Aug
BERLIN
2015-0228
FL F V P van Rolleghem and
crew
28 Aug
NURNBERG
2100-0404
FL F V P van Rolleghem and
crew
31/1 Sep
BERLIN
2001-0524
FL F V P van Rolleghem and
crew
3/4 Sep
BERLIN
FL F V P van Rolleghem and
crew
1946-0346
5/6 Sep
MANNHEIM
1937-0204
FL F V P van Rolleghem and
crew
6/7 Sep
MUNICH
1944-0335
FL F V P van Rolleghem and
crew

No 166 Squadron
23/24 Sep
MANNHEIM
1856-0156
FS C E Phelps
Sgt E P F Hillyard
FS W Mitchell
FS E D Nesbitt
Sgt R Winder
Sgt W H Clarke
Sgt H R Gibson
27/28 Sep
HANNOVER
1937-0044
SL B Pope
FS R Somerset
FO F E Claydon
Sgt E S Brown
Sgt H Fewster
FS A Barnes
Sgt D R Taylor
29 Sep
BOCHUM
1820-2306
1/Lt J C Drew USAAF
Sgt K S Lewis
FS P W Lees
PO H Ellis
Sgt S A Pett
FO H G Cook
Sgt D Lowe
1 Oct
HAGEN
1821-2346
1/Lt J C Drew and crew
2/3 Oct
MUNICH
1821-0234

1/Lt J C Drew and crew
7/8 Oct
STUTTGART
2006-0345
SL B Pope and crew
FO H Mitchell (2P)
18 Oct
HANNOVER
1711-2214
SL B Pope and crew
(SL Pope received the DFC in
December 1943)
20/21 Oct
LEIPZIG
1746-0020
FO H Mitchell
Sgt P G Edvain-Walker
Sgt F F Clarke
FO J D Maddox
WO E Merralls DFM
Sgt D G Day
Sgt C Cushing
22 Oct
KASSEL
1807-2351
FO H Mitchell and crew
(Mitchell and crew were
killed in a flying accident, 17
November 1943)
3 Nov
DUSSELDORF
1721-2156
FO R J Robinson
Sgt A R Bird
PO F F Denney
Sgt C E Clarke
Sgt D J Stoken
Sgt N O Jones
PO B O Wright
(shot-up by night fighter and
damaged)
29/30 Dec
BERLIN
1649-0151
PO J Horsley
Sgt A W E Pilgrim (FE)
Sgt W C Morgan (BA)
PO K Cornwell (N)
Sgt M Smith-Crawshaw
Sgt J R McCourt (MU)
Sgt J Davies (RG)
(engine troubles and flak
damage)

1944
28/29 Jan
BERLIN
0019-0807
FS R B Fennell
Sgt W Pettis (FE)
FS W J C Keigwin (BA)
FS J Smyth (N)
Sgt L Parker RAAF
Sgt W Jones (MU)
Sgt W J Allen RAAF
24/25 Feb
SCHWEINFURT
2010-0515
FS R V G Frandson
Sgt K L Pile
Sgt R G Leevers
Sgt J W Latham
Sgt D V Randall
Sgt H V Reed
Sgt W V Francis
15/16 Mar
STUTTGART

1900-0335
FS J Gagg
Sgt A R Branson
Sgt P A Jessop
FS H J Vinden
Sgt H M McCann
Sgt L Moran
Sgt L A Irvine
18/19 Mar
FRANKFURT
1900-0035
FS J Gagg and crew
22/23 Mar
FRANKFURT
1830-0115
FS J A Sanderson RNZAF
Sgt F J Soloman
Sgt C Farley
Sgt R G Marks
Sgt W T Violett
Sgt J Cockburn
Sgt J A W Bodsworth
24/25 Mar
BERLIN
1825-0209
FS J A Sanderson and crew
3/4 May
MAILLY-LE-CAMP
2200-0350
PO P J Wilson DFC RCAF
S H R Moncrieff (2P)
Sgt F Reed
FO P N J Noble RCAF
WO G W Knowles RCAF
Sgt P J Hardiman
FO B W G Felgatem
Sgt C G M Meadows RCAF
(Destroyed Me110 night
fighter)
21/22 May
DUISBURG
2235-0335
PO T W Boyce
Sgt J D Carter
FO M A Monks
FO A B Leonard
Sgt S Aldis
Sgt H Rothwell
Sgt I M Rose
(unable to identify target,
brought bombs back. This
crew FTR 12 June)

No 550 Squadron
12/13 Jun
GELSENKIRCHEN
2308-0319
PO C Beeson
Sgt K J R Hewlett (FE)
FS A R McQuarrie RCAF
Sgt D Neall (BA)
Sgt J K Norgate (WOP)
Sgt H S Picton (MU)
Sgt J A Trayhorn (RG)
16/17 Jun
STERKRADE
2321-0400
PO J Lord
Sgt K W C Down
FS R Sebaski RCAF
Sgt A A Vass RCAF
PO J Elliott
Sgt A J Schemberg RAAF
Sgt P J Sculley
22 Jun
PAS DE CALAIS
1415-1718

PO D A Shaw
Sgt C A Bruce
Sgt R N Harris
Sgt S Gartland
Sgt L L Llanwarne
Sgt E J Griffiths
Sgt H A Buckingham
23/24 Jun
SAINTES
2229-0645
PO D A Shaw and crew
(bombed on three engines)
24/25 Jun
V1 SITE
0134-0505
PO D A Shaw and crew
27/28 Jun
V1 SITE
0134-0518
PO D A Shaw and crew
29 Jun
DOMLEGER
1153-1501
PO D A Shaw and crew
30 Jun
OISEMONT
0558-0954
PO D A Shaw and crew
2 Jul
V1 SITE
1213-1553
PO D A Shaw and crew
5 Jul
DIJON
2115-0552
PO D A Shaw and crew
6 Jul
V1 SITE
1849-2235
PO L W Hussay RCAF
Sgt E Elliott
FO M DeGast RCAF
FO H S W Nelson RCAF
Sgt M H Collings
Sgt A G Sale
Sgt R L Holmgreen
7 Jul
CAEN
2017-2322
PO D A Shaw and crew
12/13 Jul
RAVIGNY
2121-0656
PO D A Shaw and crew
(aborted over target by
Master Bomber due to cloud)
18 Jul
SANNEVILLE
0320-0732
FO L N B Cann
Sgt C Shaw
FL K MacAleavey RCAF
Sgt N I E Ostrom
FS J Lyons
FO H Yates
Sgt J L R Remilard
(MacAleavey was navigator,
a/c captain and A Flight
Commander; he won the
DFC in August but failed to
return on 28 August)
19/20 Jul
SCHOLVEN
2328-0313
FO L N B Cann and crew
20 Jul
V1 SITE

1914-2247
FO L N B Cann and crew
Sgt R V Fisher in, FL
MacAleavey out
23/24 Jul
KIEL
2247-0402
FO L N B Cann and crew
24/25 Jul
STUTTGART
2135-0630
FO L N B Cann and crew
25/26 Jul
STUTTGART
2132-0612
PO D A Shaw and crew
28/29 Jul
STUTTGART
2128-0544
PO D A Shaw and crew
1 Aug
LE HAVRE
1808-2129
PO D A Shaw and crew
FS R Hopman (2P)
2 Aug
BELLE CROIX
1854-2148
FO L N B Cann and crew
(aborted over target by
Master Bomber due to cloud)
3 Aug
TROSSY ST MAXIM
1143-1608
PO D A Shaw and crew
3 Aug
LE HAVRE
1703-2032
PO D A Shaw and crew
4 Aug
PAUILLAC
1330-2126
PO D A Shaw and crew
5 Aug
PAUILLAC
1426-2240
FO L N B Cann and crew
8/9 Aug
FONTENAY
2111-0030
PO D A Shaw and crew
10 Aug
DUIGNY
0906-1438
PO D A Shaw and crew
12 Aug
BORDEAUX
1122-1814
PO D A Shaw and crew
(shrapnel hole through wind-
screen)
14 Aug
FONTAIN LE PAIN
1328-1725
FL R P Stone
Sgt G E White
FS C W Sayers
FS S W Holliday
Sgt D E Norgrove
Sgt L G B Wartnaby
FS F Wright
15 Aug
LE COULOT A/F
1006-1327
FL D A Shaw and crew
17/18 Aug
STETTIN

2104-0502
FL D A Shaw and crew
19 Aug
LA NIBBLE/V1 SITE
1918-2244
FO J C Cameron
Sgt D Eldrige
FO J R Rigby
Sgt G C Sutherland
FS J W White
Sgt J F Piertney
Sgt F E Popple
25/26 Aug
RUSSELSHEIM
2016-0508
FO R Purvis
Sgt G L Grant
Sgt T Stoddart
Sgt K R Scholefield
Sgt L W Guthrie
Sgt J Wright
Sgt V S B Scoble
26/27 Aug
KIEL
2013-0138
FL D A Shaw and crew
29/30 Aug
STETTIN
2117-0646
FO H Dodds
Sgt J W Brown
FS R J Moran RAAF
FO L O Browning
FS C W Bickingham
Sgt A Laidlaw
Sgt H Lewis
31 Aug
AGENVILLE/V1 SITE
2117-0646
FO H Dodds and crew
(Harry Dodds and four of this
crew failed to return 14
October, Dodds being killed)
5 Sep
LE HAVRE
1634-2012
FO H S Vaughan
Sgt R W Metcalfe
FO L C Davies
FS C Porter
Sgt T Elliott
Sgt W Watson RCAF
Sgt L D Pursar
6 Sep -
LE HAVRE
1722-2053
FL D A Shaw and crew
PO F B Ansell (2P)
FS V R Farmer in, Sgt Fisher
out
10 Sep
LE HAVRE
1654-2041
FO J Dawson
FL G T Pyke (2P)
Sgt E W C Edmunds
FS W F Wilmer (N)
FS D W Holliday RCAF
Sgt J M Palmer
Sgt J Earnshaw
FS W H Harkness
12/13 Sep
FRANKFURT
FL G T Pyke
Sgt G Iddles-White
Sgt D E H Hellings
Sgt A G Peters

Sgt L Adams
Sgt D S Eldred
Sgt A T Ellemont
16/17 Sep
STEENWIJK A/Γ
2147-0146
PO J Harris
Sgt C H Simpkins
FS J W Eppel RAAF
WO J C Conway RAAF
FS R G Bickford RAAF
Sgt W P Waddell
Sgt B S P Barby
20 Sep
SANGATTE
1520-1908
FO S H Hayter
Sgt L A Bassman
FO T Y Thomas
FO R R Bradshaw
Sgt A J Pearce
RNZAF
Sgt E M Watkins
Sgt F E Self
23 Sep
NEUSS
1847-2355
PO J Harris and crew
25 Sep
CALAIS
0717-1041
FL D A Shaw and crew
Sgt A Cross in, FS Farmer out
(aborted over target due to
weather)
26 Sep
CALAIS
1047-1411
FL D A Shaw and crew
28 Sep
CALAIS
0804-1150
FO H L Rose and crew
(abandoned again due to
weather)
3 Oct
WALCHERON
1301-1615
FL D A Shaw and crew
5 Oct
SAARBRUCKEN
1856-2040
FL D A Shaw and crew
FS W R Wilkins RAAF in, Cross
out
(aborted when Gee went
u/s)
7 Oct
EMMERICH
1159-1320
FL D A Shaw and crew
FO D M Stephens RCAF in,
Wilkins out
14 Oct
DUISBURG
0650-1118
FO J C Adams RCAF
Sgt W P Scott
Sgt Sterman
FO W R Elcoate
Sgt F Popple
PO F S Renton
Sgt K D Winstanley
14/15 Oct
DUISBURG
2221-0431
FO J C Adams and crew

(aborted)
23 Oct
ESSEN
1616-2132
FL D A Shaw and crew
FL E Keuffling in, Stephens out
25 Oct
ESSEN
1235-1719
FL D A Shaw and crew
28 Oct
COLOGNE

1319-1821
FL D A Shaw and crew
Sgt R W Harris in, FL Keuffling out
30/31 Oct
COLOGNE
1739-0004
FO F E Bond RCAF
Sgt D C Mortimer
FO H A Leigh RCAF
Sgt B Smith RCAF
FS R C Bowling RAAF

Sgt H F Sullivan RCAF
Sgt J H Udink RCAF
31 Oct
COLOGNE
1737-2326
FL D A Shaw and crew
FO E H Luder RAAF (2P)
2 Nov
DUSSELDORF
1615-2154
FL D A Shaw
FL J P Morris (2P)

Sgt A G Furber (FE)
FO A Leonard
Sgt H J Bailey
Sgt J R Byrne
Sgt G C Dennis
Sgt T J Hooper
(FL Morris and crew failed to return 7 December 1944)
(FL David Shaw completed his tour and received the DFC)

EE136
SPIRIT OF RUSSIA

Part of Contract No 69274/40, this was a Mk III Lancaster with four Merlin 28 engines built in early 1943. Once ready for service life, it was assigned to No 9 Squadron at Bardney, Lincolnshire (the same squadron with which W4964 was flying), on the last day of May. Given the squadron code letters of WR, its individual letter became 'R'. There was a good deal of pro-Russian feeling in the midwar years, and with Bomber Command too, as evinced by the naming of a number of aircraft with Russian themes. We have already seen Uncle Joe with ED611, and now this Lancaster was given the name 'Spirit of Russia' and soon the bomb symbols painted on its nose began to mount.

They were painted a little more forward of the cockpit than on some of the other Lancs, probably due to the name being in the more usual position, and were painted in rows of 10. Once the 10 rows were completed, the 11th row was started to the right of the name.

EE136's first operation was flown on 11 June 1943, a trip to Dusseldorf with Sergeant James H. S. Lyon RAAF in the captain's seat. He and his crew became the aircraft's regular team until September, and most of the crew were decorated which included the DFC for the pilot, now commissioned. Although the aircraft had a few minor problems, by and large it had become a reliable machine.

At some stage two tiny white swastikas appeared beneath the name, presumably for the aircraft's two encounters with German night fighters. The first came on 14/15 August 1943, during a sortie to Milan. At 18,500ft, to the west of Paris on the way back, a Me109 attacked from astern and below, first seen at 500yds by Sergeant C. R. Bolt the rear-gunner who ordered an immediate corkscrew. As Sergeant W. W. W. Turnbull RCAF began the manoeuvre to port, Cyril Bolt opened fire and the '109 broke away and disappeared. A second '109 then attacked from the starboard side above but was engaged by the mid-upper, Sergeant James Michael, and this too disappeared. Both air gunners later received the DFM and William Wrigley Watts Turnbull the DFC after being commissioned. Sergeant Joe Waterhouse, another crewmember, also received the DFM on completion of their tour.

On 5 September, a trip to Mannheim with Pilot Officer J. McCubbin in command, EE136 was attacked by a night fighter near the target whose fire almost destroyed the port fin, damaged the R/T and mid-upper turret, and wounding the gunner in the head and shoulder. They were at 20,000ft and the rear-gunner first

85

saw the single-engine fighter following them. The pilot was informed and he began to corkscrew but the fighter closed in and opened up with a long burst from 400yds. The mid-upper, Flight Sergeant C. J. Houbert, fired a few rounds before he was hit.

The fighter next came in from the starboard quarter, firing several bursts into the Lanc, the rear-gunner, Flight Sergeant J. L. Elliott returning fire all the time. In the final attack, Elliott's fire scored hits and the fighter burst into flames and dived under the starboard wing leaving a trail of fire behind it, seen by both the flight engineer and the bomb-aimer. EE136 was damaged considerably with too many hits even to start to count.

The damage put EE136 out of action until November, but in the meantime, McCubbin received the DFC and became tour-expired. Jim Elliott also received the DFM. Once back on operations, EE136 took part in the Battle of Berlin, going to the Big City 11 times over the winter, plus one aborted try when the aircraft lost its port inner engine due to loss of oil pressure.

Once back in harness, EE136 proved generally to be a reliable aircraft once more, and it flew on the night of D-Day, against V1 sites, during the Caen break-out with its regular skipper Pilot Officer R. C. Lake. When he finished his tour in August 1944, a variety of crews took the aircraft, but then in October EE136 left No 9 Squadron and went to No 189 Squadron, which was formed at Bardney before going to Fulbeck in November.

It was a little strange that No 9 Squadron should allow EE136 to be re-assigned, considering it had, by this time, flown 93 sorties. Perhaps the aircraft was getting too old and the chance to get a new Lancaster was too good to miss, but one has to wonder why, for generally a squadron which had a veteran aircraft on strength, especially one approaching 100 ops, knew that the publicity and morale boost it might give, out-weighed most other factors. In any case, upon completing 100 trips the CO might well be allowed to 'trade-in' the old aircraft for a new machine.

It has been recorded that EE136 flew 97 ops with No 9 Squadron but this total is open to question. Firstly, by counting up the trips listed in the Form 541 it is impossible to get to 97, so there has to be another reason. By the time EE136 was taken off ops, it had 109 bombs painted on its nose, so we know the aircraft did that many in total. By subtracting the 12 it was supposed to have flown with No 189 Squadron, from this total of 109 we are left with 97. However, the 12 is incorrect. Someone listed the trips flown with 189 – again from the Form 541s, but four other ops either list no serial number at all (2) or show EE126 (2). Because EE126 was a Lanc with No 207 Squadron, it is again a typing error for EE136.

However, EE136 went to No 189 Squadron, was coded CA and remained 'R'. Crews had come in from a number of squadrons to form the new unit, Flying Officer D. M. S. van Cuylenburg from No 44 (Rhodesia) Squadron taking EE136 on its first mission with the new formation on 1 November. The aircraft had no regular skipper while with 189, although of its 16 missions there, Flight Lieutenant E. J. Abbott (a former No 61 Squadron man) flew the aircraft on seven of them.

EE136's last sortie was to Karlsruhe on 2/3 February 1945 after which the aircraft was made Cat B. It went to No 1659 CU and then on 2 April 1946 to No 20 MU, becoming 5918M at No 1 Radar School, Cranwell. EE136 ended up at the RAF Fire School, Sutton on the Hill, in the early 1950s where the hulk was used for fire-fighting practice – an ignominious end to such a long flying career

1943
No 9 Squadron
11/12 Jun
DUSSELDORF
2345-0434
Sgt J H S Lyon RAAF
Sgt K Pack
Sgt R W Corkill
Sgt H W E Jeffrey
Sgt A Fielding
Sgt A G Denyer
Sgt G Clegg
14/15 Jun
OBERHAUSEN
2246-0332
Sgt J H S Lyon and crew
8/9 Jul
COLOGNE
2216-0414
Sgt J H S Lyon and crew
9/10 Jul
GELSENKIRCHEN
2251-0518
Sgt J H S Lyon and crew
12/13 Jul
TURIN
2227-0807
Sgt W W W Turnbull RCAF
Sgt J Wellings
Sgt J Waterhouse
Sgt J McMasters
Sgt B Owen
Sgt J Michael
Sgt C R Bolt
27/28 Jul
HAMBURG
2230-0309
FS J K Livingstone
Sgt F Parsons
Sgt F T Watson
Sgt H C Brewer
Sgt J Predergast
Sgt R N Browne
Sgt T C Taylor
29/30 Jul
HAMBURG
2300-0404
Sgt C Payne
Sgt C A Gilbert
PO K W Armstrong
Sgt S C Young
Sgt J B Robinson
Sgt N D Bennett
Sgt P A S Twinn
30/31 Jul
REMSCHEID
2212-0254
FS J H S Lyon and crew
Sgt F L Chipperfield and FS C J Houbert in;
Sgts Corkill and Denyer out

2/3 Aug
HAMBURG
2329-0439
PO K Painter
Sgt T Deacon
Sgt R C Saunders
Sgt J E Bacon
Sgt T Andrews
Sgt S P Hone
Sgt D W Angell
(jettisoned bombs due to encountering bad weather)
7/8 Aug
MILAN
2107-0531
FS J H S Lyon and crew
Sgt Denyer back, FS Houbert out
9/10 Aug
MANNHEIM
2310-0521
Sgt W W W Turnbull and crew
FS C J Houbert in, Sgt Bolt out
10/11 Aug
NURNBERG
2144-0507
FS J H S Lyon and crew
12/13 Aug
MILAN
2136-0110
Sgt W W W Turnbull and crew
(Gee went u/s; jettisoned bombs and returned early)
14/15 Aug
MILAN
2122-0610
Sgt W W W Turnbull and crew
15/16 Aug
MILAN
FS J H S Lyon and crew
2026-0415
17/18 Aug
PEENEMUNDE
2126-0407
FS J H S Lyon and crew
22/23 Aug
LEVERKUSEN
2122-0146
Sgt W W W Turnbull and crew
27/28 Aug
NURNBERG
2119-0411
FS J H S Lyon and crew
30/31 Aug
MUNCHEN-GLADBACH
0006-0508
FO W English
Sgt Mitchell
PO J E Evans
Sgt L V Fussell

Sgt L G Lane
Sgt D R Carlisle
Sgt P W Hewitt
1/2 Sep
BERLIN
2016-0343
FO W English and crew
3/4 Sep
BERLIN
2002-0425
FS J H S Lyon and crew
(Lyon was commissioned later and received the DFC in December)
5/6 Sep
MANNHEIM
2021-0259
PO J McCubbin
Sgt N D Owen
PO B J Sherry
FS K J Pagnell
FS A M Smith
FS C J Houbert
FS J L Elliott
(damaged by night fighter, mid-upper gunner wounded)
10/11 Nov
MODANE
2054-0446
PO W E Siddle
Sgt A R Wilson
FS M C Wright
FS N Machin
Sgt J W Culley
Sgt J C Parker
Sgt C C Moore
18/19 Nov
BERLIN
1741-0127
PO H Blow
FS F S Colman
Sgt S W A Hurrell
Sgt H P Smith
Sgt R O Smith
Sgt Hartley
Sgt W E Miller
22 Nov
BERLIN
1711-2346
PO W J Chambers
Sgt W E Haywood
FL J Beeston
Sgt J Hannon
Sgt Mulcock
Sgt A Steward
FS J Campbell
26/27 Nov
BERLIN
1724-0115
PO J G R Ling
Sgt L Moss
Sgt H Laws

Sgt T Fletcher
Sgt E A Gauld
Sgt E J Rush
Sgt I Prada
20 Dec
FRANKFURT
1704-2258
PO J G R Ling and crew
Sgt J N Carter in, Sgt Laws out
23/24 Dec
BERLIN
0032-0803
PO W E Siddle and crew
PO J W Hearn in, FS Wright out
29/30 Dec
BERLIN
1649-0004
PO W E Siddle and crew
FS C H Peak (2P)
Sgt S Greenwood in, PO Hearn out

1944
1/2 Jan
BERLIN
0006-0802
PO R W Mathers
Sgt A Ball
Sgt T A Cave
FO W E Pearson
Sgt J R Donaldson
PO R R Nightingale
Sgt A P Bartlett RAAF
5/6 Jan
STETTIN
0008-0930
PO W E Siddle and crew
PO E Singer (2P)
Sgt R T C Lodge in, Sgt Greenwood out
14 Jan
BRUNSWICK
1638-2202
PO W E Siddle and crew
20 Jan
BERLIN
1627-2336
FL L G Hadland
GC N C Pleasance(2P)
Sgt A W Cherrington
FO C R Brown
Sgt J Gaskell
Sgt A D Tirel
FS G C Moore
21/22 Jan
MAGDEBURG
2019-0258
FS C A Peak
Sgt E W Kindred
Sgt T W Varey

Sgt E J Wilkes
Sgt W V Torbett
Sgt J W Nelson
Sgt J Hogan
27/28 Jan
BERLIN
1734-0231
FS C A Peak and crew
(FS Peak and crew failed to
return 10 April 1944)
28/29 Jan
BERLIN
0029-0822
PO R W Mathers and crew
Sgt J Thomas and FO D A
Keeble in;
Sgt Ball and FO Pearson out
15/16 Feb
BERLIN
1745-0026
PO R W Mathers and crew
Sgt W Wilson and FS H R
Robinson in;
PO Nightingale and Sgt
Bartlett out
19/20 Feb
LEIPZIG
2327-0715
PO R W Mathers and crew
20/21 Feb
STUTTGART
0002-0716
PO R W Mathers and crew
Sgt W Bingham in, Sgt Wilson
out
24/25 Feb
SCHWEINFURT
1854-0235
PO H Forrest
Sgt A W Hutton
Sgt S Harwood
FS R D Hassell
Sgt D McCauley
Sgt F M Corssman
Sgt D B Pinchin
9/10 Mar
MARIGNANE
2043-0619
PO R W Mathers and crew
Sgt Ball back, Sgt Thomas out
15/16 Mar
STUTTGART
1917-0302
PO R W Mathers and crew
22/23 Mar
FRANKFURT
1905-0037
PO R W Mathers and crew
24/25 Mar
BERLIN
1854-0151
PO R W Mathers and crew
Sgt C W Howe in, Sgt Bell out
21/22 May
DUISBURG
2301-0341
PO R C Lake
Sgt R W Baird
FS J A Peterson RCAF
Sgt G B Watts RCAF
Sgt G E Parkinson
Sgt S G D L Major
Sgt R D Kerr RCAF
(due to a navigational error,
bombed south-west
Munchen-Gladbach)
22/23 May

BRUNSWICK
2243-0507
PO R C Lake and crew
27/28 May
NANTES
2256-0443
PO R C Lake and crew
31/1 Jun
SAUMUR
2334-0513
PO R C Lake and crew
3/4 Jun
CHERBOURG
2313-0248
PO R C Lake and crew
6/7 Jun
ARGENTAN
2321-0323
PO R C Lake and crew
8/9 Jun
RENNES
2315-0616
PO R C Lake and crew
12/13 Jun
POITIERS
2239-0456
PO R C Lake and crew
14/15 Jun
AUNAY-SUR-ODON
2234-0345
PO R C Lake and crew
15/16 Jun
CHATELLERAULT
2123-0415
PO R C Lake and crew
21/22Jun
GELSENKIRCHEN
2320-0342
PO R C Lake and crew
23/24 Jun
LIMOGES
2254-0543
PO L J Wood
Sgt M T Gordon
FS N Oates
Sgt R L Lutwyche
Sgt D C Mumford
Sgt N Hannach
Sgt J E Shuster RCAF
24/25 Jun
PROUVILLE/V1 SITE
2248-0225
PO R C Lake and crew
27/28 Jun
VITRY LE FRANCOIS
2202-0630
PO L J Wood and crew
(raid abandoned over target
by controller)
4/5 Jul
CREIL
2330-0335
PO G A Langford
Sgt C G Fenn
Sgt J L Wright
FS S M Mitchell RAAF
Sgt E Feldman
Sgt J Wright RCAF
FO G T Baseden
7/8 Jul
ST LEU D'ESSERENT
2248-0314
PO G B Scott
Sgt J E Simkins
Sgt L A Harding
Sgt L W Langley
Sgt E M Hayward

Sgt F A Saunders
Sgt L J Hambly
12/13 Jul
CULMONT-
CHALANDRY
2202-0603
FO R C Lake and crew
Sgt J R Gunnee, Sgt Baird out
15/16 Jul
NEVERS
2212-0530
FO R C Lake and crew
Sgt Baird back, Sgt Gunnee
out
17 Jul
CAEN
0410-0759
FO W J Sheppard
Sgt R Johnstone
FO J G Glashan RCAF
FS J Mulhearn
FS W J Toomey
FS W J Harris
WO B S Dean RCAF
19 Jul
THIVERNY/V1 SITE
1925-2321
FO R C Lake and crew
20/21 Jul
COURTRAI
2305-0225
FO R C Lake and crew
24/25 Jul
DONGES
2233-0355
FL E H M Relton
Sgt F Johnson
FS C H Edwards
FS J K Scott
FS C T Scott
FS D W McConville
FS W R Andrew
(this crew failed to return 13
August, Edward Relton being
killed)
28/29 Jul
STUTTGART
2202-0553
FO R C Lake and crew
30 Jul
CAHANNES
0626-1044
PO G B Scott and crew
31 Jul
JOIGNY-LA-ROCHE
1746-2253
FO W D Tweddle
Sgt C G Heath
FS E Shields
Sgt J W Singer RCAF
Sgt A Carson
Sgt J A Foot
Sgt K Mallinson
1 Aug
MONT CANDON/V1 SITE
1640-2127
FO R C Lake and crew
(raid abandoned over target
by controller due to cloud
and fog)
2 Aug
BOIS DE CASSON/
V1 SITE
1430-1912
FO R C Lake and crew
11/12 Aug
GIVORS

2046-0438
FL G C Camsell RCAF
Sgt W Andrews
Sgt P R Aslin
FO R H Thomas
Sgt D Beevers
Sgt W J Herbert RCAF
Sgt A E Boon RCAF
16 Aug
LA PALLICE
1628-2301
FO C Newton RCAF
Sgt W Gregory
Sgt P Grant
Sgt R Flynn RCAF
Sgt L G Kelly
Sgt E H Cooper RCAF
Sgt R S Stevens RCAF
18 Aug
LA PALLICE
1123-1745
FS S F Bradford
Sgt J W Williams
Sgt R W Cooper
FO A P Hull
Sgt L G Roberts
Sgt W Brand
Sgt D Winch
24 Aug
IJMUIDEN
1227-1547
FO A F Jones RAAF
Sgt A E W Biles
Sgt S Scott
FO R L Blunsden
FS R L Birch RAAF
Sg R Glover
FS J Acheson
27 Aug
BREST
1410-1846
FS A L Keeley
Sgt A E Wotherspoon
Sgt W Chorley RCAF
Sgt L W Tanner
Sgt S D Chambers
Sgt C H Cornwell
Sgt J E Johnson
26/27 Sep
KARLSRUHE
0057-0744
FO R W Ayrton RAAF
Sgt H K Huddlestone
Sgt M J Herkes
FS N Bardsley
Sgt W Scott
Sgt D K Chalcroft
Sgt J A W Davies
27/28 Sep
KAISERSLAUTERN
2144-0432
FO R W Ayrton and crew
5 Oct
WILHELMSHAVEN
0801-1230
FO E I Waters RNZAF
Sgt C Booth
Sgt R Miles
Sgt S Coxon
FS E French RNZAF
Sgt G Jones
FL W T G Gabriel
6 Oct
BREMEN
1753-2241
FO W G Rees
Sgt H Mayhew

Sgt G A Hammond
FS D A MacIntosh
Sgt R A Morrow
Sgt W L King
Sgt G M Heppell
7 Oct
FLUSHING
1150-1454
FS A L Keeley and crew
11 Oct
FLUSHING
1314-1612
FO H Anderson
Sgt W D Loakes
FS A B Vivian
Sgt E Summer
Sgt A Cornfoot
Sgt K A Ashworth
FS G M Young RCAF
15 Oct
SORPE DAM
0638-1148
FS A L Keeley and crew
19/20 Oct
NURNBERG
1725-0103
FS A L Keeley and crew
Sgt Rogers and PO E E Stevens in
Sgts Chambers and Cornwell out

No 189 Squadron
1 Nov
HOMBERG
1416-1807
FO D M S van Cuylenburg
Sgt G H Thomson

WO A C R Havers
Sgt N E Goodhand
Sgt F W Ostopovitch
Sgt A Smith
Sgt K Pursehouse
11 Nov
HARBURG
1643-2150
FL E J Abbott
Sgt H Henderson
FS J F Charlton
FS J D Rowan
Sgt W Ashford
FL W R Kennedy
Sgt J C Oberneck
16 Nov
DUREN
1236-1744
FO A D D Brian
Sgt L C Boyle
FO D W Deubert
FS C Morgan
Sgt L W H Lambert
FO J D Constable
Sgt D C Wainwright
4 Dec
HEILBRONN
1648-2332
FL E J Abbott and crew
FO J Gilmour RCAF (2P)
Sgt E White in;
FL Kennedy out
6 Dec
GIESSEN
1710-2342
FL E J Abbott and crew
8 Dec
URFT DAM

0916-1346
FO I V Seddon RAAF
FL J A Skilton
FS L A Laurence
Sgt W Helliker
Sgt A Tuthill
FS H B Hodder
Sgt G R White
17/18 Dec
MUNICH
1607-0216
FO I V Seddon and crew
18/19 Dec
GDYNIA
1726-0311
FL E J Abbott and crew
21/22 Dec
POLITZ
1652-0303
FL E J Abbott and crew
30/31 Dec
HOUFFALIZE
0358-0817
FO I V Seddon and crew
Sgt J H Sands in;
FL Skilton out

1945
4/5 Jan
ROYAN
0123-0744
FO S J Reid RCAF
Sgt F N Benson
FO T J Nelson RCAF
FO H G Harrison
Sgt R McCormack
Sgt M R Bullock RCAF
Sgt C F Caley RCAF

7/8 Jan
MUNICH
1725-0240
FO J S Fenning
Sgt A E Veitch
FO H Loggin RCAF
FS R B Revill
WO A G Denyer
FS W Langmaid
FS J Brown
13/14 Jan
POLITZ
1700-0301
FO P Glenville
Sgt F Pallister
FS L S Harper
WO C J Gallagher
Sgt J L Nolan
FO D P Hammersley
Sgt L Moore
14/15 Jan
MERSEBURG
1639-0214
FO P Glenville and crew
(this crew failed to return on 21 February; Patrick Glenville was killed)
1 Feb
SIEGEN
1611-2236
FL E J Abbott and crew
Sgt S J Joyner in;
Sgt Oberneck out
2/3 Feb
KARLSRUHE
2027-0344
FS E J Abbott and crew

EE139
PHANTOM OF THE
RUHR

Coming off Avro's production line just after EE136, this Mk III Lancaster was also destined to become a veteran of over 100 operations. It had Merlin 28 engines and was assigned to No 100 Squadron at Waltham, near Grimsby, Lincolnshire, on 31 May 1943.

It received the squadron codes of HW and initially the individual letter 'A' but in July 1943 became R-Roger. EE139 completed at least 29 sorties with No 100 Squadron, which seems correct as there is a very slight change in the design of the bomb symbols beginning with the 30th, painted on by the new squadron. The bomb symbols were in fifteens and eventually there were eight of them (120) plus one more bomb, making a total of 121 ops.

This Lanc was also given the name 'Phantom of the Ruhr' painted on its side by its first flight engineer, Sergeant Harold Bennett, just forward of the bomb-log. Above the legend a grim-looking painting of a ghoulish skeleton figure dressed in a hooded garment, reaches over some clouds while dropping one bomb and holding another. Just in front of the 'Phantom' insignia was the mustard-coloured circular gas detection patch which appeared on aircraft of No 1 Group Bomber Command.

The Phantom's first captain, Sergeant J. R. Clark, whom we have already met at the beginning of this book, also recalls the Phantom's origins:

'The painting on the nose of the aircraft was executed by "Ben" Bennett, who says that he may have had feelings of revenge after suffering frequent bombing raids as a ground engineer in Fighter Command earlier in the war. Geoff Green, our rear-gunner, adds that I was influenced by the film 'The Phantom of the Opera' showing at the time. The Grand Operatic Teutonic sagas of the British and the Germans performed nightly over the Fatherland, should have been accompanied by the music of Siegfried. I felt afterwards that something a little less ghoulish would have been more appropriate.'

The Phantom's first raid came on the night of 11/12 June 1943 and its fifth, one month later, was to Turin, denoted on the bomb-log not by a bomb but by an ice-cream cone, an unofficial marking often used in Bomber Command to show a raid on an Italian target. Thus it is that the Phantom's 12th, 15th and 16th 'bombs' were also cones – for trips to Genoa, and two more to Milan. This marking process gives us a clear indication of how some aborted ops were counted, for EE139's second op – to Oberhausen on 14/15 June – only lasted 2 hours 38 minutes, the crew coming back after the R/T failed, jettisoning their bombs into the North Sea. It was not counted.

A second abort, on 2/3 August, due to intense predicted flak, severe electrical storms and icing in the Elbe area, made getting rid of the bombs essential, the crew dumping them over Bremen, a last-resort target. When they raided Berlin, which they did on three occasions (the Phantom went there 15 times in all), the bombs on the nose were marked in red.

Ron Clark and company flew the Phantom on 24 trips (23 plus a half to Hamburg which was later upgraded), and were close to 'buying it' over Mannheim on 23/24 September. They were coned by searchlights and hit by flak which damaged the starboard elevator, another shell going through the bomb-bay and out through the top of the fuselage, narrowly missing the wireless-operator, Sergeant L. Y. 'Lish' Easby. The Lancaster went into a dive and while still held in the searchlight beams was attacked by a fighter, but the flight engineer ably assisted his pilot to pull the Lanc out of the dive. It was their last trip in the Phantom which was out of action until early November. However, both Ron Clark and Ben Bennett were decorated for the Mannheim trip with the DFC and DFM respectively. Ron Clark recalls:

'Over the target we were coned by the searchlights and this time they held on and the aircraft was badly damaged by flak and a night fighter.

'Under the most adverse conditions, with the aircraft shaking like a leaf due to aileron control damage, Ben got out his penknife, delved into the control pedestal and somehow found the right trimming wires and severed them, which stopped the severe buffeting of the wing. He was awarded a well-merited DFM for this action.

'In every direction up to 50 searchlights brazenly stared us right in the eye. I was conscious of having the personal and un-divided attention of the thousands of resolute defenders below, no doubt jubilant, with the scent of blood in their nostrils, as we struggled to control the aircraft and evade the fighter. We were to cheat them again, however, as we eventually extricated ourselves from their implacable embrace. 'Geoff Green in the rear-turret was immobilised by the G-forces imposed by the gyrating aircraft and blinded by searchlights. In his waggish way Geoff said later that at least he had enough light to finish his crossword puzzle and when Lish yelled, "Corkscrew!" from the astrodome, Geoff thought it was drinks all round at last!'

This had not been EE139's first brush with enemy aircraft. On the ground it was being directed into its dispersal on one occasion when a German intruder aircraft attacked the airfield. The airman guiding the Phantom's pilot rapidly dropped his torch and got clear, but a couple of rounds raked the Lanc.

Towards the end of November 1943, C Flight of No 100 Squadron became the new No 550 Squadron, EE139 being transferred to the new unit and re-coded BQ B-Baker. The Lanc had already done six trips to Berlin prior to leaving No 100 Squadron, and continuing with the Battle of Berlin, the Phantom went to Berlin on nine more occasions with 550. Its usual skipper with 550 was a Canadian, Flight Sergeant Vernon J. Bouchard RCAF, who flew the aircraft on 13 ops before he became tour-expired in May, commissioned and received the DFC. He would have flown more but EE139 went off for a major inspection on 15 March 1944 and when it returned, its new pilot became Flight Sergeant T. M. Shervington who also flew 13 trips in the aircraft until he and his crew failed to return on the night of D-Day (in ME556-F).

The Phantom missed being a part of D-Day because it had been damaged by a night fighter on 3/4 June, attacked by a FW190 which appeared to be a decoy for a Me110. EE139 was on its bomb-run when all this happened, but the mid-upper probably damaged the '190 while the rear-gunner returned the '110's fire, both enemy machines breaking off and not coming back.

Returning to operational status on 17 June, Pilot Officer J. C. Hutcheson now took the Phantom over during the V1-site raids and attacks in support of the Invasion. He also piloted this aircraft to Le Havre on 5 September 1944, recorded as its 100th sortie. Hutcheson was a Scot from Troon, Ayrshire, and by that date had flown 25 of the crew's 26 sorties in the Phantom and went on to do four more to complete their tour.

Hutcheson, who had been a chemist pre-war, had had his share of excitement with the Phantom. On 12/13 July they had lost an engine but carried on to the target – even dodging a night fighter. On 24/25 July he again brought the Lanc back from Germany on three engines. No doubt Hutcheson got on well with the Phantom's groundcrew who were mostly Scots; Sergeant Cuthbertson, the NCO in charge, came from Kilmarnock, her fitter was LAC R. Taylor from Dollar, near Stirling and the rigger was LAC J. Birney from Glasgow. Hutcheson's crew was the usual mix of nationalities: Smithy his navigator came from Harrow, London, the flight engineer from Liverpool, and both the bomb-aimer and wireless-operator hailed from Sydney, Australia. The rear-gunner came from Sale, Lancashire, the mid-upper – another Scot – from Dundee. Both Hutcheson and the Phantom soon received a DFC each, the latter's ribbon being painted to the right of the first row of bombs.

When Hutcheson left to become an instructor with No 17 OTU, the Phantom had a variety of skippers but none flew the aircraft regularly or permanently. Its last sortie was to the Aschaffenburg Marshalling Yards on 21 November, flown by one of the flight commanders on No 550 Squadron, Squadron Leader W. F. Caldow AFC, DFM, who had arrived on 13 November. Caldow had won the DFM following a tour with No 142 Squadron then became an instructor although he flew a few raids between tours. Finally he instructed on Lancasters with No 1656 CU receiving the AFC for this work. He would receive the DSO in 1945 while with No 100 Squadron. Willie Caldow remembers:

'I recall quite clearly the circumstances surrounding EE139's last operational sortie. I was posted to No 550 Squadron as A Flight Commander and flew the first sortie of my second tour on 18 November with my all-officer crew, all very experienced, and all with one complete tour behind them on various No 1 Group squadrons.

'Between that and my next sortie, my deputy flight commander, Flight Lieutenant George Pyke came to me and said that he was a bit concerned that EE139 wasn't behaving very well. So, with nearly 900 hours experience on Halifaxes and Lancasters (about 800 on the latter) as an instructor, I considered myself qualified to pass judgement on the aircraft so decided to take it on the next sortie. On the way to Aschaffenburg I felt very sorry for the pilots who'd been flying it and George's concern was completely justified because with a full bomb-load, EE139 wallowed around the sky. The trim had to be adjusted constantly; one of its engines kept over-heating and the aircraft was very tiring to fly. The sortie was 6 1/2 hours of extreme discomfort.

'After the trip I spoke to both the squadron engineering officer and then the squadron commander, recommending that the aircraft be taken off operational flying and replaced. Although it handled better without a bomb-load it wasn't a good aircraft of the type. Rumour has it that it was found to have a slightly twisted fuselage too.'

Leaving No 550 Squadron with 121 ops to its name, EE139 moved to No 1656 CU on 1 December, became Cat AC on 19 February 1945 and was sent to No 58 MU. Back it came to No 1656 CU on 5 June, then it was assigned to No 1660 CU on 17 December. The aircraft became Cat AC again on 12 January 1946 and was finally made Cat E and Struck off Charge on 19 February.

1943
No 100 Squadron
11/12 Jun
DUSSELDORF
2313-0438
Sgt J R Clark
Sgt H Bennett (FE)
Sgt J H Siddell (N)
Sgt D Wheeler (BA)
Sgt L Y Easby (WOP)
Sgt L R Simpson (MU)
Sgt W G Green (RG)
12/13 Jun
BOCHUM
2310-0429
Sgt J R Clark and crew
(hit by flak over target, damage to rear turret)
14/15 Jun
OBERHAUSEN
2302-0140
Sgt J R Clark and crew

(aborted due to R/T becoming u/s; jettisoned bombs in North Sea)
16/17 Jun
COLOGNE
2247-0320
Sgt J R Clark and crew
8/9 Jul
COLOGNE
2306-0413
Sgt J R Clark and crew
12/13 Jul
TURIN
2202-0853
Sgt J R Clark and crew
24/25 Jul
HAMBURG
2255-0352
Sgt J R Clark and crew
25/26 Jul
ESSEN
2237-0232

Sgt J R Clark and crew
27/28 Jul
HAMBURG
2235-0406
WO J R Clark and crew
29/30 Jul
HAMBURG
2205-0324
WO J R Clark and crew
30/31 Jul
REMSCHEID
2213-0250
WO J R Clark and crew
2/3 Aug
HAMBURG
2313-0355
WO J R Clark and crew
(jettisoned bombs over Bremen in bad weather at 0143)
7/8 Aug
GENOA

2105-0602
WO H Wright
Sgt J S Henderson
PO W Bentley
Sgt O'Dea
Sgt T S McCleod
Sgt S O Hodges
Sgt J McKean
9/10 Aug
MANNHEIM
2310-0545
FS E C Bagot
Sgt S F May
PO F Lampin
Sgt C E Webster
Sgt A Head
Sgt J J Lloyd
Sgt A Neal
10/11 Aug
NURNBERG
2206-0555
WO J R Clark and crew

12/13 Aug
MILAN
2109-0556
WO J R Clark and crew
14/15 Aug
MILAN
2107-0630
FS E C Bagot and crew
17/18 Aug
PEENEMUNDE
2108-0401
WO J R Clark and crew
FO K J Wilson in, Sgt Simpson out
22/23 Aug
LEVERKUSEN
2142-0232
WO J R Clark and crew
Sgt M R Shear in, FO Wilson out
23/24 Aug
BERLIN
2041-0313
WO J R Clark and crew
Sgt E Gordon in, Sgt Shear out
27/28 Aug
NURNBERG
2122-0013
WO J R Clark and crew
Sgt J M Rodgers in, Sgt Gordon out
(abandoned sortie near Sevenoaks after port engine seized)
30/31 Aug
MUNCHEN
GLADBACH
0025-0417
WO J R Clark and crew
Sgt McRae in, Sgt Rodgers out
31/1 Sep
BERLIN
2001-0322
WO J R Clark and crew
3/4 Sep
BERLIN
1942-0329
WO J R Clark and crew
Sgt Simpson back
5/6 Sep
MANNHEIM
1932-0155
WO J R Clark and crew
FS Hardman (2P)
6/7 Sep
MUNICH
1940-0347
WO J R Clark and crew
22/23 Sep
HANNOVER
1850-0032
WO J R Clark
Sgt R J Cook (2P)
Sgt D Brown
Sgt J Berger
Sgt V H Thompson
Sgt R Henderson
Sgt R V Wells
Sgt J Ringwood
(twice escaped from a Ju88 night fighter)
23/24 Sep
MANNHEIM
1840-0136
WO J R Clark and crew

(hit by flak on bomb-run and damaged; also attacked by a night fighter)
3 Nov
DUSSELDORF
1732-2135
WO T V Heyes
Sgt P L Ashenden
Sgt S L Emmett
Sgt W Kondra
Sgt G R Jenkins
Sgt K W Kemp
Sgt J W Nash
10/1 Nov
MODANE
2045-0437
WO G F Peasgood
Sgt S Sykes
Sgt W P Morris
Sgt G Walker
Sgt S J Richards
Sgt S O Jones
Sgt P P Clarkin
18/19 Nov
BERLIN
1714-0100
WO G W Brook
Sgt L B Martin
WO M B Wareham
Sgt W H Ferdinando
Sgt J J McAnaney
Sgt J Godsave
Sgt J Flynn
23 Nov
BERLIN
1719-2345
PO C Dripps RAAF
Sgt J C Scott (FE)
PO J E Stewart RCAF
Sgt W T Sibley RCAF
Sgt D Campbell (WOP)
(BA)
PO A A van Walwyk
Sgt D P Lawrence(RG)
(MU)

No 550 Squadron
26/27 Nov
BERLIN
1708-0106
FS V J Bouchard RCAF
Sgt R Binney
FS D H Knight RCAF
Sgt J H Knox RCAF
Sgt E J Baker
Sgt C A Rann
Sgt J J H Galvin
2 Dec
BERLIN
1704-2358
FS V J Bouchard and crew
3/4 Dec
LEIPZIG
0010-0823
FS A H Jeffries
Sgt E C W Bull
Sgt H Simpson
Sgt D S Jeffrey
Sgt S A Keirle
Sgt E Brennan
Sgt D W Whiteley
16 Dec
BERLIN
1608-2342
FS V J Bouchard and crew
23/24 Dec
BERLIN

2338-0745
FS V J Bouchard and crew
29 Dec
BERLIN
1654-2337
FS V J Bouchard and crew

1944
1/2 Jan
BERLIN
0002-0813
FS V J Bouchard and crew
14 Jan
BRUNSWICK
1652-2230
Sgt C G W Kenyon
Sgt J N Ellis
PO P C Sharp
Sgt W R J Maroney
FS C E Duncan
Sgt G K Logan
Sgt L M Collicutt
20/21 Jan
BERLIN
1615-2326
FS V J Bouchard and crew
27/28 Jan
BERLIN
1737-0142
FS V J Bouchard and crew
30 Jan
BERLIN
1704-2325
FS V J Bouchard and crew
19/20 Feb
LEIPZIG
2351-0654
FS V J Bouchard and crew
FL A D McConnell RNZAF
(2AB)
20/21 Feb
STUTTGART
2328-0707
FS V J Bouchard and crew
24/25 Feb
SCHWEINFURT
1954-0431
FS V J Bouchard and crew
25/26 Feb
AUGSBURG
1815-0245
FS V J Bouchard and crew
18/19 Apr
ROUEN
2154-0200
FS T M Shervington
Sgt A Small
FS J R Mawhinney
Sgt K R Ansell
Sgt E R Hall
Sgt A C Griffiths
Sgt R G Dennett
20/21 Apr
COLOGNE
2353-0425
FS T M Shervington and crew
22/23 Apr
DUSSELDORF
2239-0346
FS T M Shervington and crew
24/25 Apr
KARLSRUHE
2134-0407
FS T M Shervington and crew
26/27 Apr
ESSEN
2256-0350

FS T M Shervington and crew
27/28 Apr
FREIDRICHSHAVEN
2134-0301
FS T M Shervington and crew
(returned early with port outer engine u/s; jettisoned bombs into the sea)
30/1 May
MAINTENON
2135-0215
FS T M Shervington and crew
3/4 May
MAILLY-LE-CAMP
2150-0328
FS T M Shervington and crew
6/7 May
AUBIGNE-RACON
0032-0505
FS T M Shervington and crew
7/8 May
RENNES A/F
2141-0253
FS D C Barton
Sgt R O G Ashby
FO W H Twitchell RCAF
FS A H Ingram
Sgt S Sulley
Sgt S G Reeve
Sgt K Coleman
9/10 May
MARDYCK
2247-0126
Sgt H C White
Sgt D D G Pryce
FS W E Megaw RCAF
FO C Garner
Sgt F D Mason
Sgt H W Jamieson
Sgt M H A Campbell RCAF
11/12 May
HASSELT
2154-0218
FS T M Shervington and crew
(no attack due to haze; ordered to jettison bombs into the sea)
19/20 May
ORLEANS
2158-0242
FO K B Bowen-Bravery
F.Off G P Fauman(2P)
Sgt T A Thompson
PO G L Thomas
FS J H Fyfe
WO P E E F Keeley
Sgt R Blackburn
Sgt R R Thompson
21/22 May
DUISBERG
2241-0313
FS T M Shervington and crew
22/23 May
DORTMUND
2232-0142
FS T M Shervington and crew
(abandoned; intercom u/s, rear-turret u/s; trouble cleared once bombs had been jettisoned)
24/25 May
AACHEN
2354-0431
2/Lt G P Faumen USAAF
Sgt W J Killick
FS A E Stebner RCAF
FO M S Merovitz RCAF

94

Sgt P E Cooksey
Sgt J A Ringrow
Sgt W A Drake
27/28 May
AACHEN
0005-0425
FS T M Shervington and crew
3/4 Jun
WIMEREUX
2347-0305
PO N D Holdsworth
Sgt H Mimmack
Sgt V Kirby
Sgt C E Venebles
Sgt J A S Steer
Sgt W Johnston
Sgt H Granger
22 Jun
PAS DE CALAIS/V1 SITE
1411-1721
PO J C Hutcheson
Sgt S Wright (FE)
Sgt D F Smith (N)
FS J K Francis RAAF
WO W Y Smith RAAF (BA)
Sgt E C Hodgson (MU)
Sgt A F Tosh (RG)
23/24 Jun
SAINTES
2230-0557
PO J C Hutcheson and crew
24/25 Jun
V1 SITE
0128-0514
PO J C Hutcheson and crew
27/28 Jun
V1 SITE
0128-0514
PO J C Hutcheson and crew
29 Jun
DOMLEGER/V1 SITE
1206-1514
PO J C Hutcheson and crew
30 Jun
OISEMONT/V1 SITE
0557-0953
PO J J W Dawson
Sgt E W Edmunds
Sgt F W Willmer
FS K P Brady RAAF
Sgt J M Palmer
Sgt J Earnshaw
Sgt W A Harkness
2 Jul
V1 SITE
1212-1551
PO J C Hutcheson and crew
4/5 Jul
ORLEANS
2206-0407
PO J C Hutcheson and crew
5/6 Jul
DIJON
2114-0551
PO J C Hutcheson and crew
6 Jul
V1 SITE
1845-2226
PO J J W Dawson and crew
7 Jul
CAEN
1935-2330
PO J J W Dawson and crew
12/13 Jul
REVIGNY
2118-0637
PO J C Hutcheson and crew

14/15 Jul
REVIGNY
2109-0610
PO J C Hutcheson and crew
(ordered not to bomb; landed on three engines)
18 Jul
SANNERVILLE
0335-0724
PO J C Hutcheson and crew
18/19 Jul
SCHOLVEN-BUER
2252-0318
PO J C Hutcheson and crew
20 Jul
WIZERNES/V1 SITE
1925-2241
PO J C Hutcheson and crew
23/24 Jul
KIEL
2259-0354
PO J C Hutcheson and crew
24/25 Jul
STUTTGART
2122-0554
PO J C Hutcheson and crew
28/29 Jul
STUTTGART
2124-0214
PO J C Hutcheson and crew
(aborted due to air pressure problems)
30 Jul
CAHAGNES
0638-1027
PO J C Hutcheson and crew
1 Aug
LE HAVRE
1811-2150
FO L N B Cann
Sgt C Shaw
Sgt R V Fisher
Sgt N I E Ostram
FS J Lyons
FO H Yates
Sgt J L R Remillard RCAF
3 Aug
LE HAVRE
1716-2043
FL R P Stone
Sgt C E White
Sgt R F Ferry
FS E W Holliday RCAF
Sgt D E Nargrove
Sgt L G B Wartnaby
FS S Wright
4 Aug
PAUILLAC
1341-2138
WO W H S Ansell
FS G W Battersby
FO C R Cameron RCAF
FO I H R Hood
Sgt A Anderson
Sgt E W Parker
Sgt O Tabuteau
5 Aug
PAUILLAC
1435-2230
FO L W Hussey RCAF
Sgt E Elliott
FO M DeGast RCAF
FO H S W Nelson RCAF
FS F E Dawson
Sgt A G Sale
Sgt R L Holmgren RCAF

8 Aug
FONTENAY
2121-0129
FO H G Manley RAAF
Sgt R Hughes
FO R O George
FS G E Hill
FO G P Brown
Sgt G B Wood
PO J K MacDonald
(abandoned, cause unknown)
10 Aug
DUGNY
0920-1407
FO J C Hutcheson and crew
(hit by flak over target; bomb-aimer slightly wounded in leg)
11 Aug
DOUAI
1303-1742
FO J C Hutcheson and crew
FS S E Card in, FS Francis out
12/13 Aug
FALAISE
0009-0357
FO H G Manley and crew
14 Aug
FONTAIN LE PAIN
1326-1718
FO J C Hutcheson and crew
FL C W Peck in, FS Card out
17/18 Aug
STETTIN
2107-0513
Sgt G H Town
Sgt G Hope
FS D J T Slimming
FS J H Windsor
FS P D Probert
PO E C Ball
WO J Teasdale
19/20 Aug
GHENT
2216-0117
FO J C Hutcheson and crew
Sgt R R-Jones in, FL Peck out
25/26 Aug
RUSSELSHEIM
2015-0444
FO J C Hutcheson and crew
FO W F Cox in, Sgt R-Jones out
26/27 Aug
KIEL
2012-0133
FO J C Hutcheson and crew
FL C W Peck in, FO Cox out
29/30 Aug
STETTIN
2107-0555
FO J C Hutcheson and crew
Sgt A Sutherland in, FL Peck out
3 Sep
GILZE-RIJEN A/F
1558-1910
F O J C Hutcheson and crew
FO A F W Nelson in, Sgt Sutherland out
5 Sep
LE HAVRE
1604-1947
FO J C Hutcheson and crew
FO L O Browning in, FO Nelson out

6 Sep
LE HAVRE
1716-2056
FO J C Hutcheson and crew
FS W Windsor in, FO Browning out
8 Sep
LE HAVRE
0636-1025
FO J C Hutcheson and crew
FS Gartland in, FS Windsor out
(abandoned over the target by Master Bomber)
10 Sep
LE HAVRE
1653-2036
PO V P Ansell
Sgt C P Sythes
Sgt T D Rogers
Sgt L Trudgian
FS H A Elderfield
PO G J Horsfall
Sgt S J H Adams
12/13 Sep
FRANKFURT
1814-0341
PO V P Ansell and crew
(slight damage from Me109 attack; DR compass u/s after combat)
16/17 Sep
STEENWIJK A/F
2152-0126
FO H A Shenker RCAF
Sgt J Faren
Sgt G H Lennox RCAF
Sgt L M Johnson RCAF
FS R J Emmett RAAF
Sgt P A Lander RCAF
Sgt A Ingram RCAF
17 Sep
BIGGE-KERKE/BOULOGNE
1632-1923
FO H A Shenker and crew
20 Sep
SANGATTE
1525-1833
FO J C Hutcheson and crew
FS Hutcheson RCAF in, FS Gartland out
23 Sep
NEUSS
1845-2320
FO J C Hutcheson and crew
11 Oct
FT FREDERIK HENDRIK
1524-1819
FO F E Bond RCAF
Sgt C C Mortimer
FO H K Leigh RCAF
Sgt B R-Smith RCAF
FO R C Boweling RAAF
Sgt H F Sullivan RCAF
Sgt J H Udink RCAF
12 Oct
FT FREDERIK HENDRIK
0622-0945
FO F E Bond and crew
14/15 Oct
DUISBERG
2217-0428
FL M F A Maltin
Sgt G N Raynes
FO J D Nelson RCAF
FS G B McGhee RCAF
Sgt G King

Sgt V Montagne
Sgt R C Dyke
19/20 Oct
STUTTGART
2132-0429
FO K F Sidwell
Sgt J Allen
Sgt J W Hewitt
Sgt J L Banks RCAF
Sgt J F Chapman
Sgt F W E Woodley
Sgt D G Whitmarsh
25 Oct
ESSEN
1254-1748
FO W P F Daniels
Sgt J G Woodall
Sgt R G Roberts

Sgt R Wright
Sgt F R Easton RCAF
Sgt A E Baker
Sgt W F Baker
28 Oct
COLOGNE
1324-1853
FO W P F Daniels and crew
30 Oct
COLOGNE
1745-2346
FO W P F Daniels and crew
31 Oct
COLOGNE
1755-2257
FO W P F Daniels and crew
4 Nov
BOCHUM

1732-2218
FO L O Williams
Sgt W Aspinall
FS C W Jones RCAF
FS W E Reed
FS P E Binder
FS W M Johnson
FS D L Marke
6 Nov
GELSENKIRCHEN
1145-1615
FO W P F Daniels and crew
16 Nov
DUREN
1242-1750
FO E A Stevenson RCAF
Sgt W T Woodhams
FS D T Morrison RCAF

FS W W Fitch
FS E N Pearson
FS M O Olsen RCAF
Sgt C Copperthwaite
21 Nov
ASCHAFFENBURG
1531-2208
SL W F Calder AFC DFM
FO S P George DFM (FE)
FL J Cassidy (N)
FO D E Sloggett (BA)
FO D Gear (WOP)
FO C Squires (MU)
FO J H Marston (RG)

96

Above: Pilot Officer D. A. Shaw (in cockpit) and crew with ED905 showing 70 ops. The others from l to r: L. L. Llanwarne, H. A. Buckingham, C. A. Bruce, R. N. Harris and S. Gartland.

Below: Flying Officer F. V. P. van Rolleghem DSO, DFC, receives the DFC from Air Vice-Marshal E. A. B. Rice CBE, MC, AOC No 1 Group, on 17 August 1943.

Below right: PO Jock Shaw in the cockpit of ED905 showing 99 bombs and its 'new' coat of arms insignia, on the occasion of the 100th op.

Top: David 'Jock' Shaw taking ED905 off on its 100th operation, 2 November 1944.

Above: 'Spirit of Russia', EE136 WS-R of No 9 Squadron, is believed to have completed over 90 sorties between June 1943 and October 1944, bringing its score to 109 by February 1945 with No 189 Squadron.

Right: Some of Ron Clark's crew with the post office ladies: Geoff Green, Clark, lady, Les Simpson, Ben Bennett, lady, Lish Easby.

Right: Ron Clark DFC of No 100 Squadron, who took EE139 'Phantom of the Ruhr' on its first sortie, 11/12 June 1943.

Below: EE136's insignia and score-board, showing two swastikas for night kills.

Above: 'The Phantom', now with No 550 Squadron, with 95 bombs recorded (in rows of 13), and a DFC ribbon. Note the four ice-cream cones for Italian raids.

Opposite page, top: EE139 with the Hutcheson crew about to go on the 100th op on 5 September 1944. Note how the bomb symbols have not yet caught up with the actual number of sorties flown.

Right: 'Phantom' — BQ B-Baker — at North Killingholme in 1944.

Left: Squadron Leader Willie Calder AFC, DFM, took EE139 on its 121st and last sortie on 21 November 1944, then had to 'sideline' the aircraft.

Below: 'Mickey the Moocher', EE176, carried out 128 ops with Nos 97 and 61 Squadrons. Here, 83 bombs can be seen in an unsual pattern together with the nose artwork.

Above: 'Just Jane' was JB138 of No 61 Squadron, with a possible 123 ops.

Right: No 100 Squadron's 'Take it Easy', JB603, failed to return from its 111th operation, on 5/6 January 1945.

Below: 'Easy's' crew on three Berlin raids. From l to r: Tom Hayes DFC, AFC, Peter Ashenden DFC, Sid Emmett DFM & Bar, William Kondra DFM, Glynn Jenkins, Ken Kemp DFM, and Jock Ross.

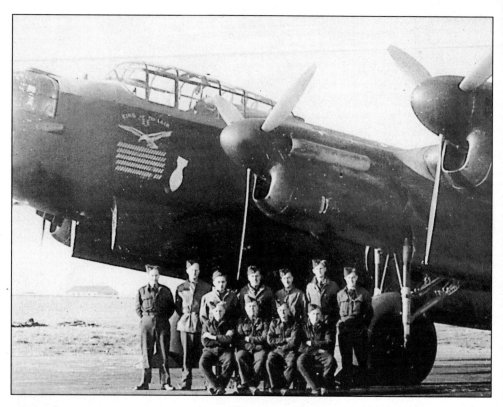

Above: The big bomb for the 100th sortie of No 106 Squadron's JB663 'King of the Air', on 4 November 1944.

Below: No 15 Squadron's LL806, LS-J, in flight.

Opposite page, top: LL806's crew prior to the 100th op on 5 January 1945. From l to r: groundcrew man, Ken Dorsett, R. Hopper-Cuthbert, Don Inglis, Sid Lewis; in doorway and on ladder: Bob Heatley, George Charlton, and Wally Lake.

Opposite page, bottom: With 134 bombing raids, three 'Manna' and three 'Exodus' trips, LL806's scoreboard looks impressive.

Top left: Doug Hunt took LL806 on two of its sorties. From l to r: Pat Russell, Paddy Kirrane, Flight Lieutenant Hunt, George Pitkin and John Shepherd.

Bottom left: PO-D LL843 of No 467 (RAAF) Squadron also served with No 61 Squadron, achieving a grand total of 118 ops.

Above: Pilot Officer J. L. Sayers RAAF (far right) took LL843 on 20 trips. With him are Flying Officers E. W. Warner and B. P. Barry while with No 617 Squadron, the latter also his rear-gunner on No 467 (RAAF) Squadron.

Centre right: 'Jig' was LL885 and served with No 622 and later No 44 (Rhodesia) Squadrons, notching up 113 ops.

Right: Jack Lunn (3rd from left rear row) and crew took LL885 on its first mission on 30/31 March 1944 and went on to fly a further 12 ops in the aircraft.

Top left: J-Jig is waved off on another sortie.

Left: LM227 of No 576 Squadron, about to take off for its first bombing raid on 4/5 July 1944.

Above: Some of No 576 Squadron's groundcrew. From l to r: Bill Honeywell (rigger), Johnny Cook (engine fitter), Norman Hall (engines), Norman Bryan (engine fitter to LM227 and the nose artist), Paddy Bennet (rigger); in front: Ronnie Brett (engine fitter).

Right: LM227 pictured after 24 ops and one night kill. Note the 21st trip is recorded as a key on a parachute — a mining trip.

Above: The Guilfoyle crew. From l to r: Pete Dodwell, Jack Powell, George Tabner, Harry Guilfoyle, S. C. Wilkins, N. S. Cassidy and Nick Hawrelenko.

Below: The Till crew. From l to r: Kevin Oliver, John Shorthouse, Derek Till, Geoff Gripps, Charles Bray, Bob Hamilton, D. E. Holland.

Above: LM227's 100th op was a 'Manna' trip flown on 8 May 1945, piloted by Flying Officer B. Simpson (centre standing). To his right are T. E. Melocke and J. Ellis, navigator and engineer.

Right: Beer mugs and a barrel were painted on LM550, its motto being 'Let's Have Another'. The aircraft ended the war with 118 mugs!

Above: LM550 on the morning of 13 March 1945, noting 100 ops 'not out'; although the accuracy of the date is suspect. Standing left to right: FL J. Trusler (passenger), FO B. F. Rea-Taylor (BA), Sgt W. D. Thomson (FE), FL H. W. Langford (P), FO D. S. McDonald (N), Sgt D. W. Hallam (MU), Sgt K. A. Hawkins (RG), Sgt T. E. Jones (WOP)

Left: Bill Capper and crew. Clockwise: Capper, Graham Bale, Len Sparvell, Tom Morris and George Luckraft.

EE176
MICKEY THE MOOCHER

A Manchester-built Lancaster Mk III fitted with Merlin 28 engines, EE176 was part of contract No B69274/40 and came off the production line in the spring of 1943 after which it was assigned to No 7 Squadron at Oakington, Cambridgeshire, on 11 June. Within 10 days it was re-assigned to No 97 (Straits Settlements) Squadron at nearby Bourn, although the squadron's Flights were detached to three other bases, C Flight being at Oakington, where EE176 had joined it.

With No 97 Squadron it carried the unit code letters OF. Its individual letter was 'N', possibly a legacy from No 7 Squadron, but by early August this had been changed to 'O'. EE176's first trip was to Cologne on 3/4 July, and once more we see the not unusual sequence of a new aeroplane being used by various crews for the first few weeks, before a regular crew becomes evident – in this case Flying Officer D. Moodie. However, after 15 trips, which included the famous raid on Peenemunde, three to Hamburg, one to Milan and two to Berlin, EE176 was moved again, this time to No 61 Squadron at Syerston, Nottinghamshire, on 20 September. Here it became QR-M.

Aircraft lettered 'M' were usually known as Mike or Mother, but one of EE176's crews called it Mickey and on its nose was painted Walt Disney's Mickey Mouse character, walking along pulling a bomb-trolley on which sat a bomb. Beneath this was the name 'Mickey the Moocher', an obvious parody of the popular song of the time 'Millie the Moocher'. Mickey is walking towards a sign-post upon which was written '3 Reich' and 'Berlin'. Bomb symbols began to appear in rows of 10 behind the Mickey character, and after five such rows, a second column began aft of the first, although, oddly, after the first 10, the second row only had eight bombs, the next three having just five. There is a suggestion that the Lanc eventually had 115 marked on its nose. How, or if, its accredited 128 ops were shown is not known.

With No 61 Squadron Mickey became the regular aircraft for Flight Lieutenant J. E. R. Williams and crew, who took it to Germany 13 times between October 1943 and February 1944, out of its first 28 sorties with the squadron. During this period Mickey went to Berlin 15 times, although it had to abort the mission on 1/2 January 1944 when the starboard outer engine suddenly caught fire. Williams then went to No 617 Squadron and two captains, Pilot Officer J. A. Forrest RAAF, and the flight commander, Squadron Leader S. J. Beard DFC began to fly Mickey.

On the fateful Nurnberg raid of 30/31 March 1944 under the captaincy of Forrest, the aircraft was flown well north of track owing to wrong winds being broadcast and with petrol getting low, the skipper decided to home on the 'B' Lattice Line. Near Hannover they ran into hail and sleet storms. Reaching the North Sea, Forrest headed for Coningsby only to meet violent electrical storms off the Norfolk coast. The Lanc was hit by lightning, the shock passing right through the machine, stunning and temporarily blinding Forrest who lost control. Believing they had already crossed the coast he ordered the crew to bale out and EE176 dived earthwards. The WOP and the rear-gunner were the only two crew members who recovered sufficiently to do so, but then, when just 1,000ft from the ground – actually the sea – Forrest regained control and quickly landed at the nearest base he could locate. Although an immediate search was made, once they realised they had still been over the sea, neither of the two men who had baled out were found. Forrest and the rest of his crew were to die in action on another mission shortly after this.

In May, Pilot Officer D. E. White RCAF and Beard shared most of Mickey's ops, Beard flying the aircraft on D-Day. Mickey flew all through the summer of 1944, attacking V1 sites and communications points, Beard taking the Lanc on the Caen raid that heralded the Allied break-out of Normandy. The aircraft had very few aborts due to mechanical problems, and by August was fast approaching its 100th trip, which appears to have been flown on 12/13 August by Delbert White, to Russelsheim. In all White flew 28 trips in Mickey and won the DFC; Sidney Beard flew 17, receiving a bar to his DFC.

Mickey encountered a night fighter going to Rheydt on 19/20 September. It opened fire at 600yds, closing to 200yds from the starboard quarter. Both gunners opened fire too and the fighter broke away to port and was not seen again.

Flying Officer Norman Hoad, who had flown several trips in another 100-plus veteran on No 61 Squadron – ED860 – took Mickey to the Dortmund-Ems Canal on 23/24 September, noting: 'Bombed from 7,800ft to attain accuracy needed to hit this precision target'. He also had to take evasive action from a Ju88 which attempted an attack, although the enemy fighter did not get into a firing position. This was EE176's 110th op. Norman Hoad and his crew failed to return from Brunswick in Lancaster ME595 on 14/15 October, Hoad ending up as a prisoner of war.

Mickey's last combat took place on a trip to Brunswick on 14/15 October. Nearing the target area the mid-upper spotted a fighter approaching from the port quarter above which it then appeared to side-slip into position behind them. He ordered the pilot to corkscrew as he opened fire, while also giving the rear-gunner the fighter's position and who, upon seeing it too, also opened fire. The fighter dived quickly away, the mid-upper giving it a final burst as it disappeared out of range.

EE176 left the squadron at the end of November, going to No 1653 CU where it was marked H4-X but became Cat AC on 21 April 1945. In May it became 5260M at BOAC Whitchurch and was converted to an instructional airframe.

It is not certain just how many ops EE176 flew. After Norman Hoad noted the 23/24 September trip being the 110th, the aircraft went on nine further sorties, although the last two were frustrated by orders not to bomb, both because of cloud covering the target. Thus it seems that EE176 made 119 trips although, as stated earlier, the Lanc seems to have had 115 bombs on its nose when it left the squadron, while other records show the aircraft as flying 128 trips. The figure of 119 seems about right.

1943
No 97 Squadron
3/4 Jul
COLOGNE
2245-0340
FL J H Sauvage
Sgt W G Waller (FE)
FO H A Hitchcock (N)
PO F Burbridge (BA)
FS E Wheeler (WOP)
FO J E Blair (MU)
Sgt G W Wood (RG)
24/25 Jul
HAMBURG
2209-0357
FL J H Sauvage and crew
27/28 Jul
HAMBURG
2243-0354
FS I Baker
Sgt W Vaughan
PO C W Webb
Sgt Davis
Sgt J Richards
Sgt Lowden
Sgt P Edwards
2/3 Aug
HAMBURG
2317-0445
Sgt C S Chatten
Sgt C Baumber
Sgt L R Armitage
FO Webb
Sgt W A Reffin
FS J R Kraemer RAAF
Sgt D V Smith
(this crew were attacked by a German Me410 intruder on 23 August, Chatten being wounded and his Australian mid-upper killed)
10/11 Aug
NURNBERG
2227-0533
FO D Moodie
Sgt L E Melbourne
Sgt J T Bundle
Sgt N E Clausen
Sgt T E Stamp
Sgt L A Drummond
Sgt F A Hughes
12/13 Aug
MILAN
2109-0507
SL J M Garlick
Sgt J M Anderson
Sgt A G Boyd
Sgt E O Charlton
Sgt J Kenny
Sgt T N Ward

Sgt F Edwards
17/18 Aug
PEENEMUNDE
2101-0430
FO W Richen
Sgt G Winter
Sgt H W Watts
Sgt E H Pack
Sgt J Wrigley
Sgt R W Lowe
Sgt F C Nordhoff
22/23 Aug
LEVERKUSEN
2137-0227
FO D Moodie and crew
23/24 Aug
BERLIN
2051-0351
FO D Moodie and crew
27/28 Aug
NURNBERG
2132-0441
FO D Moodie and crew
30/31 Aug
MUNCHEN
GLADBACH
0044-0429
FO D Moodie and crew
3/4 Sep
BERLIN
2026-0436
FO D Moodie and crew
5/6 Sep
MANNHEIM
2005-0222
FO D Moodie and crew
6/7 Sep
MUNICH
2000-0343
FO K M Steven
Sgt A C East
Sgt S B Stevenson
PO R R Brown
Sgt W C Gadsby
Sgt L A Drummond
Sgt K D Newman
15/16 Sep
MONTLUCON
2114-0220
2/Lt J E Russell
Sgt J H Lazenby
2/Lt R Wright
FS L W Golden
Sgt J P Dow
Sgt E W Bark
FS R H Marston
No 61 Squadron
4/5 Oct
FRANKFURT
1850-0101
FL J E R Williams

Sgt W Beach (FE)
FO A J Talbot (N)
PO A J Walker (BA)
Sgt A E Potter (WOP)
Sgt R Blagdon (MU)
Sgt K S Jewell (RG)
7/8 Oct
STUTTGART
2014-0236
FL N D Webb
Sgt J W Brown
Sgt P S Watkins
Sgt J Bailey
FS C J Colingwood
FS L J Powell
Sgt D J Chapman
8/9 Oct
HANNOVER
2250-0411
FL J E R Williams and crew
Sgt J Soilleux and FS A G Leslie in;
Sgt Beach and PO Walker out
18 Oct
HANNOVER
1736-2345
FL J E R Williams and crew
20/21 Oct
LEIPZIG
1700-0037
FL J E R Williams and crew
22/23Oct
KASSEL
1811-0013
FL J E R Willaims and crew
3/4 Nov
DUSSELDORF
1717-2156
PO R A Walker
Sgt G M Ward (2N)
Sgt H E Houldsworth
FS N J Cornell
PO J Wells
Sgt R C Bailey
Sgt C R Taylor
Sgt D R Kelly
10/11 Nov
MODAN
2053-0515
Sgt C J Gray
Sgt W J McCulloch
Sgt Ward
Sgt R Jones
Sgt L E Jackson
Sgt W M Morrison
Sgt D R Hay
18/19 Nov
BERLIN
1719-0040
FL N D Webb and crew

(FL Webb and crew failed to return 24/25 February 1944)
23/24 Nov
BERLIN
1723-0040
FL J E R Williams and crew
26/27 Nov
BERLIN
1719-0118
FL J E R Williams and crew
2/3 Dec
BERLIN
1653-2359
FL J E R Williams and crew
3/4 Dec
LEIPZIG
0018-0746
PO R H Todd
Sgt S Robson
FO J Hodgkinson
Sgt V R DuVall
FS W Housley
Sgt M McCloskey
Sgt J Cartwright
16/17 Dec
BERLIN
1619-0011
Sgt C J Gray and crew
20 Dec
FRANKFURT
1720-2321
Sgt L Cannon
Sgt K H Dean
FS G J Hull
Sgt R Stones
Sgt H Wyrill
Sgt H Sherliker
FS J M Green
23/24 Dec
BERLIN
2355-0803
SL S J Beard DFC
FS J K Burnside
FL J R Anderson
WO D C Davies
Sgt A E Wood
PO J C Hodgkins
WO J Graham
29/30 Dec
BERLIN
1647-2359
FL J E R Williams and crew

1944
1/2 Jan
BERLIN
2234-0226
FL J E R Williams and crew
(returned early; starboard outer caught fire)

2/3 Jan
BERLIN
2324-0713
FL J E R Williams and crew
5/6 Jan
STETTIN
1650-2242
PO F J Nixon
Sgt W Craig
Sgt J W Devenish
FS A T Garnett
Sgt J E Chapman
Sgt H F Bore
Sgt H W Pain
20 Jan
BERLIN
1624-2324
FL J E R Williams and crew
21/22 Jan
BERLIN
2039-0308
PO H H Farmiloe
Sgt G N Whitley
Sgt T K Telfer
Sgt K Vowe
Sgt E A Davidson
Sgt H J Newey
Sgt R Noble
27/28 Jan
BERLIN
1735-0202
PO H Wallis
Sgt T F Preston
FS L E Tozer
Sgt A Pardoe
FS K Sims
Sgt D J Brewer
FS E Bremner
28/29 Jan
BERLIN
0021-0739
FL J E R Williams and crew
30 Jan
BERLIN
1725-2345
PO F P Moroney
Sgt G N Whitley
FO M Jenkins
Sgt J Davies
Sgt D Dushman
Sgt C A Mills
FS J Bell
15/16 Feb
BERLIN
1723-0028
FL J E R Williams and crew
19/20 Feb
LEIPZIG
2351-0721
SL S J Beard and crew
24/25 Feb
SCHWEINFURT
2029-0432
SL S J Beard and crew
15/16 Mar
STUTTGART
1916-0248
PO L Cannon and crew
(PO Les Cannon and crew
failed to return 18/19 March,
all but one being killed)
18/19 Mar
FRANKFURT
1900-0124
FS D Newman
Sgt R Jones
Sgt C Rattner

Sgt E Outram
Sgt R Taylor
PO E Alston
Sgt R C Gardener
22/23 Mar
FRANKFURT
1856-0102
PO J A Forrest RAAF
Sgt A H Davies
Sgt J Wood
Sgt D C Newman
Sgt L Darben
FS H W Pronger RAAF
Sgt J Macfie
24/25 Mar
BERLIN
1857-0153
SL S J Beard and crew
26/27 Mar
ESSEN
1943-0153
PO J A Forrest and crew
30/31 Mar
NURNBERG
2203-0600
PO J A Forrest and crew
(hit by lightning; WOP and
MU baled out over the sea
and were lost)
5/6 Apr
TOULOUSE
2034-0444
SL S J Beard and crew
FL J Breakey (3AG, Sqn
Gunnery Leader)
10/11 Apr
TOURS
2215-0429
SL S J Beard and crew
(SL Beard acted as 1st
Deputy Leader)
11/12 Apr
AACHEN
2038-0057
FS J Kramer
Sgt N H Shergold
Sgt R W Burkwood
Sgt C W Greenaway
Sgt P Donoghue
Sgt A N Savery
Sgt R F Coleman
18/19 Apr
JUVISY
2059-0135
PO J A Forrest and crew
FS E J Kemish and Sgt C C
Scrimshaw in;
FL J Breakey (Sqn Gunnery
Leader)
20/21 Apr
PARIS
2255-0345
SL S J Beard and crew
22/23 Apr
BRUNSWICK
2305-0533
SL S J Beard and crew
26/27 Apr
SCHWEINFURT
2121-0615
SL S J Beard and crew
Sgt C W Saunders in, FS
Burnside out
29/30 Apr
ST MEDARD
2214-0547
SL S J Beard and crew

FL I M Pettidrew in, Sgt
Saunders out;
Sgt C C Scrimshaw (3AG)
1/2 May
MALINES
2202-0345
FS J G Gibbard RAAF
Sgt R F Saunders
Sgt C Michael
FS E J Roberts
Sgt F Gibbons
Sgt T M Blackie
Sgt C J Jackson
Sgt C C Scrimshaw (3AG)
6/7 May
LOUAILLES
0045-0552
FS J G Gibbard and crew
Sgt H S Rosher (3AG)
8/9 May
BREST A/F
2124-0245
PO F Norton and crew
Sgt K Chapman
FS W H Webb
FS K W Clement
Sgt H J Eldrett
FS J H Tollitt
FS S Calver
10/11 May
LILLE
2156-0115
PO W North
FS H Crowley
Sgt L Morton
FS N E Jarvis RAAF
Sgt G I Monteigh
Sgt D A Bartlett
Sgt P E O'Shea
19/20 May
TOURS
2207-0442
PO D E White RCAF
Sgt J A Lyon
WO M B Blackwood RCAF
FO P O Points
FS A F Harrow
Sgt D E Gibb RCAF
Sgt L C Anderson RCAF
22/23 May
BRUNSWICK
2230-0515
PO D E White and crew
27/28 May
NANTES
2228-0444
PO D E White and crew
28/29 May
ST MARTIN
2244-0205
PO D E White and crew
31/1 Jun
SAUMUR
2341-0454
SL S J Beard and crew
3/4 Jun
FERME D'URVILLE
2309-0239
SL S J Beard and crew
Sgt P F Lewis in FL Pettigrew
out
5/6 Jun
ST PIERRE
0250-0702
SL S J Beard and crew
FL Pettigrew in, Sgt Lewis out

6/7 Jun
ARGENTAN
2336-0336
SL S J Beard and crew
Sgt B F Rowland in, FL
Pettigrew out
10/11 Jun
ORLEANS
2245-0353
PO D E White and crew
12/13 Jun
POITIERS
2241-0435
PO D E White and crew
14/15 Jun
AUNAY-SUR-ODON
2230-0307
PO D E White and crew
15/16 Jun
CHATELLERAULT
2127-0405
PO D E White and crew
19/20 Jun
WATTEN/V1 SITE
2218-0010
SL S J Beard and crew
Sgt L Morton in, Sgt Rowland
out
(most aircraft recalled due to
cloud)
21/22 Jun
GELSENKIRCHEN
2316-0429
PO D E White and crew
23/24 Jun
LIMOGES
2245-0519
PO D E White and crew
27/28 Jun
VITRY
2207-0522
PO D E White and crew
29 Jun
BEAUVOIR/V1 SITE
1250-1545
PO D E White and crew
4/5 Jul
ST LEU D'ESSERENT
2303-0318
PO D E White and crew
7/8Jul
ST LEU D'ESSERENT
2217-0250
PO D E White and crew
12/13 Jul
CULMONT
2203-0531
PO D E White and crew
14/15 Jul
VILLENEUVE
2206-0451
PO D E White and crew
15/16 Jul
NEVERS
2224-0520
PO D E White and crew
18 Jul
CAEN
0337-0707
SL S J Beard and crew
18/19 Jul
REVIGNY
2308-0418
PO D E White and crew
20/21 Jul
COURTRAI
2301-0209

PO D E White and crew
24/25 Jul
DONGES
2225-0356
FO G M Taylor RCF
Sgt G Gaunt
Sgt A F Nixon
FO J Meek
Sgt S Adair
Sgt J H Jebb
FS H S Grahn
25 Jul
ST CYR
1750-2201
FO R F Heath
Sgt P Davis
FO A A Lorton
FO F J Kelly RCAF
Sgt C E Ruane
Sgt E D Fisher
Sgt J Ryan
28/29 Jul
STUTTGART
2200-0543
FO R J King
Sgt W N Heritage
FO M C Macfarlane
FO C H Oliver RCAF
FO R R Collard
Sgt F W Futter
Sgt E G Cutting
30 Jul
CAHAGNES
0556-1045
FO R J King and crew
(did not bomb, all aircraft
aborted on instructions from
controller)
31 Jul
RILLY-LA-MONTAGE
1720-2220
PO D E White and crew
1 Aug
MONT CONDON/V1 SITE
1645-2113
PO D E White and crew
(all aborted due to fog)
2 Aug
BOIS DE CASSAN
1440-1920
PO D E White and crew
3 Aug
TROSSY ST MAXIM
1140-1612
SL S J Beard and crew
Sgt M Kelly in, Sgt Morton
out;
FSs C C Scrimshaw and R R

Langley AGs
5 Aug
ST LEU D'ESSERENT
1055-1530
PO D E White and crew
7/8 Aug
SECQUIVILLE
2115-0105
PO D E White and crew
(ordered not to bomb by
controller; dropped bombs in
Channel)
9/10 Aug
CHATELLERAULT
2027-0250
PO D E White and crew
11 Aug
BORDEAUX
1200-1937
PO D E White and crew
12/13 Aug
RUSSELSHEIM
2134-0303
PO D E White and crew
14 Aug
BREST
1734-2229
PO D E White and crew
15 Aug
GILZE RIJEN A/F
1002-1343
FO J S Cooksey
Sgt D E Rose
Sgt E P Meaker
Sgt T Stephenson
Sgt A R Marshall
Sgt S J Scarrett
Sgt N W Elliott
5 Sep
BREST
1558-2037
FO H L Inniss
Sgt H T Steele
FO J T O Dickinson
FO J R Morrison RAAF
Sgt K F Moseley
Sgt J A Aldridge
FS R R Langley
10 Sep
LE HAVRE
1507-1916
FO J F Boland RAAF
Sgt L Muir
Sgt J Boon
Sgt J A Fitzgerald
Sgt L E Welch
Sgt H Cave
Sgt J L Jones

11 Sep
LE HAVRE
0358-0906
FO J F Boland and crew
11/12 Sep
DARMSTADT
2056-0257
FO M L Hunt RNZAF
Sgt P Askew
FS A T Harvey RNZAF
FS A T Bell RAAF
Sgt R Nesbit
Sgt S H Jobson
Sgt J W H Harrison
17 Sep
BOULOGNE
0755-1113
SL H W Horsley
PO L A Cawthorne
FO J C Webber
FO J P Wheeler
FS G Twyneham
Sgt H W Jennings
Sgt R T Hiskisson
18 Sep
BREMERHAVEN
1821-2318
FO S E Miller
Sgt G McChrystal
FO H A Hunt
Sgt R T Galloway
FS D C Mummery RAAF
Sgt L A Hay
Sgt J E Norcutt
19/20 Sep
RHEYDT
1855-2358
FO S E Miller and crew
(night fighter combat)
23/24 Sep
DORTMUND-EMS CANAL
1859-0012
FO N E Hoad
Sgt C S Webb
FO K W Ball
FO W H Pullin
Sgt C P Boyd
Sgt N England
Sgt C V Embury
26/27 Sep
KARLSRUHE
0100-0716
FL H B Grynkiewicz
Sgt W M Ratcliffe
Sgt J L Jones
Sgt E W Gibb
Sgt E J Day
Sgt G E G Walter

Sgt E W Browne
27/28 Sep
KAISERSLAUTERN
FO M L Hunt and crew
2150-0410
5 Oct
WILHELMSHAVEN
0750-1315
FO C A Donnelly
Sgt A H Steers
FL J H Vincent
FS F D Green
Sgt R G D Brook
Sgt L Ayres
Sgt T J Kerrigan
6 Oct
BREMEN
1743-2215
FO M L Hunt and crew
14/15 Oct
BRUNSWICK
2257-0622
FO F A Mourtiz RAAF
Sgt A J Leith
FS L A Cooper
FO P M R Smith
FS D C Blomfield
Sgt A G B Bass
Sgt D C Cluett
19/20 Oct
NURNBERG
1730-0051
FO F A Mouritz and crew
23 Oct
FLUSHING
1439-1735
FO F A Mouritz and crew
28/29 Oct
BERGEN
2210-0544
FO T Bain
Sgt P G Gee
Sgt H G Harding
Sgt T E Morgan
WO N G McKenzie
Sgt J H Hodgkins
Sgt J Casey
(aborted due to cloud over
target)
6 Nov
GRAVENHORST
1635-2228
FO F A Mouritz and crew
(aborted due to ground
smoke)

JB138
JUST JANE

Part of an order for 550 Lancasters placed with A. V. Roe in 1941, JB138 rolled off the production line in the early summer of 1943 and was assigned to No 61 Squadron at Syerston on 22 August. Given the squadron letters of QR, its individual letter of 'J' eventually helped shape the aeroplane's name of 'Just Jane'. Shape is right, for the artist who painted the name and motif on the Lanc's nose produced a very buxom nude stretched out on a bomb; undoubtedly inspired by the popular *Daily Mirror* newspaper cartoon character Jane.

Unlike most Lancaster bomb tallies, Jane's was not marked in neat rows of the more usual 10 but in rows of 30. The photograph – believed to have been taken in about July 1944, and reproduced here – shows the aircraft with 64 bombs marked, and it went on to make its 100th trip in early October. Jane has been variously credited with 113 ops or even as high as 123, although records from the Form 541 would indicate something around 120, so the higher figure may be correct; and 113 was either a typing or a purely arithmetical error.

Jane's first operation was to Nurnberg on 27/28 August 1943, just in time to fly during the Battle of Berlin that autumn and winter. The aircraft went to the Big City eight times but its only problem was the trip on 1/2 January 1944. Jane's pilot, Howard 'Tommy' Farmiloe, had to land at the US 8th Air Force base at Molesworth when fuel ran low. Edward Davidson recalls this sortie and the crew's first on 29/30 December:

'On 29/30 December we bombed on the red TI's from 22,000ft, the glow of the fires being seen through the clouds for a considerable time as we headed for home. This was our first operation.

'It was back to Berlin on 1/2 January. Tommy reported seeing route markers at the pinpoint, and we later bombed on three Wanganui flares from 22,000ft. Our Electrical Direct Circuit unit was overcharging and failed entirely on leaving the target, which put the rear-turret u/s. Then the main petrol feed line became partly blocked causing the starboard inner engine to overheat. The Gee also failed.

'We landed at the American base at Molesworth and while Tommy was whisked off to the Officers Mess, the rest of us NCOs had to spend the night sleeping in armchairs in a room, which we didn't find amusing. Their "class" system was far worse than ours ever was. The Americans were fascinated by our Lancaster, but at least they put a guard on it overnight.'

102

Howard Farmiloe and crew crash-landed another Lanc at Little Snoring coming back from Berlin on 24/25 March. He received an immediate DSO, while Ed Davidson and the bomb-aimer received DFMs; the navigator got a DFC.

Jane was engaged by night fighters on several occasions, the first on the night of 29/30 September, by no fewer than three fighters right in the target area of Bochum. Over Hannover on 8/9 October, flying under a half-moon at 19,000ft, when 20 miles south of Bremen, a fighter was seen by the rear-gunner making for another Lanc to starboard and opened fire, although no results were seen. Reaching the target they were attacked again by a FW190 which came in from the starboard beam. The rear-gunner again opened up, the '190 breaking away to port as the Lanc did a diving turn to starboard. Hits were observed by both gunners and the '190 was claimed as damaged.

Early in 1944, with the squadron CO Wing Commander R. N. Stidolph in command, Jane was attacked by a Me210 on the Stettin raid of 4/5 January. Monica picked up the approaching bogey but neither gunner saw it until it began firing, closing in very fast. Both gunners fired back and saw the Messerschmitt break to port and climb, the gunners continuing to fire as it went. It was then seen to dip a wing in preparation for another attack but then smoke could be seen pouring from its starboard engine as it then turned away and went down. It was claimed as destroyed. The '210 had, however, scored hits on the Lanc – a large hole had been blasted in the port fin and a smaller hole in the elevator. The port outer engine was set on fire and the port wing holed, while cannon shell splinters holed the port-side fuselage and mid-upper turret, fortunately without injuring the gunner, Sergeant E. A. Gardener. The rear-turret, powered as it was by the now feathered engine, became u/s. The gunner was Pilot Officer J. H. Pullman who was on his second tour.

Wing Commander Stidolph also discovered the Gee, visual Monica and R/T all u/s and to add to his problems, the aircraft became unmanageable on three engines, so they had to return on just the two inboard motors at maximum boost and revs, and with the starboard-outer throttled right back. They landed at Matlaske with just 8 1/2 gallons of petrol left after nearly 10 hours flying.

After this night's damage, Jane was declared Cat AC for almost a month, but was back on duty by 5 February. Its next major problem might well have been fatal, for on a daylight sortie to Duren on 16 November the aircraft was hit by a falling bomb and lost an engine! The Lanc was quickly repaired at No 54 MU and was back by 2 December. However, the Form 541 notes Jane as flying three operations in late November which must be suspect. However, these entries were more than likely clerical errors, Jane's number being repeated without checking what actual aircraft was being flown.

Jane's more-or-less regular skipper was Flying Officer Norman F. Turner DFM (he had won the DFM with the squadron back in 1942) who did some 18 trips in the aircraft, receiving the DFC. Then came Pilot Officer Frank Norton who did over 20. Some of Jane's captains were men who had flown another No 61 Squadron veteran, EE176 – notably Squadron Leader S. J. Beard, Flight Lieutenant D. E. White and Flying Officer H. L. Inniss.

Jane operated on both nights of D-Day, despite a lucky escape a few nights earlier just after midnight on 1 June. The crew had just set course when the bomber was hit by lightning. The machine stalled and as the port wing dropped, the aircraft went into a partial spin. Pilot Officer D. C. Freeman recovered and flew back to base, setting course again but the aircraft's stability was very bad so they were forced to abort.

The aircraft's last combat occurred on 11/12 September 1944 while returning from Darmstadt. A Me109 came in from behind shortly after leaving the target area, the rear-gunner saw it and started firing, followed by the mid-upper. As the pilot put the Lanc into a corkscrew, strikes were seen on the fuselage and engine of the '109, which was then claimed as damaged. Jane's pilot on this night had been Flying Officer A. P. Greenfield RAAF, who was to do most of his tour in Jane.

Jane's last trip was made on 14/15 January 1945 and on 2 February the aircraft was assigned to No 5 LFS. On 22 April it went to No 4 School of Technical Training where it became 5224M and was converted to an instructional airframe. JB138 was finally Struck off Charge on 16 October 1946.

1943
No 61 Squadron
27/28 Aug
NURNBERG
2050-0435
FO N F Turner DFM
Sgt G A Turnbull (FE)
Sgt J Barr (N)
Sgt R Freeth (BA)
WO R J Russell (WOP)
PO M Root-Reid (MU)
Sgt E A Walker (RG)
30/31 Aug
MUNCHEN-
GLADBACH
0002-0519
FO N F Turner and crew
31/1 Sep
BERLIN
2018-0406
FO N F Turner and crew
3/4 Sep
BERLIN
2002-0436
PO P H Todd
Sgt S Robson
PO J Hodgkinson
Sgt V R Duvall
Sgt W Housley
Sgt Pattrick
Sgt J Cartwright
6/7 Sep
MUNICH
2008-0344
PO H Wallis
Sgt T A Cooksey

Sgt L Tozer
Sgt A Holbrook
Sgt K Sims
Sgt D T Brewer
Sgt Sgt E Bremne
22/23 Sep
HANNOVER
1858-0054
PO R A Walker
Sgt H E Houldsworth
Sgt N J Cornell
PO J Wells
Sgt R C Bailey
Sgt C R Taylor
Sgt D R Kelly
23/24 Sep
MANNHEIM
1907-0150
PO R A Walker and crew
27/28Sep
HANNOVER
1912-0044
FO N F Turner and crew
29 Sep
BOCHUM
1844-2324
PO E J Nixon
Sgt W Craig
FS J W Devenish
Sgt A T Garrett
Sgt J E Chapman
Sgt W Leary
Sgt H F Bore
(engaged by three enemy fighters in the target area)

1/2 Oct
HAGAN
1814-2337
FO N F Turner and crew
WO M R Braines in, PO
Root-Reid out
2/3 Oct
MUNICH
1825-0235
SL E C Benjamin
PO J J Stephenson
FO H L Hewitt
Sgt F W Stead
Sgt H D Dinsdall
Sgt J Frawley
PO M Root-Reid
3/4 Oct
KASSEL
1853-0046
FO J E Williams
Sgt G N Whitley
PO A J Talbot
PO A J Walker
Sgt A E Potter
Sgt R Blagdon
Sgt K S Jewell
4/5 Oct
FRANKFURT
1828-0029
FO N F Turner and crew
Sgt R S Brown in PO
Root-Reid out
7/8 Oct
STUTTGART
2018-0240
FO N F Turner and crew

PO S R Hughes in, Sgt Brown
out
8/9 Oct
HANNOVER
2247-0404
FO K R Ames
Sgt V R Biggerstaff
PO A J Wright
Sgt T H Savage
FS H Glasby
Sgt S P Nicholas
Sgt R S Parle
(rear-gunner damaged a
German night fighter)
18 Oct
HANNOVER
1710-2220
SL E C Benjamin and crew
Sgt A B Woodvine and FS D
Grice in;
PO J J Stephenson and PO
Root-Reid out
20/21 Oct
LEIPZIG
1702-0041
FO N F Turner and crew
PO H H Farmiloe (2P)
22 Oct
KASSEL
1809-2102
FO N F Turner and crew
(ASI failed, returned early;
jettisoned 'cookie' over North
Sea)
3 Nov
DUSSELDORF

1700-2131
WC R N Stidolph
Sgt H J Anthony
FO J H Dyer
FO G F Aley
Sgt J D Barnes-Moss
Sgt E A Gardener
PO J H Pullman
10/11 Nov
MODANE
2101-0455
PO F J Nixon and crew
18/19 Nov
BERL!N
1737-0055
FO N F Turner and crew
FO G E Sharpe (2P)
FS J S Cook and Sgt G A
Davey in;
Sgt Freeth and PO Root-Reid
out
22 Nov
BERLIN
1623-2315
SL E C Benjamin and crew
(SL Benjamin tour-expired 5
December)
2 Dec
BERLIN
1655-2335
FO N F Turner and crew
20 Dec
FRANKFURT
1712-2000
SL S J Beard DFC
FS J K Burnside
FL J R Anderson
WO D C Davies
Sgt J D Barnes-Moss
PO J C Hodgkins
WO J Graham
(returned early when starboard
inner failed, jettisoned 'cookie')
29/30 Dec
BERLIN
1712-0004
PO H H Farmiloe
Sgt G A Jerry (FE)
Sgt T K Telfer (N)
Sgt K Vowe (BA)
Sgt E A Davidson
Sgt H Sherlinker (MU)
Sgt R Noble (RG)

1944
1/2 Jan
BERLIN
0008-0830
PO H H Farmiloe and crew
Sgt W M Morrison in, Sgt
Sherlinker out
5/6 Jan
STETTIN
0002-0955
WC R N Stidolph and crew
(attacked by Me210 at
0358hrs and badly damaged
after bombing)
1/2 Mar
STUTTGART
2324-0712
FL N F Turner and crew
FS C J Woolnough (2P)
FS R A Bunyan and Sgt H J
Silzer RCAF in;
WO Russell and Sgt Davey
out

10/11 Mar
CHATEAUROUX
1958-0133
FL N F Turner and crew
FL G L Dunstone and PO
Davey in;
FSs Bunyan and Silzer out
WO W Mainwaring (3AG)
(also attacked airfield SE of
target with machine-gun fire
from zero feet)
15/16 Mar
STUTTGART
1904-0237
FL N F Turner and crew
FO J G Cox (2P)
24/25 Mar
BERLIN
1840-0109
FL N F Turner and crew
26/27 Mar
ESSEN
2006-0038
FL G A Berry RCAF
Sgt F C Astell
FS A G Williams
FO R T Reid RCAF
Sgt E F Sutton
Sgt L G Brand
Sgt R W Levett
30/31 Mar
NURNBERG
2201-0503
PO F Norton
Sgt K Chapman
FS W Webb
FS K Clement
FS H Eldrett
FS J Tollitt
FS S Calver
5/6 Apr
TOULOUSE
2037-0403
PO F Norton and crew
FS N Audley (3AG)
(front escape hatch blown off
on bomb run and had to
make second run)
10/11 Apr
TOURS
2231-0422
FS J Kramer RCAF
Sgt N H Shergold
Sgt R W Burkwood
Sgt C W Greenaway
Sgt P Donaghue
Sgt A N Avery
Sgt R F Coleman
11/12 Apr
AACHEN
2046-0039
FL G A Berry and crew
(starboard inner u/s over target;
returned on three engines)
18/19 Apr
JUVISY
2057-0111
FL N F Turner and crew
20/21 Apr
PARIS
2305-0327
FL N F Turner and crew
FS O Baldwin in PO Davey
out
22/23 Apr
BRUNSWICK
2324-0538

PO F Norton and crew
PO G A Davey (3AG)
24/25 Apr
MUNICH
2057-0610
PO F Norton and crew
26/27 Apr
SCHWEINFURT
2148-0613
FO D Paul
Sgt J Bosworth
PO H Griffin
Sgt J Miller
Sgt F R Brazzier
Sgt S Billington
Sgt P McGierney
28/29 Apr
ST MEDARD-EN-JALLES
2250-0616
PO F Norton and crew
Sgt C C Scrimshaw (3AG)
(ordered not to bomb due to
smoke and haze, bombs jetti-
soned over sea)
1/2 May
ST MARTIN-DU-
TOUCH
2126-0530
PO F Norton and crew
3/4 May
MAILLEY-LE-CAMP
2150-0339
PO F Norton and crew
FS R Coxon (3AG)
6/7 May
LOUAILLES
0034-0535
PO F Norton and crew
FS C Baldwin (3AG)
24/25 May
EINDHOVEN
2305-0122
PO F Norton and crew
(sortie aborted; port inner
u/s)
27/28 May
NANTES
2240-0355
PO F Norton and crew
(FL Turner tour-expired early
June, to No 17 OTU)
31/1 Jun
SAUMUR
2328-0350
PO D C Freeman
Sgt P E Cook
Sgt E J Coe
FS J W Morris
Sgt J H Heasman
Sgt D Gordon
Sgt L C Whitehead
(hit by lightning, returned to
base)
3/4 Jun
FERME D'URVILLE
2315-0301
PO D C Freeman and crew
5/6 Jun
ST PIERRE-DU-MONT
0241-0648
PO D C Freeman and crew
6/7 Jun
ARGENTAN
2357-0343
PO R H Passant RAAF
Sgt T E Hunt
Sgt P Uren

Sgt A M Frew
FS G E Nash RAAF
Sgt C Howard
Sgt D Copson
8/9 Jun
RENNES
PO W North
2316-0645
Sgt L Morton
FS H Crawley
FS N E Jarvis
Sgt G I Monteith
Sgt D A Bartlett
Sgt E W Ravenhill
10/11 Jun
ORLEANS
2218-0345
PO F Norton and crew
12/13 Jun
POITIERS
2215-0434
FL F Norton and crew
14/15 Jun
AUNAY
2212-0250
FL F Norton and crew
FO H L Inniss (2P)
19/20 Jun
WATTEN/V1 SITE
2215-0117
FL F Norton and crew
(aborted after recall signal
due to poor weather)
21/22 Jun
GELSENKIRCHEN
2304-0205
FL F Norton and crew
(aborted when starboard
inner engine caught fire)
23/24 Jun
LIMOGES
2222-0505
FL F Norton and crew
FS C Baldwin in, FS Calver
out
24/25 Jun
PROUVILLE/V1 SIT
2252-0203
FO H L Inniss
Sgt H T Ansell
FO J T Dickinson
FO J R Morrison
Sgt K F Moseley
Sgt J Aldridge
Sgt R R Langley
27/28 Jun
VITRY
2204-0514
FL F Norton and crew
FS J Slome in, Sgt Chapman
out
29 Jun
BEAUVOIR/V1 SITE
1153-1522
FL F Norton and crew
Sgt G A Jerry in, FS Slome out
4/5 Jul
ST LEU D'ESSERENT
2309-0325
FL F Norton and crew
7/8 Jul
ST LEU D'ESSERENT
2221-0248
FL F Norton and crew
(returned early, port inner
engine u/s)

12/13 Jul
CULMONT
2153-0616
FL F Norton and crew
15/16 Jul
NEVERS
2219-0523
FS K J Burnside in, Sgt Jerry out
18 Jul
CAEN
0353-0749
FL F Norton and crew
Sgt Jerry and FS Calver back
18/19 Jul
REVIGNY
2303-0411
FS S Parker
FS J K Burnside
Sgt H J Richardson
Sgt J H Wellens
Sgt K G West
Sgt K Smith
Sgt C H Stothard
20/21 Jul
COURTRAI
2326-0242
FL F Norton and crew
FS J K Burnside in, Sgt Jerry out
24/25 Jul
DONGES
2219-0350
FO R F J Heath
Sgt P Davis
FO F J Kelly
FO A A Lorton
Sgt C F Ruane
Sgt E D Fisher
FS R Nicholson
26/27 Jul
GIVORS
2113-0635
FL F Norton and crew
FS R G Poulter in, FS Burnside out
(FL Norton tour-expired early August, posted to No 1660 CU on 28 August)
30 Jul
CAHAGNES
0617-1134
FO R F J Heath and crew
(raid aborted on orders from controller)
31 Jul
JOIGNY-LA-ROCHE
1741-2252
FO R F J Heath and crew
FS J Ryan in, FS Nicholson out
1 Aug
SIRACOURT/V1 SITE
1509-1926
FO R F J Heath and crew
(raid aborted on orders from controller - fog)
2 Aug
BOIS DE CASSAN
1440-1920
FO R F J Heath and crew
3 Aug
TROSSY ST MAXIM
1208-1625
FO R F J Heath and crew

5 Aug
ST LEU D'ESSERENT
1108-1515
FO H L Inniss and crew
7/8 Aug
SECQUIVILLE
2116-0134
FL F Norton and crew
Sgt C C Scrimshaw (3AG)
(ordered to abort on run-up to target)
9/10 Aug
CHATELLERAULT
2028-0245
FL F Norton and crew
Sgt Scrimshaw (3AG)
11/12 Aug
GIVORS
2040-0436
FL F Norton and crew
FS J F Boland RAAF (2P)
12/13 Aug
RUSSELSHEIM
2128-0303
WO A P Greenfield RAAF
Sgt F F Fraser
Sgt W J A Gibb
Sgt W J Haddon
WO V P Smith
Sgt J P King
Sgt S D P Goodey
13 Aug
BORDEAUX
1631-2341
FO J S Cooksey
Sgt E D Rose
Sgt L P Meaker
Sgt T Stephenon
Sgt A R Marshall
Sgt S J Scarrett
Sgt N W Elliott
14 Aug
BREST
1744-2239
WO A P Greenfield and crew
15 Aug
GILZE RIJEN A/F
1010-1347
WO A P Greenfield and crew
16/17 Aug
STETTIN
2102-0526
FS J E Boland
Sgt J Boon
Sgt L Miller
Sgt J A Fitzgerald
Sgt L E Welch
Sgt H Cave
Sgt J L Jones
18 Aug
BORDEAUX
1126-1554
WO A P Greenfield
Sgt F F Fisher
FO A A W Larton
FO F J Kelly
FS A C R Brydges
FS J Ryan
Sgt S D Fisher
19 Aug
LA PALLICE
0524-1152
FS J E Boland and crew
25/26 Aug
DARMSTADT
2022-0508
FL D E White DFC RCAF

Sgt A Lyon
PO M B Blackwood
FO A E Jones
WO H F Harron
FS L E Gibb
FS L G Anderson
26/27 Aug
KONIGSBERG
2006-0702
FO H L Inniss and crew
31 Aug
ROLLENCOURT
1609-2000
Sgt M L Hunt RNZAF
Sgt F Askew
FS A T Harvey
FS A T Bell
Sgt R Nesbitt
Sgt S H H Jobson
Sgt J W Harrison
5 Sep
BREST
1604-2038
FO A P Greenfield and crew
10 Sep
LE HAVRE
1513-1852
FO A P Greenfield and crew
11 Sep
LE HAVRE
0413-0950
FO N P Blain RCAF
Sgt F T Nicholls
Sgt R E Fulcher RCAF
FO K S Porter
Sgt K C Alder
Sgt A E Smith
FS D G Clement RCAF
11/12 Sep
2043-0300
DARMSTADT
FO A P Greenfield and crew
(damaged Me109 in fighter attack)
17 Sep
BOULOGNE
0743-1120
FO A P Greenfield and crew
18 Sep
BREMERHAVEN
1805-2302
FO A P Greenfield and crew
19/20 Sep
RHEYDT
1831-2233
FO A P Greenfield and crew
(sortie aborted, R/T u/s)
23/24 Sep
MUNSTER
1904-0043
FO A P Greenfield and crew
26/27 Sep
KARLSRUHE
0054-0724
FO A P Greenfield and crew
27/28 Sep
KAISERSLAUTERN
2233-0414
FO A P Greenfield and crew
(whole bomb-load hung-up)
5 Oct
WILHELMSHAVEN
0738-1326
FO A R Goodbrand
Sgt F C Sayer
FO S A Reeves
Sgt W Devane

Sgt E A Tyler
Sgt A H Siddons
Sgt P D Tointon
6 Oct
BREMEN
1753-2219
FO C A Donnelly RAAF
Sgt A H Steers
FL J H Vincent
FS Green
Sgt R G D Brock
Sgt L Ayres
Sgt T J Kerrigan
11 Oct
FLUSHING
1337-1713
FO C A Donnelly and crew
14/15 Oct
BRUNSWICK
2233-0155
FO A P Greenfield and crew
(sortie aborted due to pilot being ill; jettisoned 'cookie')
19/20 Oct
NURNBERG
1730-0055
FO R Edwards
Sgt J R Owen
Sgt A S Madden
Sgt A Grant
FS C Brookes
Sgt J Clarke
Sgt J Rogers
23 Oct
FLUSHING
1436-1749
FO A P Greenfield and crew
28/29 Oct
BERGEN
2221-0500
FO A P Greenfield and crew
1 Nov
HOMBERG
1340-1833
SL H W Horsley
FS L Morton
FS S Fleet
FS R J Sawyer
FS J Chapman
FS H A Grahn RCAF
Sgt R Hoskinson
2/3 Nov
DUSSELDORF
1629-2144
FO R A Lushey RAAF
Sgt G Goodier
Sgt J W A Brewster
Sgt L Brewin
Sgt H A Parsons
Sgt R L Humphries
Sgt W A Fox
4 Nov
DORTMUND-EMS CANAL
1743-2222
FO J F Swales
Sgt A J M Davies
FO C H Saunders
Sgt R Taylor
FS D M Easton
Sgt T Torney
FS H P L Hardy
6 Nov
GRAVENHORST CANAL
1648-2232
SL H W Horsley and crew

(did not bomb, brought bombs back)
11/12 Nov
HARBURG
1613-2155
FO H R Smith
Sgt R Harris
Sgt R A Williams
Sgt M Gibson
FS R Schmidt RAAF
FO G R Bennett RCAF
FO G Bobenic RCAF
16Nov
DUREN
1255-1745
FO H R Smith and crew
(hit by falling bomb)
21 Nov
DORTMUND-EMS CANAL
1747-2312
FL A P Greenfield and crew
22/23 Nov
TRONDHEIM
1539-0144

FO E W Hutchins and crew (ordered to abort, bombs jettisoned)
26/27 Nov
MUNICH
2328-0905
FO B S Tasker
Sgt T McKnight
FO E Walker
FS W P R Boobyer
WO N G McKenzie
Sgt S W Herring
Sgt V A Edwards

1945
1 Jan
GRAVENHORST CANAL
1637-2105
FO R A Mouritz
Sgt A J Leith
FS L A Cooper
FO P M R Smith
FS D C Bloomfield
Sgt A Bass

Sgt D C Cluett
4/5 Jan
ROYAN
0045-0749
FL A P Greenfield and crew
FS G A Robson in, Sgt Haddon out
5/6 Jan
HOUFFALIZE
0054-0538
FO R Edwards
Sgt J R Owen
Sgt A G Madden
FS A Grant
Sgt D S Tandy
Sgt A McQuilkin
Sgt J Rogers
7/8 Jan
MUNICH
1649-0307
FL A P Greenfield and crew
Sgt L Muir in, Sgt Sayer out
13 Jan
POLITZ
1643-1847

FO E W Hutchins
Sgt E Cleave
Sgt N C Shepherd
FS G F Burmaster
Sgt P W Scoley
FS A G Brookes
FS W G Webb
(aborted when mid-upper turret went u/s)
14/15 Jan
MERSEBERG
1609-0223
FO E W Hutchins and crew
(the three raids noted as being flown by JB138 in November 1944 when the aircraft record card notes it was under repair, were 21/22 Dortmund-Ems Canal – FL Greenfield and crew; 22/23 Trondheim – FO E W Hutchins and crew; 26/27 Munich – FO B S Tasker and crew)

JB603
TAKE IT EASY

This Mk III Lancaster came off the production line in the autumn of 1943, part of contract No 1807 built at Manchester, with four Merlin 38 engines. Assigned to No 100 Squadron on 3 November at Waltham, near Grimsby, Lincolnshire, it was given the squadron code letters of HW and the individual letter E-Easy.

From this it was named 'Take it Easy', the legend appearing below a large flying bird painted beneath the cockpit. Ahead of this, rows of bomb symbols in tens began to appear regularly, the bombs being slightly angled rather than the more normal vertical ones on other aircraft.

Easy arrived on the squadron just as the Battle of Berlin was warming up, and in fact, of its first 15 raids, no fewer than 12 were to the Big City. These also brought moments of excitement. On Easy's second trip heavy flak scored a hit in the mainplane whilst flying over Hannover, then on the fifth trip its crew had a scrap with a night fighter which the gunners claimed as destroyed. Then on the first raid of 1944 the inner starboard petrol tank and inner engine cowling were holed by machine-gun fire, believed to have come from another Lancaster.

A period of mechanical and crew problems in March 1944 led to three aborts in succession but thereafter followed a trouble-free month, until a flak shell went through the starboard wing between the inner engine and the fuselage on a trip to Friedrichshaven on 27/28 April. The fuel had to be emptied from No 1 tank in that wing but the sortie was completed.

Easy's first regular skipper was Flight Sergeant T. F. Cook who arrived on 7 January (he was later commissioned), and whose rear-gunner, Flying Officer H. G. D. Pawsey was on his second tour. Terry Cook did some 22 ops in Easy and won the DFC, his flight engineer, Harry Widdup, navigator Eric W. Norman and bomb-aimer H. P. Peachey collecting DFCs, as did Harold Pawsey too.

As JB603 was one of three Lancaster veterans to operate with No 100 Squadron, it was almost inevitable that crews who flew in Easy would also have flown in the others. So occasionally familiar names are in the crew lists. Pilot Officer D. W. Lee was one pilot that flew Easy a great deal after Cook became tour-expired. In fact he did 20 trips in the aircraft, including the last op of his tour, on 5/6 October 1944, which was about Easy's 85th.

During this period Easy had flown twice on D-Day, operated over V1 sites and supported the armies in France following the invasion, and all with a minumum of fuss and breakdowns. The only problems seemed to come from flak, with the

Lanc being hit over Essen in daylight on 25 October, which damaged its wings, tailplane and bomb-doors, and again on a daylight to Dortmund on 11/12 November.

Easy's last regular captain was Flying Officer L. T. Harris, who did 18 ops in the aircraft between September and December 1944. Again it is not possible to be exact on when Easy flew its 100th op, although it had to be about mid-November, but a couple of mechanical aborts confuse the picture somewhat.

However, by early 1945 Easy was into its second hundred, some records noting that the aircraft flew a total of 112 ops, while the squadron diary notes that it failed to return from Hannover on 5/6 January, while on its 111th sortie. Easy was flown on this fateful occasion by Flying Officer Reg Barker and crew, all but one of whom were killed and buried in Gorssel Parish Cemetery, near Deventer, Holland. No 1 Group Headquarters had received the crew's target attacked message at 2206 hours – then silence.

1943
No 100 Squadron
18/19 Nov
BERLIN
1726-0111
PO F H Tritton
Sgt K Wright (FE)
WO A Selman (N)
Sgt G Hopkins (BA)
Sgt H O Hinderwell
PO T D Seager (MU)
Sgt J C Knox (RG)
22 Nov
BERLIN
1648-2307
PO F H Tritton and crew
(damaged by flak over
Hannover)
23 Nov
BERLIN
1744-2332
FO F H Tritton and crew
26/27 Nov
BERLIN
1655-0055
PO F H Tritton and crew
2/3 Dec
BERLIN
FS C W Henderson
1644-0037
PO C N Waite
FO J M Ogilvie
Sgt C H Hendry
Sgt N Bowman
Sgt R D Stoneman
Sgt D Sissons
(gunners shot down
twin-engined night fighter)
3/4 Dec
LEIPZIG
0009-0758
FS E E Tunstall
Sgt J Wunderley
Sgt J C Sharp
PO J A Honey
Sgt J J Green
Sgt W Essar
Sgt R Allison

16 Dec
BERLIN
1617-2334
WO T V Hayes
Sgt P L Ashenden (FE)
Sgt S E Emmett (N)
Sgt W Kondra RCAF
Sgt G R Jenkins(WOP)
(BA)
Sgt K W Kemp (MU)
Sgt J S Ross (RG)
20 Dec
FRANKFURT
1708-2233
WO T V Hayes and crew
23/24 Dec
BERLIN
2330-0739
FS J A Crabtree
Sgt R T Davies
FO M O Rees
Sgt J W Knight
Sgt F Holm
Sgt J J Whelan
Sgt T S Sanders
29 Dec
BERLIN
1650-2314
WO T V Hayes and crew

1944
1/2 Jan
BERLIN
WO T V Hayes and crew
(shot-up, possibly by another
Lancaster)
21/22 Jan
MAGDEBERG
2023-0018
FO D F Gillam
Sgt K Talbot
FS K Drury
Sgt D C Gemmell
Sgt W Moffatt
Sgt H R Crompton
Sgt G H Warren
(aborted when Gee went u/s
and the H2S was intermit-
tently u/s)

27/28 Jan
BERLIN
1756-0210
Sgt K W Evans RAAF
Sgt J J Lapes
Sgt P Atha
Sgt D Francis
Sgt J Armstrong
Sgt S C Brookes
Sgt F Whitehouse
30 Jan
BERLIN
1729-2344
FS T F Cook
Sgt H Windup
FO E W Norman
Sgt J C Stewart RCAF
Sgt H P Peachey
Sgt A K Burchell
FO H G D Pawsey
15/16 Feb
BERLIN
1735-0005
FS T F Cook and crew
19/20 Feb
LEIPZIG
2350-0710
FS T F Cook and crew
Sgt J Hewitson in, Sgt
Burchell out
(port outer engine u/s, possi-
bly due to flak on home jour-
ney)
24/25 Feb
SCHWEINFURT
1810-0210
FS T F Cook
Sgt E R Belbin (2P)
Sgt R A Cassell
FO E W Norman
FS K G Wilde
Sgt F T Baldwin
Sgt E T Duckett
Sgt J R Truman
1/2 Mar
STUTTGART
2330-0740
FS T F Cook and crew
(main oil pipe for rear-turret

burst after crossing French
coast)
15/16 Mar
STUTTGART
1855-2205
FS T F Cook and crew
(aborted due to rear-gunner's
electrical suit failing)
18/19 Mar
FRANKFURT
1905-0100
FS T F Cook
Sgt R W Pryce
FO E W Norman
FO J Spector RCAF
Sgt A J Alcott
Sgt F Parrish
FO H G D Pawsey
22/23 Mar
FRANKFURT
1915-0055
FO E L Eames
Sgt R W Pryce
FS D S Kirkwood
RCAF
FO J Spector RCAF
Sgt A J Alcott
Sgt F Parrish
Sgt C Bird
24/25 Mar
BERLIN
1900-0210
FS J Littlewood
Sgt C McCartney
FS D A Tovell RCAF
Sgt J G Hughes
Sgt R W Gilbey
Sgt S C Smith
Sgt J Taylor
26/27 Mar
ESSEN
2015-0115
FO E L Eames and crew
30/31 Mar
NURNBERG
2220-0525
FS D T Fairbairn
Sgt H R Tufton
FO F Tovery

109

FO J M Wilder
Sgt L Gibbons
Sgt G Tunstall
Sgt J G Wookey
9/10 Apr
MINING/GDYNIA
2135-0555
FS T F Cook and crew
Sgt J G Wookey in, FO Pawsey out
10/11 Apr
AULNOYE
2330-0450
FS T F Cook and crew
11/12 Apr
AACHEN
PO E D King
Sgt R W Bland
FO L H Salt
Sgt F W Cheetham
Sgt S Beardsall
Sgt D Ralston
Sgt D W Young
18/19 Apr
MINING/BALTIC
2105-0355
FS T F Cook and crew
20/21 Apr
COLOGNE
2335-0359
FS T F Cook and crew
22/23 Apr
DUSSELDORF
2230-0309
FS T F Cook and crew
24/25 Apr
KARLSRUHE
2209-0432
FS T F Cook and crew
PO R P Anderson in, Sgt Widdup out
26/27 Apr
ESSEN
2300-0310
FS T F Cook and crew
Sgt Widdup back
27/28 Apr
FRIEDRICHSHAVEN
2200-0610
FS T F Cook
PO J A Orr (2P)
Sgt H Widdup
FO E W Norman
Sgt J M Campbell
Sgt W J Strange
Sgt H W K Welsby
FO H G D Pawsey
(damaged by flak)
3/4 May
MAILLY-LE-CAMP
2155-0320
FO E L Eames and crew
7/8 May
BRUZ
2140-0250
PO J D Rees
Sgt M J Dunphy
FS J Amory
FO E Jackson
Sgt A Palmer
Sgt T L Daly
Sgt V E Locke
9/10 May
MERVILLE
2155-0105
PO T F Cook and crew
PO T G Page RNZAF (2P)

Sgt D Henderson (FE)
Sgt D M Jones (2AB)
Sgt R Watson (RG)
10/11 May
DIEPPE
2235-0140
FL E L Eames and crew
FO T Slater in, FO Spector out
21/22 May
DUISBURG
2245-0305
PO T F Cook and crew
22/23 May
DORTMUND
2240-0250
PO T F Cook and crew
24/25 May
LE CLIPON
2235-0105
PO T F Cook and crew
FS R W W Pye (MU)
27/28 May
MERVILLE
2345-0323
PO T F Cook and crew
28/29 May
EU, NR CAYEAUX
2224-0200
PO W Kay
Sgt H Dale
Sgt F H Fulsher
FO J Frink
Sgt E Harrop
Sgt T E Sharpley
Sgt W E Struck
31/1 Jun
TERGNIER
0001-0421
PO W Kay and crew
Sgt W Everitt in, Sgt Struck out
2/3 Jun
BERNAVAL
2335-0310
PO T F Cook and crew
5/6 Jun
ST MARTIN DE VARREVILLE
2135-0145
PO T F Cook and crew
6/7 Jun
VIRE
2155-0305
PO T F Cook and crew
7/8 Jun
FORET DE CERISY
2320-0400
PO J H Shaw
FL C N Waite
Sgt J G Locke
FS B W Young
Sgt W J Jones
Sgt J G Gorman
Sgt W Everitt
10/11 Jun
ACHERES
2305-0400
PO A J Orr
Sgt H Gibson
Sgt W Beet
Sgt J M Campbell
Sgt W J Strange
Sgt H W K Welsby
Sgt C H Barron
14/15 Jun
LE HAVRE
2045-0015

PO J H Shaw and crew
16/17 Jun
DOMLEGER/V1 SITE
0015-0405
FO D S Milne
Sgt W H Lanning
Sgt R A D Newman
FO R B Hutchinson
Sgt B Nundy
Sgt H Taylor
Sgt K Yuelett
7 Jul
CAEN
1950-2335
FO C M Stuart RCAF
Sgt H Prince
FO R H Rix
Sgt P Burnett
Sgt P W T Dunn
FO J F Insell
Sgt K S Kowal
12/13 Jul
TOURS
2130-0400
FO C M Stuart and crew
14/15 Jul
REVIGNY
2125-0500
FO D W Lee RNZAF
Sgt F B Whitehouse (FE)
FS K G Braithwaite
Sgt E Atherton (BA)
Sgt R Parkin (WOP)
Sgt L W Histed (MU)
Sgt R Spencer (RG)
(sortie abandoned at 0156 hours)
18 Jul
SANNERVILLE/CAEN
0355-0740
PO D W Lee and crew
20/21 Jul
COURTRAI
0010-0310
PO D W Lee and crew
23/24 Jul
KIEL
2305-0340
PO D W Lee and crew
25 Jul
COQUEREAUX/V1 SITE
0705-1050
PO K I Cole RAAF
Sgt F Ritchie
Sgt J Jefferies
Sgt A Paterson
Sgt A Plastow
Sgt K J Hamilton
Sgt G V George
25/26 Jul
STUTTGART
2130-0545
PO J A Orr RNZAF
Sgt H Gibson
FS W Boot
FS J M Campbell
Sgt W J Strange
Sgt H W K Welsby
Sgt L Cohen
28/29 Jul
STUTTGART
2130-0535
PO D W Lee and crew
30 Jul
VILLERS-BOCAGE/
CAUMONT
0630-1030

PO D W Lee and crew
31/1 Aug
FORET DE NIEPPE
2155-0055
PO D W Lee and crew but with
FO L S Bell RCAF (2P)
Sgt T A Kewley (FE)
FO R Watson (BA)
Sgt E A Pocock (WOP)
Sgt T Brennan (MU)
3 Aug
TROSSY ST MAXIM
1135-1600
PO D W Lee and crew
4 Aug
PAUILLAC
PO D W Lee and crew
1325-2125
5 Aug
PAUILLAC
1430-2245
WO W T Ramsden
Sgt E G Stubbings
Sgt S T Howard
FO R P Simpson
Sgt R M Chestnutt
Sgt R J Williams
Sgt N Crompton
7/8 Aug
FONTENAY-LE-MARMION
2120-0125
FO L S Bell RCAF
Sgt T A Kewley
PO T M Shewring
FO R Watson
Sgt E A Pocock
Sgt T Brennan
Sgt C Barker
10 Aug
VINCLY/V1 SITE
1030-1420
FS D W McKenzie RNZAF
Sgt R McLelland
Sgt F R Ford
FS W J D Allen RNZAF
WO H E Thornley
FS J F Malvern
Sgt C W Anderson RCAF
(did not bomb due to cloud over target)
11 Aug
DOUAI
1340-1735
FO L S Bell and crew
14 Aug
FALAISE
1210-1600
FO L S Bell and crew
15 Aug
VOLKEL A/F
1020-1340
FS D W McKenzie and crew
16/17 Aug
STETTIN
2123-0520
WO J Thompson RAAF
Sgt H R Tufton
FO J Honeyman
FO S C Hatfield
FS J R Lyons
FS E K Lindorff
FS J Roberts
18/19 Aug
RIEME
2220-0120

PO D W Lee and crew
WO M S Paff in, FS
Braithwaite out
25/26 Aug
RUSSELSHEIM
2005-0440
PO D W Lee and crew
PO T Batley (2P)
FS Braithwaite back; Sgt D
Clay (2BA)
26/27 Aug
KIEL
2015-0120
PO D W Lee and crew
29/30 Aug
STETTIN
2125-0545
PO D W Lee and crew
31 Aug
RAIMBERT
1235-1545
PO D W Lee and crew
3 Sep
GILZE RIJEN A/F
1610-2000
PO L T Harris
Sgt A F Lambert
Sgt H G Roberts
FS C H Smith
FS B E Lawton RAAF
Sgt G S Gilbert
Sgt C A Daventry-Bull
5 Sep
LE HAVRE
1615-1930
PO D W Lee and crew
6 Sep
LE HAVRE
1715-2035
PO D W Lee and crew
8 Sep
LE HAVRE
0710-1045
PO D W Lee and crew
(abandoned due to low
cloud over target)
10 Sep
LE HAVRE
1650-2015
PO D W Lee and crew
12/13 Sep
FRANKFURT
1825-0130
FO C M Stuart and crew
20 Sep
CALAIS
1524-1921
FO L T Harris and crew
23/24 Sep
NEUSS
1850-0010
FO L T Harris and crew
25 Sep
CALAIS
0735-1100

FO L T Harris and crew
(abandoned due to low
cloud over target)
26 Sep
CAP GRIS NEZ
FO L T Harris and crew
FS S Kowal in, Sgt
Daventry-Bull out
27 Sep
CALAIS
0830-1220
FO L T Harris and crew
Sgt A G Tipple in, FS Kowal
out
5/6 Oct
SAARBRUCKEN
1845-0020
PO D W Lee and crew
7 Oct
EMMERICH
1210-1610
FO L T Harris and crew
Sgt Daventry-Bull back, Sgt
Tipple out
14 Oct
DUISBURG
0650-1055
FO F O Griffiths
Sgt A Dawson
FS A D Cozens
FS M A Kravenchuck
Sgt E McGuire
Sgt J I Morgan
Sgt C Nelson
14/15 Oct
DUISBURG
0025-0545
FO F O Griffiths and crew
PO D M Ward (2P)
FO F C Squires (2AB)
19/20 Oct
STUTTGART
1645-0000
FO L T Harris and crew
23 Oct
ESSEN
1620-2150
FO L T Harris and crew
25 Oct
ESSEN
1250-1735
FO L T Harris and crew
(damaged by flak)
30 Oct
COLOGNE
1735-2325
FO L T Harris and crew
31 Oct
COLOGNE
1750-2305
FO L T Harris and crew
2 Nov
DUSSELDORF
1615-2130
FO L T Harris and crew

FO W O Nobes RCAF (2P)
Sgt N L Warren RCAF (2BA)
4 Nov
BOCHUM
1745-2220
FO L T Harris and crew
6 Nov
GELSENKIRCHEN
1155-1640
FO C O P Smith RCAF
Sgt J Kieran
Sgt F G Dean RCAF
Sgt J Newlan RCAF
FS C Jones RCAF
Sgt G H Booth RCAF
Sgt A W Jenkins RCAF
6 Nov
WANNE EICKEL
0815-1300
FO C O P Smith and crew
11 Nov
DORTMUND
1635-2125
FO R T Hoyle
Sgt P T Bickley
PO G S Charles
Sgt R A Ward
Sgt A E Law
PO L A Hoptroff
Sgt R Marshall
16 Nov
DUREN
1250-1800
FO L T Harris and crew
21 Nov
ASCHAFFENBURG
1540-2220
FO W O Nobes RCAF
Sgt J G Kerr
FO J H Kimpton
Sgt N L Warren RCAF
Sgt R Doherty
Sgt C L Taylor
Sgt L A Schofield
27 Nov
FREIBURG
1610-2250
PO D N Shrimpton
Sgt C W Overton
Sgt J B Booth
Sgt R Livingstone
Sgt H J Cleghorn
Sgt R A Tilly
Sgt C Reilly
29 Nov
DORTMUND
1210-1610
FO L T Harris and crew
(abandoned sortie at 1422
hours; starboard outer
engine cut and unable to
maintain height)
3 Dec
URFT DAM
0755-1220

FO L T Harris and crew
(sortie abandoned as
instructed by controller due
to cloud over target)
6/7 Dec
MERSEBURG
1640-0040
FO C O P Smith and crew
12 Dec
ESSEN
1615-2200
FO L T Harris and crew
15 Dec
LUDWIGSHAVEN
1440-2110
FO L T Harris and crew
17 Dec
ULM
1515-2300
FO L T Harris and crew
21 Dec
BONN
1455-2055
FO L T Harris and crew
28 Dec
MUNCHEN
GLADBACH
FO J A Scholey RCAF
Sgt G Grundy
FO D H Lennox RCAF
1545-2120
FO H F Amies RCAF
FO A A Templeton
FS W L O'Shea RCAF
FS J A E Willis RCAF
29 Dec
GELSENKIRCHEN
1510-2115
FO J A Scholey and crew

1945
2 Jan
NURNBERG
1520-2340
WO W H Evans
Sgt J J Paxton
Sgt J L Pearson
Sgt A W Dack
Sgt J G Sutherland
Sgt K W J Hodges
Sgt F Burdett
5/6 Jan
HANNOVER
1910-
FO R Barker
Sgt A S Gordon
FS F S Elliott
FS A A Law
PO J M C Wilson
Sgt E Gillen
Sgt B G Aldred
(Failed to return)

111

JB663
KING OF THE AIR

Named 'King of the Air', JB663 came off Avro's Woodford production line with four Merlin 38 engines in the autumn of 1943. This Mk III was then assigned to No 106 Squadron based at Syerston, Nottinghamshire, on 15 November, where it was marked with the squadron codes of ZN and given the individual aircraft letter A-Able.

The Battle of Berlin was on when LB663 began ops and the first trip on 26/27 November was to the Big City, followed by eight more, plus a spoof raid to Magdeburg when the Main Force hit Berlin again on 21/22 January 1944.

JB663 was flown by senior and junior ranks alike, including the Squadron Commander, Wing Commander R. E. Baxter DFC, the Flight Commander, Squadron Leader A. R. Dunn – the latter 10 times – and later his replacement, Squadron Leader E. Sprawson. Eric Sprawson flew this aircraft once only, but he had an unusual experience a month afterwards when he was shot down on the night of D-Day and he and his bomb-aimer evaded and reached Allied positions in Normandy the following month. Albert Dunn and Sprawson each won the DFC.

One of JB663's more regular skippers was Pilot Officer B. F. Durrant, who first flew in April and then completed 28 of his 34 trips in this Lanc by the end of July. Brian Durrant received the DFC. Another senior airman to fly JB663 was the Station Commander, Group Captain W. N. McKechnie, holder of the George Cross who, of course, had no need to fly ops at all. Unhappily, William Neil McKechnie was killed in action flying in another Lanc on 29 August 1944.

JB663's next regular captain was Flight Lieutenant S. H. Jones, who took the aircraft on eight trips, including one with a BBC reporter on 18 August 1944, but Jones failed to return in ND868 on 23 September. JB663 then had various skippers until late in its career. Flying Officer L. P. Bence RAAF took over on 11 November and flew the Lanc on its last 11 trips, although two of these were abandoned by the bombing controller while a third had to be aborted when the starboard inner engine developed a glycol leak.

On occasion No 106 Squadron acted as Pathfinders and, on 10/11 March 1944, Wing Commander Baxter flew in JB663 (captained by Albert Dunn) acting as Leader and Pathfinder Controller in an attack on the Michelin factory at Clermont-Ferrand in France by 33 Lancs from No 5 Group. On the night of 23/24 March, JB663 dropped flares continually to illuminate No 617 Squadron's attack on an aero-engine factory at Lyons. Flight Lieutenant D. U. Gibbs was the pilot, but when all the flares had gone down there were none left for his own attack. Gibbs dropped two green target indicators from low-level to help guide 617 further but the target could not be located so 617's leader gave the order to abandon the raid.

JB663 flew on both the night ops of D-Day with Brian Durrant in command, then against several V1 sites during June and transport centres in July. The Lanc's groundcrew consisted of Sergeant G. W. Vidgen, Corporal W. Black, Leading Aircraftsmen L. A. Smith, J. Picker and A. Hirst, and Aircraftsman First Class R. J. Roberts.

Following its earlier Pathfinding role, the squadron finally became part of PFF in September 1944 and the squadron crews were given the opportunity of volunteering for PFF work (which usually meant a longer tour of ops) or they could go to other Main Force squadrons. Flight Lieutenant Jones was among those who stayed but failed to return before the month was out as stated above, but he appears to have survived.

JB663's 100th mission was recorded as 4 November 1944, a raid on the Dortmund-Ems Canal. The following morning a huge bomb was painted on the aircraft's nose next to the 99 bombs already there, in five neat, slanting rows of 20 and one of 19 bombs. Above this impressive scoreboard with wings outstretched was an eagle, looking much the same as the RAF eagle, and above this were the words 'King of the Air'.

In all JB663 flew 111 operational sorties, and statistics compiled on this particular aircraft noted that it had flown 985 hours, covered 150,000 miles and carried over 600 tons of bombs and incendiaries to targets in Germany and France. Its final trip was a seven-hour sortie to a synthetic oil plant at Brux in Western Czechoslovakia.

JB663 became Cat B on 11 April 1945 at No 24 MU, whence the aircraft had gone after its final trip. It was eventually at No 15 MU in August 1946 and was Struck off Charge on 26 October of that year.

1943
No 106 Squadron
26/27 Nov
BERLIN
1730-0120
FL M I Boyle
Sgt A T Cox (FE)
Sgt H Dixon (N)
FS D R Waddell (BA)
Sgt M Webb (WOP)
Sgt J H Higdon (MU)
Sgt M P Butler (RG)
3/4 Dec
LEIPZIG
0001-0805
SL A R Dunn
Sgt T E Eddowes
Sgt F Mycoe
FL R L Wake
FS E G King
Sgt R Nightingale
Sgt D Pinckard
16/17 Dec
BERLIN

1625-0020
FL M I Boyle and crew
20 Dec
FRANKFURT
1705-2320
FL M I Boyle and crew
Sgt R Edwards in, FS King out
23/24 Dec
BERLIN
2335-0735
SL A R Dunn and crew
Sgt E Fortune in, FL Wake out
29/30 Dec
BERLIN
1620-0020
FO J B Latham
Sgt H Weinow
FL W Williamson
FO R T Martins
Sgt T E Witts
Sgt D J Crowley
Sgt H Burn

1944
1/2 Jan
BERLIN
2359-0740
SL A R Dunn and crew
5/6 Jan
STETTIN
0010-0850
FO J B Latham and crew
14/15 Jan
BRUNSWICK
1650-2210
FO W B Jardine
Sgt J P Olive
FO G H Wright
Sgt A Dunae
Sgt J R Whitehead
Sgt J J Phillips
FO J W Paige
20 Jan
BERLIN
1635-2345
FS G S Milne RCAF
Sgt P J Butcher

FS J B Bevan
FS S J Halvorsen
Sgt M J Kimber RCAF
Sgt S S Harris
FS G Whittaker
21/22 Jan
MAGDEBURG
2000-0425
PO D U Gibbs
Sgt W R Mason
FS R Appleyard
FO D L Cramp
Sgt H Stubbs
Sgt J E Charnock
Sgt R F Birch
27/28 Jan
BERLIN
1750-0155
FO E R F Leggett
Sgt E F Windeatt
FS A G Mearns
FO F B Chubb
Sgt T H Jones
PO S W Payne

Sgt J C Harrison
29/30 Jan
BERLIN
0050-0520
FS G S Milne and crew
(sortie aborted due to engine trouble)
30/31 Jan
BERLIN
1715-0005
FO E R F Leggett
FOs A E Bristow and A I Johnson in;
FS Mearns and FO Chubb out
15/16 Feb
BERLIN
1710-0040
SL A R Dunn and crew
19/20 Feb
LEIPZIG
2345-0700
SL A R Dunn and crew
FO E L Sharp (2N)
20/21 Feb
STUTTGART
2355-0450
FO E R Penman
Sgt R N Johnson
FO E L Sharp
FO E O Aaron RCAF
Sgt S R Patti
Sgt R F Stubelt RCAF
Sgt J A Roberts
(sortie abandoned, pitot-head cover not removed)
24/25 Feb
SCHWEINFURT
1830-0230
WC R E Baxter DFC
Sgt T E Eddowes
FS F Mysoe
Sgt E Fortune
WO E G King
Sgt R Nightingale
FS D Pinckard
25/26 Feb
AUGSBURG
1830-0215
SL A R Dunn and crew
1/2 Mar
STUTTGART
2310-0720
SL A R Dunn and crew
10/11 Mar
CHATEAUROUX/DEOLS
2000-0140
SL A R Dunn and crew
WC R Baxter flying as raid controller
15/16 Mar
METZ
2100-0155
SL A R Dunn and crew
16/17 Mar
CLERMONT-
1915-0220
SL A R Dunn and crew
FERRAND
18/19 Mar
BERGERAC
1940-0215
FL D U Gibbs and crew
Sgt T Monteith in, Sgt Charnock out
20/21 Mar
ANGOULEME

1910-0135
FL D U Gibbs and crew
Sgt M P Monton in, Sgt Monteith out
22/23 Mar
FRANKFURT
1905-0045
FO E R Penman and crew
23/24 Mar
LYONS
1925-0225
FL D U Gibbs and crew
25/26 Mar
LYONS
1930-0305
FL D U Gibbs and crew
FS R J Smith and Sgt E Long in;
Sgt Monton and Sgt Birch out
26/27 Mar
ESSEN
1950-0305
FL D U Gibbs and crew
Sgt Monton back, Sgt Long out
(crew's last trip of first tour)
5/6 Apr
TOULOUSE
2040-0410
PO B F Durrant
Sgt F R Broad
Sgt J C Pittaway
FS A Buchanan RCAF
Sgt N H Jones
Sgt W Martin
Sgt K N Warwick
9/10 Apr
MINING/'PRIVET'
2120-0615
FO E R Penman and crew
10/11 Apr
TOURS
2235-0400
FO E R Penman and crew
(FO Penman and crew failed to return 7/8 May)
18/19 Apr
JUVISY
2100-0135
PO B F Durrant and crew
20/21 Apr
LA CHAPELLE
2310-0405
PO B F Durrant and crew
22/23 Apr
BRUNSWICK
2325-0240
PO B F Durrant and crew
(sortie abandoned due to engine failure)
24/25 Apr
MUNICH
2055-0655
PO B F Durrant and crew
26/27 Apr
SCHWEINFURT
2130-0620
PO B F Durrant and crew
1/2 May
TOULOUSE
2135-0530
SL E Sprawson
FL A W Williams (2P)
Sgt K Anderton
FO R R C Barker
FO E L Hogg
Sgt W D Low

Sgt E J Wiggins
FO P S Arnold
3/4 May
MAILLY-LE-CAMP
2205-0325
FL S J Houlden
Sgt R H Cosens
WO T H White RAAF
FS K T Millikan RAAF
FS H Pringle RAAF
Sgt R C Hulme
FS S N Kelly
7/8 May
SALBRIS
2205-0405
FS P C Browne
Sgt G A Gray
FS R Carmichael RCAF
Sgt W J Markey
Sgt C Tate
Sgt E A Stead
Sgt W A Greenwood
9/10 May
GENNEVILLIERS
2220-0235
PO B F Durrant and crew
11/12 May
BOURG-LEOPOLD
2215-0145
PO B F Durrant and crew
FO P H George in, Sgt Pittaway out
19/20 May
TOURS
2215-0335
PO B F Durrant and crew
FS A R Kitto (2P)
Sgt Pittaway back
5/6 Jun
ST PIERRE DU MONT
0235-0645
PO B F Durrant and crew
6/7 Jun
CAEN
0035-0500
PO B F Durrant and crew
8/9 Jun
RENNES
2310-0345
PO B F Durrant and crew
10/11 Jun
ORLEANS
2200-0405
PO B F Durrant and crew
FO L C W Boivin (2P)
12/13 Jun
POITIERS
2230-0420
PO B F Durrant and crew
14/15 Jun
AUNAY-SUR-ODON
2230-0235
PO B F Durrant and crew
15/16 Jun
CHATELLERAULT
2130-0345
PO S M Wright RAAF
Sgt W S McPhail
Sgt H Smith
Sgt W R Knaggs
FS L McGregor RAAF
Sgt A T Clarke
FS W Beutel RAAF
19/20 Jun
WATTEN/V1 SITE
2250-0110
PO S M Wright and crew

(raid abandoned due to weather)
21/22 Jun
GELSENKIRCHEN
2315-0330
PO S M Wright and crew
24/25 Jun
POMMERVAL/V1 SITE
2220-0150
PO B F Durrant and crew
27/28 Jun
VITRY
2155-0515
PO B F Durrant and crew
29 Jun
BEAUVOIR/V1 SITE
1220-1530
FO G S Mather RCAF
Sgt J L Lucas
PO A E Power
FO J S Kingston RCAF
Sgt W Stewart
Sgt W A Waldren RCAF
Sgt J Crawford RCAF
(raid abandoned due to cloud over target)
(Mather failed to return 7 July)
4/5 Jul
ST LEU D'ESSERENT
2315-0310
PO B F Durrant and crew
Sgt E C Bumford (2P)
7/8 Jul
ST LEU D'ESSERENT
2240-0255
PO B F Durrant and crew
12/13 Jul
CULMONT
2150-0550
FO B F Durrant and crew
FO H E Sayeau RCAF (2P)
14/15 Jul
VILLENEUVE
2215-0445
FO B F Durrant and crew
15/16 Jul
NEVERS
2200-0515
FO B F Durrant and crew
19 Jul
THIVERNY/V1 SITE
1925-2315
FO B F Durrant and crew
20/21 Jul
COURTRAI
2310-0205
FO B F Durrant and crew
23/24 Jul
KIEL
2255-0335
FO B F Durrant and crew
24/25 Jul
STUTTGART
2145-0535
FO B F Durrant and crew
25 Jul
ST CYR
1750-2135
FO B F Durrant and crew
26/27 Jul
GIVORS
2120-0600
FO B F Durrant and crew
28/29 Jul
STUTTGART
2200-0600

FL S H Jones
Sgt D Levene
FO C G Bryan RCAF
Sgt H I Shepherd RCAF
Sgt R H Julien
Sgt J F W Clarke RCAF
Sgt K A McLaughlin RCAF
30 Jul
CAHAGNES
0600-1045
FL S H Jones and crew
(raid abandoned by controller)
31 Jul
RILLY-LA-MONTAGNE
1725-2215
GC W N McKechnie
GCFO B R Markes (2P)
Sgt R Howarth (FE)
SL G Crowe DFC (N)
FO K Pender RCAF (BA)
Sgt J C Burns (2N)
Sgt H H Beards (WOP)
Sgt J A Monk (MU)
Sgt P Baldwin (RG)
2 Aug
TROSSY-ST-MAXIM
1420-1850
FO P C Brown
Sgt G A Gray
WO R E Carmichael
FS W J Markey RCAF
Sgt C Tate
Sgt E A Stead
Sgt W A Greenwood
3 Aug
TROSSY-ST-MAXIM
1145-1600
FO P C Brown and crew
(aircraft hit by flak)
5 Aug
ST LEU D'ESSERENT
1040-1525
FO P C Brown and crew
6 Aug
LORIENT
1735-2250
FO P C Brown and crew
7/8 Aug
SECQUEVILLE
2140-0050
FO P C Brown and crew
9/10 Aug
CHATELLERAULT
2055-0300
FO P C Brown and crew
11 Aug
BORDEAUX
FO1155-1920
P C Brown and crew
12/13 Aug
BRUNSWICK
2120-0250
FL S H Jones and crew
14 Aug
QUESNAY
1205-1550
FO H E Sayeau RCAF
Sgt S A Coucill
FS M H Moore RCAF
FS L P Mason RCAF
Sgt A W Stewart
FS A P Fontaine RCAF
FS M Waite RCAF
15 Aug
GILZE-RIJEN A/F
1005-1340

Sgt J R Ford
Sgt W A Ambrose
Sgt R C Fondt
Sgt A T Worthington
Sgt P J O'Brien RNZAF
Sgt F S Williams
Sgt J B Dandy
16/17 Aug
MINING/STETTIN
2130-0455
FL S H Jones and crew
18 Aug
FORET DE L'ISLE ADAM
1130-1610
FLS H Jones and crew
Mr I Wilson, BBC reporter
25/26 Aug
DARMSTADT
2050-0515
FL S H Jones and crew
26/27 Aug
KONIGSBERG
2015-0615
FL S H Jones and crew
29/30 Aug
KONIGSBERG
2040-0650
FL S H Jones and crew
(FL Jones failed to return 23
September)
31 Aug
AUCHY-LES-HESDIN
1535-2000
FS W E Brunton RNZAF
Sgt W Dyson
Sgt H H Harris
FS W S Sutherland RAAF
Sgt F N Evans
Sgt M J Mehan
Sgt A Gray
3 Sep
DEELEN A/F
1535-1915
FO B R Marks
Sgt R S Howarth
Sgt J C Burns
FO K M Render RCAF
Sgt H H Beard
Sgt J A Monk
Sgt B Baldwin
10 Sep
MUNCHEN-GLADBACH
0245-0710
FO R G Waterfall
Sgt D D Jones
Sgt E Kindley
Sgt G Dixon
Sgt F J Swindly
Sgt J J Campbell
Sgt S E Cunningham
11 Sep
LE HAVRE
0605-0905
FO A M Dow
Sgt J Chamberlain
FO R A Muddle
FO H A Orrell
FO G R Willey
Sgt H F Brittain
Sgt W E McNeill
11/12 Sep
DARMSTADT
2105-0235
FO A M Dow and crew
12/13 Sep
STUTTGART
1910-0150

FO R B Sexton
Sgt R C Aird
PO I G Martin
FO H J Milne
Sgt R Jeavons
Sgt W R Orr
Sgt R Burtenshaw
17 Sep
BOULOGNE
0820-1105
FO W D Kelley RAAF
Sgt J Howarth
Sgt F J Turkentine
Sgt A James
FS J H Grubb
Sgt F Cawlishaw
Sgt R F Dyson
18 Sep
BREMERHAVEN
1830-2315
FO W D Kelley and crew
19 Sep
RHEYDT
1900-2350
FO W D Kelley and crew
23 Sep
DORTMUND-EMS CANAL
1915-2115
FO W D Kelley and crew
26/27 Sep
KARLSRUHE
0100-0725
Lt P A Becker SAAF
Sgt R C Osman
FS W M Ching RNZAF
Sgt V P Tomei
Sgt C G Lees
Sgt N T Deacon
Sgt A R Roselt
27/28 Sep
KAISERSLAUTERN
2215-0400
Lt P A Becker and crew
5 Oct
WILHELMSHAVEN
0755-1305
FO W D Kelley and crew
6 Oct
BREMEN
1740-2205
FO K R Simpson
Sgt R C Witcombe
FO A Stott
FS G H Lloyd
FO W D Mitchell
Sgt N D Reynolds
Sgt C Powell
14/15 Oct
BRUNSWICK
2255-0555
FO A M Dow and crew
19/20 Oct
NURNBERG
1740-0050
FO A M Dow and crew
24/25 Oct
MINING/ 'SILVERTHORNE'
1720-0050
FO E Barratt
Sgt J F Emerson
Sgt A Berry
Sgt E C Towle
FS C P Calvert
Sgt R E Day
FS W D Lloyd
30 Oct
WALCHEREN

1040-1325
FL A C Barden
Sgt S Routh
Sgt J N W King
FO F S Rowbory
Sgt J F Richards
PO A C R Udson
Sgt J W Jones
1 Nov
HOMBERG
1350-1805
FO G F Laidlaw RAAF
Sgt G A Cryer
FO C S Oliver
FS C W Kinneen
FS R Beardsley
PO W P O'Shaunessy
Sgt K J Mitchell
2 Nov
DUSSELDORF
1700-2150
FO G F Laidlaw and crew
4 Nov
LADBERGEN
1750-2205
FO F E Day
WO N K Whitby DFM
FO A J Henington
FO H Jones RNZAF
Sgt J O'Dennell
Sgt E Harrison-Owen
Sgt R J Robertson
6 Nov
DORTMUND-EMS CANAL
1630-2150
FO G F Laidlaw and crew
(raid abandoned by controller due to markers having
difficulty finding target)
11 Nov
HARBURG
1630-2210
FO L P Bence RAAF
Sgt F Raine
FS R E Tolley
Sgt T Donovan
Sgt R M Taylor
Sgt R S Marchant
Sgt K F Judd
16 Nov
DUREN
1300-1805
FO L P Bence and crew
21/22 Nov
LADBERGEN
1740-2330
FO L P Bence and crew
22/23 Nov
TRONDHEIM
1605-0315
FO L P Bence and crew
(raid abandoned by controller; target covered by
smoke screen)
26/27 Nov
MUNICH
0005-0915
FO L P Bence and crew
30/31 Dec
HOUFFALIZE
0235-0645
FO L P Bence and crew
(sortie abandoned, target not
seen)

1945
1 Jan
GRAVENHORST
1700-2355
FO L P Bence and crew
4/5 Jan
ROYAN

0025-0750
FO L P Bence and crew
5/6 Jan
HOUFFALIZE
0025-0325
FO L P Bence and crew
(returned early due to glycol

leak to starboard inner
engine)
7/8 Jan
MUNICH
1640-0200
FO L P Bence and crew

16/17 Jan
BRUX
1755-0300
FO L P Bence and crew

LL806
JIG

One of the lesser-known of the top half-dozen century-plus Lancasters, LL806 was part of contract No 239 awarded to Avro and built under licence by Armstrong-Whitworth. Produced as a Mk I, it came from the factory in late 1943 equipped with four Merlin 24 engines and on 22 April 1944 it was assigned to No 15 Squadron at Mildenhall, Suffolk, where No 15 Squadron had not long converted to Lancasters from Short Stirlings.

The squadron's code letters LS were painted on LL806's fuselage as well as the individual letter 'J', to become J-Jig. The aircraft carried no nose-artwork, but when it completed sorties, a regular tally of bomb symbols began to appear beneath the cockpit, in rows of 10. After eight such rows, a second column was started and by January 1945, it had topped the 100 mark.

In all, 134 bombs were painted on Jig, all straightforward, with no colour-coding for day or night trips. At the end of April 1945, Jig, along with other heavy bombers, was used to make food supply drops to starving Dutch civilians, a task which was known as Operation 'Manna' – a true Manna from heaven as far as the Dutch were concerned. Although these sorties had the tacit agreement of the occupying German forces, there was still an element of risk by some determined Nazi, or a trigger-happy soldier, so these missions were counted.

Jig did three such trips and then did three 'Exodus' trips, bringing released Allied prisoners of war back to Britain from the Continent. These last six sorties were marked up on Jig's nose as three food-flour sacks, and three running stick-men, making 140 operations in all. Although the aircraft completed 134 operational war sorties, plus six of these other trips, it could be argued that it did 140 war sorties, especially as other Lancasters were credited with these 'Manna' and 'Exodus' sorties, bringing some of them into the 100 op-plus category by counting them. Thus Jig could stand next to ED888 as jointly holding the record of 140 operations.

Returning now to the opening of Jig's operational career, LL806 replaced the previous 'J' (LL801) which had been lost on 27/28 April 1944, against Wilhelmshaven. LL806's first sortie was flown on 1/2 May against the railway stores and repair depot at Chambly, France, although the aircraft had a 1,000-pounder bomb hang-up. Jig's first pilot was Pilot Officer Mervyn J. Sparks, from Christchurch, New Zealand, who would survive the war as a flight lieutenant, DFC; his navigator Lancelot Elias and bomb-aimer Edward Spannier RCAF also won DFCs.

Sparks went on to fly 18 ops in Jig, although he missed the D-Day sorties with the aircraft. A few days after D-Day, on 10/11 June, Jig's gunners shot down a Me109 night fighter. Jig was fortunate to have few aborts due to mechanical problems and continued to plod on, flying regularly over the summer and autumn of 1944, including such famous sorties as Caen on 18 July and Falaise on 12 August. The aircraft's 100th op came on 5 January 1945 with a raid on Ludwigshaven.

One of Jig's crews that became more-or-less regular in the autumn of 1944 was that of Flying Officer R. H. Hopper-Cuthbert RAAF, who were to do some 18 trips in this Lanc. Sergeant Ken Dorsett was the rear-gunner and recalls:

'I cannot give a "Dambuster"-like image of J-Jig, she was just an old lady who did her bit. I had done seven operations as a spare gunner when I met Hopper and crew, who came rear-gunner-less from LFS. We looked over this rather tired old Lanc and you could almost feel the thoughts going through our heads – will we or won't we finish this war together?

'Well, much to our relief we did. J-Jig turned out to be a good old lady. She took a few chunks of flak, had her rivets strained to breaking point by "corkscrews", and the landings (not quite Hopper's speciality) must have played merry Hell with her undercarriage. Nevertheless she got us through with the aid of Hopper (despite his landings) and Lady Luck sitting on our shoulders.

'She even finished with us on a long haul to Königsberg, now known as Kalinengrad, just to prove she could do it. Other than that, just an old bomber in the stream that made the 100.'

Hopper-Cuthbert received the DFC and Ken Dorsett the DFM at the end of their tour.

Jig's last bombing raid was to Bremen on 22 April 1945, then came those six 'Manna' and 'Exodus' trips already mentioned above. In June and July came some Ruhr trips, flights taking some of the squadron's ground personnel – both men and women – on daylight sight-seeing sorties to devastated German cities for them to see the results of Bomber Command's war effort in which they had played their part.

Jig remained with the squadron until finally Struck off Charge on 6 December 1945.

1944	**8/9 May**	**19/20 May**	2350-0349
No 15 Squadron	CAP GRIS NEZ	LE MANS	FL S Fisher
1/2 May	2238-0039	2224-0312	FO G Tipping (N)
CHAMBLY	PO M J Sparks and crew	PO M J Sparks and crew	FL F Wright DFC (WOP)
2215-0200	**10/11 May**	FS S Mellors (2P 90 Sqdn)	FO J Wastenays (BA)
PO M J Sparks RNZAF	COURTRAI	**21/22 May**	Sgt N Berryman (FE)
FO L B A Elias (N)	2201-0041	DUISBURG	Sgt E Tiplady (MU)
Sgt J Tapping (WOP)	PO M J Sparks and crew	PO M J Sparks and crew	FL W Bossom (RG)
FS E G Spannier RCAF (BA)	**11/12 May**	2235-0330	**27/28 May**
Sgt P Hartshorn (FE)	LOUVAIN	FO W Leslie (2P)	BOULOGNE
Sgt J Aimesbury (MU)	2234-0119	**24/25 May**	0016-0243
Sgt N Freeman (RG)	PO M J Sparks and crew	AACHEN	FS W Ferguson

Sgt B Bond
PO H Thorp
FO B Harper RCAF
Sgt C Stewart
Sgt W Swent
Sgt W Poole
28/29 May
ANGERS
1809-0148
FL S Fisher
FS F Grimshaw
Sgt J Crew
FO J Wastenays RCAF
FS N Berryman
Sgt E Tiplady
FS G Allen
30/31 May
BOULOGNE
2302-0058
PO M J Sparks and crew
31/1 Jun
TRAPPES
2355-0501
FO W Leslie
FO F Frudd RCAF
Sgt J Rozier
FO E McNiece
Sgt W Gundry
Sgt E North
Sgt D Findlay RCAF
2 Jun
WISSANT
0112-0310
FS W Ferguson and crew
3 Jun
CALAIS
0040-0240
FL J P Bell
FO E Leah
PO G Bovett
FO W Henshall RCAF
Sgt E Ward
Sgt Barkshire
Sgt G Morrison
5/6 Jun
CAEN/OUISTREHAM
0327-0704
FS W Ferguson and crew
6/7 Jun
LISIEUX
2349-0323
FL J P Ball and crew
7/8 Jun
MASSY-PALAISEAU
0045-0426
FL M J Sparks and crew
8/9 Jun
FOUGERES
2134-0221
FL M J Sparks and crew
10/11 Jun
DREUX
2310-0325
FL B Payne
Sgt P Hemming
Sgt D Brady
FO C Allan RCAF
Sgt F Lambert
Sgt T Blackburn
Sgt R Butterworth
(gunners claimed
Me109 night fighter
shot down)
12/13 Jun
GELSENKIRCHEN
2303-0308
FL M J Sparks and crew

17/18 Jun
MONTDIDIER
0041-0434
FL M J Sparks and crew
21 Jun
DOMLEGER/V1 SITE
1804-2039
FL M J Sparks and crew
30 Jun
VILLERS-BOCAGE
1753-2126
FL M J Sparks and crew
2 Jul
BEAUVOIR/V1 SITE
1234-1625
FL M J Sparks and crew
5/6 Jul
WIZERNES/V1 SITE
2250-0135
FS A Barford
Sgt F Bolan RCAF
Sgt W Parks
FS P Whitehouse
Sgt E Thomas
Sgt W Day
FS E Marshall
7/8 Jul
VAIRES
2245-0328
FL B Payne and crew
9 Jul
LINZEUX/V1 SITE
1218-1614
FL M J Sparks and crew
10 Jul
NUCOURT/V1 SITE
0330-0737
FL M J Sparks and crew
12 Jul
VAIRES
1735-2221
FL B Payne and crew
(brought bombs back due to
cloud over target)
18 Jul
CAEN
0356-0749
FL W Bell DFC
PO A Hayden
FS P Sweetman
WO F Oakes
Sgt M Feit
Sgt J Brennan
Sgt T Brookfield
18/19 Jul
AULNOYE
2224-0144
FL W Bell and crew
Sgt T Pavey in, Sgt Feit out
(sortie abandoned when
starboard outer failed)
23 Jul
MONTCANDON
0633-1015
FL M J Sparks and crew
23/24 Jul
KIEL
2253-0357
FS W Mason RAAF
Sgt A Douglas RAAF
FS P Elgar RAAF
Sgt D Brown
Sgt H Leigh
Sgt N Bibby
Sgt M Hathaway
24/25 Jul
STUTTGART

2150-0519
FL M Johnston RNZAF
FS E King
FS J Paine
FS C Morris
Sgt E Marsh
Sgt A Hartley
FS N Barker
25/26 Jul
STUTTGART
2140-0558
FO S Stewart
FS J Burrett
FS N Herbert
FS W Turner RAAF
Sgt J Bax
FS D McFadden RAAF
FS K Girle RAAF
28/29 Jul
STUTTGART
2144-0538
FO H Kelly
FS J Mason
FS W Watts RAAF
FS A O'Sullivan RAAF
Sgt J Brown
Sgt R Tweddle
Sgt J Bardsley
30 Jul
AMAYE-SUR-SEULLES
0523-0925
FL B Payne and crew
1 Aug
COULON VILLERS/V1 SITE
1906-2158
FL W Leslie and crew
(sortie abandoned due to fog)
5 Aug
BASSENS
1427-2250
FL W Leslie and crew
7/8 Aug
ROCQUECOURT
2140-0135
FL W Leslie and crew
8/9 Aug
FORET DE LEUCHEAUX
2207-0117
FS L Marshall
Sgt R Bates
Sgt R Knight
FO L Ford RNZAF
Sgt D Kenny
Sgt J Kay
Sgt H Jackson
9/10 Aug
FORT D'ENGLOS/V1 SITE
2146-0028
FO W Moran
Sgt B Rennie
FS C Dyer
Sgt K Logan
Sgt W Brockett
Sgt G Donaldson
Sgt R Faint
11 Aug
LENS
1349-1730
FO H Cato RAAF
Sgt D Lee RAAF
FS J Hance RAAF
Sgt T Priddle
Sgt H Hounsome
Sgt K Knok
Sgt W Henderson
12/13 Aug
FALAISE

0039-0357
WO A McDougall
FS R Keen
WO D Moore
Sgt I Howitt
Sgt K McKie
Sgt J McNee
Sgt T Hunter
14 Aug
ST QUENTIN
1341-1808
WO A McDougall and crew
15 Aug
ST TROND A/F
0955-1354
FO R Jennings RAAF
Sgt J Fawcett
FS G Johnson
FO J Watts
Sgt J Biddle
FS G Tregoing
FS R Banks RAAF
16/17 Aug
STETTIN
2103-0529
FL W Leslie and crew
18/19 Aug
BREMEN
2138-0309
FL W Leslie and crew
26/27 Aug
KIEL
2017-0212
FL W Leslie and crew
29/30 Aug
STETTIN
2055-0700
FL W Leslie and crew
31 Aug
PONT REMY
1549-1953
FL W Leslie and crew
3 Sep
EINDHOVEN A/F
1521-1802
WO A McDougall and crew
5 Sep
LE HAVRE
1643-2053
FL W Leslie and crew
6 Sep
LE HAVRE
1616-2016
FO R Jennings and crew
8 Sep
LE HAVRE
FO R Jennings and crew
0536-0901
17 Sep
BOULOGNE
1022-1335
FO R Jennings and crew
20 Sep
CALAIS
1456-1818
FO R Jennings and crew
23 Sep
NEUSS
1852-2325
FO R Jennings and crew
24 Sep
CALAIS
1652-1927
FL H Cato and crew
27 Sep
CALAIS
0728-1054

FO R Marsh
FS V Stuckey RAAF
FS R Murray RAAF
FO S Tudor-Lee
Sgt J Swainston
Sgt L Garrett
Sgt J Wade
28 Sep
CALAIS
0714-1034
FO R Jennings and crew
Sgt J Munro in, Sgt Biddle out
5/6 Oct
SAARBRUCKEN
1905-0112
FL P Percy
FO W Shakespeare
FO A A Aleandri
FO T R Palmer
WO P Shields
FS D King
FS S Mackie
6 Oct
DORTMUND
1632-2250
FL P Percy and crew
7 Oct
KLEVE
1109-1546
FL P Percy and crew
14 Oct
DUISBURG
0652-1132
FL H Cato and crew
14/15 Oct
DUISBURG
2303-0343
FL H Cato and crew
15 Oct
WILHELMSHAVEN
1738-2145
FO R Hopper-Cuthbert RAAF
FS W Lake
FS R Heatley
Sgt G Charlton
Sgt S Lewis
Sgt D Inglis
Sgt K T Dorsett
18 Oct
BONN
0856-1318
FL H Cato and crew
19/20 Oct
STUTTGART
2157-0404
FO D Kelly RAAF
FS J Bishop
FO R Johnson
FO C Clay
Sgt J Taylor
Sgt T Pownall
Sgt C Rhodes
22 Oct
NEUSS
1332-1750
FO B Jones
FS L Beren
FS G Sim RAAF
FS R Warman
Sgt D Lord
Sgt F Atkinson
Sgt P Acton
23 Oct
ESSEN
1632-2204
FS L Hastings RAAF
FO R Smith RAAF

FS V Pearce
PO H Burns
Sgt J Munro
Sgt G Malyon
Sgt D McFadden
25 Oct
ESSEN
1212-1710
FL P Percy and crew
28 Oct
FLUSHING
0844-1107
FO R Marsh and crew
15 Nov
DORTMUND
1213-1721
FL I Buchanan RAAF
FO J Varey
WO A Kimber RAAF
FO S Hawkins
FS J Crosbie
FS A Helyer
Sgt T Field
16 Nov
HEINSBERG
1312-1806
FO R Hopper-Cuthbert and
crew
20 Nov
HOMBERG
1214-1503
FO R Hopper-Cuthbert and
crew
(aborted with electrical failure
and the Gee set caught fire)
21 Nov
HOMBERG
1223-1655
FO R Hopper-Cuthbert
23 Nov
GELSENKIRCHEN
1224-1728
FO R Hopper-Cuthbert
26 Nov
FULDA
0753-1351
FO R Hopper-Cuthbert
27 Nov
COLOGNE
1210-1700
FO R Hopper-Cuthbert
29 Nov
NEUSS
0254-0737
FO C Noble RAAF
PO R McQuaid
WO F Watson
FO C Bender RAAF
Sgt G Fox
Sgt L Brown
Sgt A Adams
30 Nov
BOTROP
1035-1447
FO C Noble and crew
2 Dec
DORTMUND
1305-1732
FO L Marriott
FS E Lumsden RAAF
FS D Woon RAAF
FS C Dane
Sgt J Wyllie
Sgt P Kite
Sgt J Ferbrache
4 Dec
OBERHAUSEN

1150-1618
FO R Hopper-Cuthbert and
crew
5 Dec
SCHWAMMENAUEL DAM
0907-1339
PO J Slaughter RAAF
Sgt H Bradbrook
FS B Philpot RAAF
Sgt J Seel
Sgt R Fearn
Sgt H Hill
Sgt G Lock
(abandoned, target covered
with cloud)
6/7 Dec
MERSEBERG
1653-0021
FS B Giles RAAF
Sgt B Cooper
FS Henry
Sgt R Bonner
Sgt A Campbell
Sgt H Bosworth
Sgt E Chettoe
8 Dec
DUISBURG
0844-1310
FS B Giles and crew
11 Dec
OSTERFELD
0828-1300
FO N Clayton
Sgt J Graham
Sgt L Brown
FO K Robertson
FS E Prewers
Sgt K Hardy
Sgt A Fletcher
12 Dec
WITTEN
1113-1610
FO N Clayton and crew
16 Dec
SIEGEN
1131-1715
FO R Hopper-Cuthbert and
crew
19 Dec
TRIER
1243-1748
FO R Hopper-Cuthbert and
crew
21 Dec
TRIER
1221-1729
FO R Hopper-Cuthbert and
crew
23 Dec
TRIER
1120-1633
FO R Hopper-Cuthbert and
crew
24 Dec
BONN
1513-2023
FO S Bignell RAAF
FS J Lacey RAAF
FS C Russell
Sgt D Jones
Sgt F Keeble-Buckle
FS T Thoroughgood
FS W Wilkie
28 Dec
COLOGNE
1212-1659
FO N Burns

Sgt S Duke
Sgt J Nicholson
Sgt E Doble
Sgt S Franks
Sgt E Davis
Sgt M Giddings
29 Dec
COBLENZ
1216-1713
FO R Hopper-Cuthbert and
crew
31 Dec
VOHWINKEL
1148-1635
FO R Hopper-Cuthbert and
crew

1945
2 Jan
NURNBERG
FO R Hopper-Cuthbert and
crew
1518-2300
3 Jan
DORTMUND
1258-1748
FO S Bignall and crew
5 Jan
LUDWIGSHAVEN
1127-1754
FO R Hopper-Cuthbert and
crew
6 Jan
NEUSS
1544-2047
FO S Bignall and crew
7/8 Jan
MUNICH
1825-0237
FO L Gray RAAF
FL S Tinkler
FS D Archibald
FS G R Dykstra
Sgt R Rawson
Sgt J H Cooper
Sgt H C Ferguson
11 Jan
KREFELD
1146-1656
FO R Hopper-Cuthbert and
crew
15 Jan
ENKERSCHWICK
F1155-1654
O R Hopper-Cuthbert and
crew
16/17 Jan
WANNE-EICKEL
2308-0432
FL C Hagues
FS V S Flower
WO D Leahy RAAF
FS D Green
FS C Houlgrave
Sgt J Byrne
Sgt A Welsey
22 Jan
DUISBURG
1657-2205
PO I McHardy
FS A Bolton
FS D Buchanan
FS J Surridge
Sgt K Ashbolt
FS J Hall
Sgt F Borrell

28 Jan
COLOGNE
1010-1601
FO L Gray and crew
FS D Metcalf RAAF in, FS
Archibald out
29 Jan
KREFELD
1022-1548
FS V Tenger RAAF
FO E James RAAF
FS A Turner RAAF
FS G Hay
Sgt E Fulton
Sgt K Brown
Sgt B Conroy
1 Feb
MUNCHEN GLADBACH
1312-1842
FO D Hunt RAAF
Sgt J Shepherd
FS P Smeeton RAAF
Sgt D A Russell
Sgt R Rawson
Sgt J Cooper
Sgt H Ferguson
3 Feb
DORTMUND
1621-2159
FS V Tenger and crew
14/15 Feb
CHEMNITZ
2010-0409
FO N Burns and crew
22 Feb
BUER
1301-1757
FO N Burns and crew
23 Feb
GELSENKIRCHEN
1135-1703
FS A Meikle
FS F Jacobs
Sgt J Whitehouse
Sgt E Utting
Sgt J Palmer
Sgt W Nunn
Sgt D Cherry
25 Feb
KAMEN

0931-1513
FS A Meikle and crew
26 Feb
DORTMUND
1036-1610
FO D Hunt and crew
27 Feb
GELSENKIRCHEN
1120-1704
FS A Meikle and crew
28 Feb
GELSENKIRCHEN
0852-1401
FO C Ayres RAAF
FS E Risley
WO P Grease
FO W Forsyth
Sgt D Keichley
FO A Chambers
FO P Girardot
1 Mar
KAMEN
1149-1729
FS A Meikle and crew
2 Mar
COLOGNE
1251-1818
FO C Ayres and crew
5 Mar
GELSENKIRCHEN
1028-1555
FS L Baxendale
FS M Vaughan
Sgt A Fisher
FS A Wright
Sgt L Taylor
Sgt R Kerwan
Sgt S Salter
7/8 Mar
DESSAU
1649-0219
FS A Wright RAAF
FO R Rees RAAF
FS J Walden
FS G Gibson RAAF
FS F Williams
FS J Smith
Sgt L Harries
9 Mar
DETTELN

1029-1549
FS A MacDonald RAAF
Sgt W Roberts
Sgt W Orford
Sgt R Davis
Sgt A Maw
Sgt N Farmer
Sgt A Cooper
10 Mar
GELSENKIRCHEN
1218-1721
FS A Wright and crew
11 Mar
ESSEN
1115-1707
FS A Wright and crew
12 Mar
DORTMUND
1248-1911
WO V Tenger and crew
FS G Naldrett in, FS Hay out
14 Mar
DETTELN
1320-1828
FS A Meikle and crew
22 Mar
BOCHOLT
1054-1611
FS W Sievers RAAF
Sgt A Clarkstone
Sgt D King
FO J Hunter
Sgt T Blenkain
Sgt R Williamson
Sgt H Wells
27 Mar
ALTENBOGGE/HAMM
1040-1545
WO V Tenger and crew
FS G Fuller in, FS Naldrett out
4/5 Apr
MERSEBERG
1827-0317
WO W Woodman
Sgt G Relf
Sgt P Gennoy
FO J Moffatt RCAF
FS J Freemantle
Sgt T Hammond
Sgt T Dickinson

5/6 Apr
KIEL
1947-0201
FS W Sievers and crew
13/14 Apr
KIEL
2010-0210
FS W Sievers and crew
14/15 Apr
POTSDAM
1824-0323
FS W Sievers and crew
18 Apr
HELIGOLAND
1014-1502
FS W Sievers and crew
22 Apr
BREMEN
1531-2056
FS W Sievers and crew
FO C Taylor in, Sgt Blenkiron
out
30 Apr
'MANNA'/ROTTERDAM
1638-1922
FO L Baxendale and crew
2 May
'MANNA'/THE HAGUE
1104-1334
FS W Sievers and crew
7 May
'MANNA'/THE HAGUE
1110-1345
FO W Woodman and crew
11 May
'EXODUS'/JUVINCOURT
0922-1453
WO V Tenger and crew
FS G Fuller in, FS Hay out
17 May
'EXODUS'/JUVINCOURT
1115-1652
FO W Woodman and crew
24 May
'EXODUS'/JUVINCOURT
1414-1930
FO W Woodman and crew

LL843
POD

Built under contract from Avro as part of order No 239/C4, by Messrs Armstrong Whitworth at Coventry, LL843 was a Mk I with four Merlin 24 engines. On 28 February 1944 it was assigned to No 467 (RAAF) Squadron, based at Waddington, Lincolnshire, and given the squadron codes of PO with the individual identification letter D.

Operations began for LL843 on 9/10 March in the experienced hands of the B Flight Commander, Squadron Leader Arthur William Doubleday DFC, RAAF, flying what was described as a 'Special Job' – a No 5 Group attack on an aircraft factory at Marignane, near Marseilles. As can be imagined, it was a long sortie, LL843 taking 9 hours 37 minutes to complete. As well as the Lanc's seven-man crew, it also carried Group Captain S. C. 'Sammy' Elworthy DSO, DFC, AFC, Waddington's Station Commander, as an eighth man. (He was later to become Marshal of the RAF, Lord Elworthy KG.)

On 5/6 April, during a raid on an aircraft factory at Toulouse, LL843, with Squadron Leader Doubleday once again in command, acted as the Windfinder aircraft reporting back the wind strengths to Group HQ. A few nights later, on the 18/19th, Flight Lieutenant J. A. Colpus and crew had a 1,000lb bomb hang-up – and then fall – onto the closed bomb-doors, but it was later jettisoned without further problems.

Pilot Officer J. L. Sayers RAAF flew LL843 a number of times during May, and his gunners had a combat with a Me110 on the 22/23rd, LL843 then suffering damage from a flak hit in a heavy searchlight belt running west to east from Emden. Sayers' gunners had another fight on 21/22 June, two enemy fighters having a go at them, but one was attacked and seen to go down with its port engine on fire; the Lanc suffered minimal damage. Sayers and some of his crew later operated with No 617 Squadron, Sayers having won the DFC.

LL843 saw operations during D-Day, and against enemy transport centres and V1 sites, and then the Caen support operation in July which proved to be the aircraft's last sortie with the Australians. The next day LL843 was assigned to No 61 Squadron at Skellingthorpe, Lincolnshire, and had the new squadron codes QR painted on its sides. The Lanc also also met up with an old friend who had flown it on a few occasions – A. W. Doubleday, now a Wing Commander and CO of No 61 Squadron, who was about to receive the DSO.

Operations slowly mounted until, on 12/13 August 1944 during a raid on Russelsheim, an attack by two night fighters and a flak hit put LL843 out of action for several weeks while it went back to Avro's for repair and a refit. The Lanc was back in late September and during the winter, settled down with two particular skippers, Flying Officer W. G. Corewyn and Flying Officer J. S. Cooksey.

LL843 completed its 100th op during March 1945 and is reported to have flown a total of 118 by the war's end. Whether any 'Manna' or 'Exodus' sorties are included is not known as no record of these sorties were kept in the Form 541. The aircraft also suffered a couple more flak hits - on 2 August 1944 and 27 March 1945. During a sortie to Karlsruhe on 2/3 February 1945, the Lanc was unable to reach the target on time due to an iced-up ASI and an unexpected cloud base being encountered. Just five minutes from the bomb-run the Master Bomber ordered 'Cease bombing', so the crew had to jettison the 'cookie' and then cut a corner off the return route so as to rejoin the bomber stream.

On 21 May 1945 LL843 was sent to No 1659 CU and then in early September it went to No 279 (ASR) Squadron but three weeks later the aircraft was transferred to No 20 MU. LL843 was finally sold to Messrs Cooley and Co, at Hounslow, Middlesex, for scrap, on 7 May 1947.

1944
No 467 (RAAF) Squadron
9/10 Mar
MARIGNANE
2037-0614
SL A W Doubleday DFC RAAF
Sgt J Slome
FO F J Nugent
FL R B Murphy
FL B Sinnamon
FO S B Gray-Buchanan
PO A A Taylor
GC S C Elworthy DSO DFC AFC
15/16 Mar
STUTTGART
1913-0310
FL A B Simpson
PO S W Archer
FS K W Manson
PO R C Watts
FS L T Watson
FS C A Campbell
FS H Thompson
18/19 Mar
STUTTGART
1902-0057
FL A B Simpson and crew
POs T H Ronaldson (WOP) and R A Weeden (RG) in
22/23 Mar
FRANKFURT
1904-0018
SL A W Doubleday and crew
FO H C Ricketts (AB) and PO G G Johnson (WOP) in
26/27 Mar
ESSEN
1944-0045
SL A W Doubleday and crew
FL D A G Andrews (WOP) in; PO Johnson out

30/31 Mar
NURNBERG
2144-0524
SL A W Doubleday and crew
FO G G Abbott (BA) in; FO Ricketts out
5/6 Apr
TOULOUSE
2027-0403
SL A W Doubleday and crew
FO Ricketts back
10/11 Apr
TOURS
2239-0435
SL A W Doubleday and crew
18/19 Apr
JUVISY
2047-0115
FL J A Colpus RAAF
Sgt K Smith
FS S T Bridgewater
FO D J Stevens
Sgt P Macdonald
WO R H Mark
Sgt E A Rutt
20/21 Apr
LA CHAPELLE
2251-0318
FL J A Colpus and crew
22/23 Apr
BRUNSWICK
2311-0501
FL J A Colpus and crew
24/25 Apr
MUNICH
2112-0634
FL J A Colpus and crew
1/2 May
TOULOUSE
2141-0543
PO S Johns RAAF
Sgt D K J Phillips

FS M J O'Leary
WO C E Langston
FS B P Molloy
FS E D Dale
FS J J Fallon
3/4 May
MAILLY-LE-CAMP
2152-0318
PO S Johns and crew
8/9 May
BREST
2140-0210
FL J A Colpus and crew
10/11 May
LILLE
2201-0054
FL J A Colpus and crew
11/12 May
BOURG-LEOPOLD
2231-0150
PO J L Sayers RAAF
Sgt G D Colquhoun (FE)
FS A G Weaver
FO E G Strem
Sgt F H Howkins
Sgt R P Kent
FS B P Barry
(as soon as bomb-doors opened, 7 x 500lb bombs went down; then general order received to stop bombing; later jettisoned 'cookie')
19/20 May
TOURS
2208-0349
PO J L Sayers and crew
22/23 May
BRUNSWICK
2245-0447
PO J L Sayers and crew
Sgt V L Johnson in; Sgt Colquhoun out

(combat with Me110; also damaged by flak hit)
24/25 May
EINDHOVEN A/F
2307-0213-
PO J L Sayers and crew
(force ordered not to bomb by controller; bombs jettisoned)
27/28 May
NANTES
2257-0346
PO J L Sayers and crew
31/1 Jun
SAUMUR
2356-0446
PO J L Sayers and crew
Sgt Colquhoun back;
Sgt Johnson out
3/4 Jun
FERME D'URVILLE
2319-0235
PO J L Sayers and crew
5/6 Jun
ST PIERRE-DU-MONT
0258-0655
PO J L Sayers and crew
6/7 Jun
ARGENTAN
PO J L Sayers and crew
2357-0319
8/9 Jun
RENNES
2303-0422
PO J L Sayers and crew
10/11 Jun
ORLEANS
2225-0335
PO J L Sayers and crew
12/13 Jun
POITIERS
2222-0444

PO P W Ryan RAAF
Sgt G A Hays
FS V E Cockroft
WO C C Jones
FS L H Porritt
FS W D D Killworth
FS J P Steffan
14/15 Jun
AUNAY-SUR-ODON
2242-0302
PO P W Ryan and crew
15/16 Jun
CHATELLERAULT
2142-0346
PO P W Ryan and crew
19/20 Jun
WATTEN/V1 SITE
2254-0128
PO P W Ryan and crew
(raid abandoned by con-
troller due to weather)
21/22 Jun
GELSENKIRCHEN
2309-0340
PO J L Sayers and crew
(two combats with night
fighters, one shot down;
aircraft slightly damaged)
23/24 Jun
LIMOGES
2234-0506
PO J L Sayers and crew
FS M G Johnson (2P)
24/25 Jun
PROUVILLE/V1 SITE
2250-0208
PO J L Sayers and crew
27/28 Jun
VITRY
2153-0505
PO J L Sayers and crew
29 Jun
BEAUVOIR/V1 SITE
1203-1529
PO J L Sayers and crew
4/5 Jul
ST LEU D'ESSERENT
2308-0319
PO J L Sayers and crew
7/8 Jul
ST LEU D'ESSERENT
2235-0250
PO J L Sayers and crew
12/13 Jul
CULMONT/CHALINDREY
2153-0638
FO S C Carey RAAF
Sgt G G C Morrison
FS G E Fisher
FS L E Formby
FS W H Brown
FO L Anchem
Sgt A G McCoy
14/15 Jul
VILLENEUVE
2221-0428
FO J L Sayers and crew
FO A R Dyer (2P)
15/16 Jul
NEVERS
2222-0506
FO J L Sayers and crew
18 Jul
CAEN
0344-0741
SL L C Deignan
FS L S Smith

WO H R Goodwin
FS C Dean
FO A W Allison
PO L G Burden
PO R W Wishart
FS R J Mellowship (2P)

No 61 Squadron
20/21 Jul
COURTRAI
2316-0206
FL S Parker
Sgt J A Palin (FE)
Sgt H J Richardson (N)
Sgt J H Wellens (BA)
Sgt K G West (WOP)
Sgt K Smith (MU)
Sgt C E Stothard
23/24 Jul
KIEL
2255-0337
FO D G Bates
Sgt G Farrow
FS A N Hughes
FO G O Cameron
Sgt F W Cotton
PO J Fletcher
Sgt D Hancock
24/25 Jul
STUTTGART
2150-0539
FL S Parker and crew
Sgt R S Steele in;
Sgt Palin out
25 Jul
ST CYR
1711-2136
WC A W Doubleday DFC
RAAF
FS J Slome
PO J F Mills
FO F J Nugent RAAF
FS A C R Brydges
FO R K W Clover
FS C Scrimshaw
26/27 Jul
GIVORS
2126-0601
WC A W Doubleday and crew
28/29 Jul
STUTTGART
2211-0541
FO C N Hill
Sgt C L Goss
FO D Hayley
FO C H Machin
Sgt M R Vagnolini
Sgt G R Mildren
Sgt J V O'Neill
30 Jul
CAHAGNES
0608-1041
WC A W Doubleday and crew
(raid aborted over target
by Master Bomber due to
cloud)
31 Jul
JOIGNY-LA-ROCHE
1732-2303
FO H Brooker
Sgt W J Morgan
FO K Brown
Sgt D J Hector
Sgt L B Smith
Sgt W P Hunter
Sgt A D'Arcy
1 Aug

SIRACOURT/V1 SITE
1655-2135
FO H Brooker and crew
(raid aborted by controller
due to ground haze)
2 Aug
BOIS DE CASSAN/V1 SITE
1450-1910
FO H Brooker and crew
(aircraft hit by flak)
5 Aug
ST LEU D'ESSERENT
1030-1545
WC A W Doubleday and crew
FL G L P Dunstone in;
FS Brydges out
6 Aug
BOIS DE CASSAN/V1 SITE
0945-1430
FO H Brooker and crew
(raid aborted by controller)
7/8 Aug
SECQUEVILLE
2105-0056
WC A W Doubleday and crew
Sgt L Chapman in;
FL Dunstone out
9/10 Aug
CHATELLERAULT
2058-0322
FO A D D Brian
Sgt L C Doyle
FO L W Deubert
FO J D Constable
FS C Morgan
Sgt D W H Lambert
Sgt D Wainwright
11/12 Aug
GIVORS
2047-0450
WO A P Greenfield
Sgt F F Fraser
Sgt W J A Gibbs
Sgt W J Haddon
WO V P Smith
Sgt J P King
Sgt S D P Goodey
12/13 Aug
RUSSELSHEIM
2138-0320
FP L A Davies
Sgt J E Jolly
Sgt B Webster
Sgt L G Simpson
FS E Oddy
Sgt L A Williams
Sgt J Watson
(attacked by two fighters and
damaged by flak)
5 Oct
WILHELMSHAVEN
0741-1301
FO N T Collins
Sgt W F Lake
Sgt S F Heavan
FO E R Bloomfield
FS R W Platt
Sgt D J Everson
Sgt W J Scott
6 Oct
BREMEN
1746-2222
Sgt D E Rose
FS E P Meaker
FS T Stephenson
FS S J Scarrett

Sgt A R Marshall
Sgt N W Elliott
7 Oct
FLUSHING
1222-1514
FO D W Scholes
Sgt C J Foreman
FS D J Murray
Sgt R Mayall
FS J J Gardner
FS J W Jackman
FS G Allen
11 Oct
FLUSHING
1328-1656
FL H B Grynkiewicz
Sgt W M Ratcliffe
Sgt J L Jones
Sgt E J Day
Sgt H H Davie
Sgt G E Gwalter
Sgt K W Browne
14/15 Oct
BRUNSWICK
2255-0555
FO W G Corewyn
Sgt P R Earl
Sgt R C Battersby
Sgt E J Boaks
Sgt S J James
Sgt J Douglas
Sgt R Richardson
19/20 Oct
NURNBERG
1734-2131
FO W G Corewyn and crew
(sortie aborted, port inner
engine failed north of Paris)
23 Oct
FLUSHING
1442-1707
FO J S Cooksey and crew
28/29 Oct
BERGEN
2234-0506
FO J S Cooksey and crew
(raid abandoned over target,
which could not be identi-
field due to cloud)
1 Nov
HOMBERG
1344-1837
FO J S Cooksey and crew
(target again not identified so
did not bomb; slight damage
by flak)
2/3 Nov
DUSSELDORF
1638-2150
FO J S Cooksey and crew
4/5 Nov
DORTMUND-EMS CANAL
1740-2153
FO W G Corewyn and crew
6/7 Nov
GRAVENHORST
1636-2230
FO J S Cooksey and crew
(bombs jettisoned on instructions from controller)
11/12 Nov
HARBURG
1627-2145
FO J S Cooksey and crew
(bombs failed to release,
'cookie' later jettisoned)
16 Nov

DUREN
1248-1734
FO W G Corewyn and crew
21 Nov
DORTMUND-EMS CANAL
1725-2323
FO W G Corewyn and crew
22/23 Nov
TRONDHEIM
1541-0258
FO W G Corewyn and crew
(raid abandoned by controller over target due to smoke-screen)
26/27 Nov
MUNICH
2344-0914
FO J S Cooksey and crew
4 Dec
HEILBRONN
1642-2313
FO J S Cooksey and crew
6 Dec
GEISSEN
1646-2331
FO J S Cooksey and crew
8 Dec
URFT DAM
0836-1321
FO F S Farren
Sgt N F Howard
FS T L Benson
Sgt J Sinclair
Sgt W S Tandy
Sgt A Lockett
Sgt N Peckham
17/18 Dec
MUNICH
1639-0201
FO W G Corewyn and crew
18/19 Dec
GDYNIA
1717-0253
FO W G Corewyn and crew
21/22 Dec
POLITZ
1621-0245
FO W G Corewyn and crew
27 Dec
RHEYDT
1229-1708
FO J P Friend RAAF
Sgt D M Bremner
Sgt N T Nuttall
Sg P Sears
FS G A Robson
Sgt C J Yates
Sgt C J Bell
30/31 Dec
HOUFFALIZE
0210-0732
FO W G Corewyn and crew

1945
1 Jan
GRAVENHORST
1630-2328
FO W G Corewyn and crew
4/5 Jan
ROYAN
0112-0740

FO P B Shaw
Sgt W E Higgins
FS F B Robinson
FO S Burns
FS E Stafford
Sgt E Robertson
FS T V G Dearing
5/6 Jan
HOUFFALIZE
0100-0605
FO W G Corewyn and crew
WOs J W Jones and G G
Donald in;
Sgts Earl and Richardson out
7/8 Jan
MUNICH
1654-0213
FO W G Corewyn and crew
Sgt P R Ball in;
WO Jones out
14 Jan
MERSEBERG
1623-2054
FO J S Cooksey and crew
(sortie aborted, rear-gunner sick)
16/17 Jan
BRUX
1758-0318
FO J S Cooksey and crew
1 Feb
SIEGEN
1541-2215
FO J S Cooksey and crew
2/3 Feb
KARLSRUHE
2000-0334
FO J S Cooksey and crew
(failed to reach target in time)
7/8 Feb
LADBERGEN
2105-0259
FO J S Cooksey and crew
8/9 Feb
POLITZ
1634-0208
FO J S Cooksey and crew
(met heavy flak over Sweden!; first aircraft to bomb)
13/14 Feb
DRESDEN
1757-0410
FO R J Palmer
Sgt B Webster
FS J E Jolly
FS L G Simpson
FS E Oddy
Sgt Weaver
Sgt J Watson
14/15 Feb
ROSITZ
1654-0211
FO R J Palmer
Sgt J Monoghan
Sgt R R Grant
FS A C Shelstad
Sgt M T Plant
FO A Dunn
FS A L Knoke

19/20 Feb
BOHLEN
2332-0812
FO R J Palmer and above crew
20/21 Feb
MITTELAND CANAL
2158-0400
FO R J Palmer and crew
24 Feb
DORTMUND-EMS CANAL
1401-1907
FO E Roocroft
Sgt Hodges
Sgt N Fellows
Sgt T L Hargreaves
Sgt F Stanney
Sgt L F Aitken
Sgt D Harvey
(unable to identify target due to cloud; did not bomb)
3/4 Mar
DORTMUND-EMS CANAL
1835-0011
FO E Roocroft and crew
5/6 Mar
BOHLEN
1718-0242
FO J S Cooksey and crew
6/7 Mar
SASSNITZ
1813-0338
FO J S Cooksey and crew
7/8 Mar
HAMBURG
1753-0048
FO Ainsworth
FS Mills
FO Breakwell
FO Merritt
FS Snelling
Sgt Kitching
Sgt Lancaster
11 Mar
ESSEN
1226-1731
FO Ainsworth and crew
12 Mar
DORTMUND
1309-1857
FO J P Friend and crew
FL J B H Billam in;
Sgt Nuttall out
14/15 Mar
LUTZKENDORF
1655-0215
FO J P Friend and crew
16/17 Mar
WURZBURG
1732-0129
FO J P Friend and crew
20/21 Mar
BOHLEN
2348-0804
SL I J Fadden
WO J W Jones
FL W A B Martin
PO H W Knight
FO W R Brown
WO S H A Neill
WO G G Donald

22 Mar
BREMEN
1128-1619
FL D G G Phillips
FS S B Watson
FS Woodward
FO J Munn
FS W A Green
Sgt G E King
Sgt A Robinson
23/24 Mar
WESEL
1931-0103
FO W J Lambert
Sgt L Allward
PO R H Pawley
FO J A Ross
FS G F Cartwright
FS L A Sandiford
Sgt Hanna
27 Mar
FARGE
1033-1518
FO E Roocroft and crew
(starboard outer engine hit by heavy flak)
4 Apr
NORDHAUSEN
0558-1218
FL L A Davies and crew
6 Apr
IJMUIDEN
0845-1215
WO P K Morrison
FS G W Lilley
FS W Haughney
FS R G P Snelling
FS H H Neilson
Sgt G E King
WO C J Renaud
9 Apr
HAMBURG
1452-1942
WO P K Morrison and crew
16/17 Apr
PILSEN
2344-0720
WO P K Morrison and crew
18/19 Apr
KOMOTAU
2315-0752
FO H S Beckett
FS D M Moinalty
FO D T Mead
PO J D Ilott
FS N Jackson
Sgt K L Dean
Sgt R H Carr
23 Apr
FLENSBURG
1520-2055
FO H S Beckett and crew
25/26 Apr
TONSBERG
2013-0316
FO H S Beckett and crew

LL885
JIG

A Mk I built under licence by Messrs Armstrong Whitworth at Coventry, LL885 rolled off their production line in early 1944, fitted with four Merlin 24 engines. On 23 March is was assigned to No 622 Squadron at Mildenhall, Suffolk, where the squadron code letters GI were applied to its fuselage sides, together with the individual identification letter J-Jig. (Oddly enough GI-J is JIG spelt backwards!)

Jig's first sortie was flown on the fateful Nurnberg raid of 30/31 March, which resulted in 97 RAF bombers failing to return. Jig might well have been one of them, for it was struck by a falling incendiary over the target which cracked the main spar, but Pilot Officer J. M. Lunn brought the aircraft and his crew safely home.

Jig was not fit again until May but then slogged away solidly during the pre-invasion build-up and then flew on both D-Day nights and then on into the V1 site and communication targets period, and also flew on the Caen raid on 18 July. Jack Lunn completed his tour of ops in July and received the DFC, having flown Jig 15 times.

Following the aircraft's hesitant start, Jig did not in fact sustain any further serious damage until the night of 28/9 July. On its way to Stuttgart, Jig had reached the Orleans area when a night fighter attacked from out of the darkness. The pilot, Flight Lieutenant R. F. Allen, ordered the crew to put on parachutes, but the mid-upper gunner must have misinterpreted this precautionary instruction for he immediately clipped on his parachute and baled out. Losing the fighter, the damaged Lancaster was turned for home and the bombs were jettisoned 'live'.

Following repairs, Jig was back on ops by 25 August, but on the second raid after its return, the Lanc again received the attentions of a night fighter when going to Kiel on the 26/27th. This time Jig was not only damaged again but the rear-gunner, Sergeant Percy Stanley Withers, was killed in his turret. Jig was out of action until 20 September.

It was 40 minutes before midnight under a quarter-moon, while at 12,000ft that the pilot first saw the enemy fighter after a fighter-flare had been dropped. The fighter, identified as a FW190, came in on the starboard bow and then attempted to follow the Lanc round as its bullets were passing ahead of the bomber. It succeeded and proceeded to rake the port side fuselage, the '190 being virtually on its back by this time, with the Lanc doing a corkscrew.

Both gunners returned fire, the rear-gunner swinging his turret at the first warning and was hit in the back. Although wounded, the gunner swung his turret to the other side and was then hit again, this time in the chest. Once the fighter had gone, no word came from the rear turret, so the bomb-aimer went aft to find Withers dead. The bomb-aimer had first to chop away the rear turret doors because the lock had been destroyed.

Jig then had a long period of relative safety on raids, although it did collect a hole in the port wing on a daylight trip to Osterfeld on 11 December. On 1 February 1945 a sortie to Munchen-Gladbach had to be abandoned when the starboard outer engine feathered of its own accord and would not stay un-feathered. The bombs were jettisoned in the Frankfurt area and Jig was turned for home.

Jig's more prolific captains were Flying Officer H. P. Peck, who flew 21 trips in the aircraft and won the DFC, and Flight Lieutenant N. Jordan RCAF, who did 13. Another Canadian, Flying Officer B. Morrison RCAF, the pilot who had the engine un-feather itself as mentioned above, went on to fly 14 trips in Jig, including its 98th and 99th raids on 4 and 5 March 1945. (He also flew a 'Manna' trip to the Hague carrying as passenger, Lieutenant Colonel Morgan, the US Military Attaché.) However, the 100th sortie was flown by Flying Officer C. B. Moore RCAF on 6 March, a daylight op against an oil refinery at Salzbergen.

March saw some flak damage on the 14th and 21st but Jig carried on to complete 114 operations, then flew six 'Manna' and three 'Exodus' trips, to make a possible 123 missions in all.

Jig went to No 44 (Rhodesia) Squadron on 27 August 1945, then to No 39 MU on 3 January 1946. The aircraft was finally Struck off Charge on 4 March 1947.

1944
No 622 Squadron
30/31 Mar
NURNBERG
2225-0555
PO J M Lunn
FS H A Trennery (N)
Sgt W E Lister (WOP)
FS P J Halloran RCAF(BA)
Sgt W Wallis (FE)
Sgt J W Farrow (MU)
Sgt K C Hughes (RG)
1/2 May
CHAMBLY
2240-0225
PO A R Taylor
PO E J Insall RNZAF
FS R E Johnston
Sgt F Harriott
Sgt L H Gregson
Sgt L S Shaw
FS G Hutchinson
7/8 May
CHATEAU BOURGET
0055-0545
PO J M Lunn and crew
8/9 May
CAP GRIS NEZ
2235-0035
PO J M Lunn and crew
10/11 May
COURTRAI
2205-0030
FL R G Godfrey RAAF
PO A H Stewart

PO F S Sewell RNZAF
FO C D Chirighin
Sgt A W Ryder
FS W Ross
FS G E G Gardner RNZAF
11/12 May
LOUVAIN
2250-0130
PO J M Lunn and crew
FL Doonan (2P)
19/20 May
LE MANS
2200-0130
FL R G Godfrey and crew
(Richard Godfrey and crew
failed to return 7 June;
Godfrey was killed)
21/22 May
DUISBURG
2250-0255
PO J M Lunn and crew
22/23 May
DORTMUND
2310-0240
PO J M Lunn and crew
24/25 May
AACHEN
0015-0415
PO J M Lunn and crew
27/28 May
MINING/GIRONDE
2245-0425
PO J M Lunn and crew
28/2 May
ANGERS

1845-0215
PO E H Cawsey
FS E Panton
Sgt I Waters
FO G Hayter
Sgt F N Poynter
Sgt W E Mayes
Sgt C Pratt
31/1 Jun
TRAPPES
2340-0440
FO A Smith
FS J Chigwidden
Sgt P Brandon
FO A Montgomery
Sgt R Lewis
Sgt A O'Connor
Sgt J Spencer
5/6 Jun
OUISTREHAM
0315-0630
PO J M Lunn and crew
6/7 Jun
LISIEUX
0010-0350
PO A B Robbins RAAF
FS E Sutton
Sgt A Eldridge
FS G Wachter RAAF
Sgt A Morgenstern
Sgt J A Mann
Sgt G R Wilby
8/9 Jun
FOUGERES
2155-0250

PO A B Robbins and crew
12/13 Jun
GELSENKIRCHEN
2300-0300
PO J E A Pyle
FS P A McGibbon
Sgt E Crowther
FS L Tomlinson
Sgt A H Hall
Sgt W H Pool
Sgt J L Spaven
13/14 Jun
MINING/BREST
2240-0330
PO J M Lunn and crew
14/15 Jun
LE HAVRE
2300-0250
PO J M Lunn and crew
15/16 Jun
VALENCIENNES
2305-0215
FL R W Trenouth
FS S West
FS W A Atkins
FO B L Good
Sgt R B Francis
Sgt C C Pulman
FS D C Harvey
17/18 Jun
MONTDIDIER
0110-0450
PO J M Lunn and crew
23/24 Jun
L'HEY/V1 SITE

2300-0145
PO J M Lunn and crew
30 Jun
VILLERS-BOCAGE
1825-2115
PO J M Lunn and crew
2 Jul
BEAUVOIR/V1 SITE
1215-1605
FL R G Allen
FS W A Bishop
Sgt J Paton
FO C D J Pennington
Sgt J Barker
FL D B Mason
PO J T Gray
9 Jul
LINZEUX/V1 SITE
1210-1555
PO J M Lunn and crew
10 Jul
NUCOURT/V1 SITE
0355-0800
FO J W Stratton
FS M L Reilly RNZAF
WO J McGuinness
FO E F Thurston RNZAF
Sgt H H Clifton
FS H G Summerton RAAF
Sgt R N Pittaway
17 Jul
VAIRES/V1 SITE
1050-1420
FL R W Trenouth and crew
18 Jul
CAEN
0335-0725
FO N V Gill
FS E Featherstonehough
FS R I Smith
FS J R Short
Sgt K J Humphries
Sgt H Harris
Sgt H P R Russell
18/19 Jul
AULNOYE
2240-0230
PO A B Robbins and crew
FS D Smith (2P)
20/21 Jul
HOMBERG
2310-0255
PO A B Robbins and crew
23 Jul
MONT CONDON
PO A B Robbins and crew
0645-1030
24 Jul
PROUVILLE
0920-1305
PO A B Robbins and crew
24/25 Jul
STUTTGART
2150-0535
FO A T Wheate
FS W Leonard
FO D C Chapman
FO D W Findlay RNZAF
Sgt R Kestrell
Sgt H Elliott
Sgt R J Thornton
25/26 Jul
STUTTGART
2120-0355
PO A B Robbins and crew
28/29 Jul
STUTTGART

2155-0255
FL R G Allen and crew
(mid-upper gunner baled
out; aircraft damaged)
25/26 Aug
RUSSELSHEIM
2035-0525
PO A B Robbins and crew
26/27 Aug
KIEL
2005-0141
FO A H Thompson RAAF
FS W S Ward RAAF
FS R W Aland RAAF
FS R J Dilley RAAF
Sgt K J Bouldton
Sgt D Smith
Sgt B S Withers
(rear gunner killed by night
fighter)
20 Sep
CALAIS
1430-1740
FO H P Peck
FS J W Barchard
FO R T Cargill
FO A G Long
Sgt R C Bowyer
Sgt J Rumsey
Sgt D C Pudney
23/24 Sep
NEUSS
1920-0015
FO H P Peck and crew
24 Sep
CALAIS
1710-1930
FL J A Brignell
FO K C Lewis RAAF
FS J Harris
FO M J McDonnell
Sgt J Irving
Sgt M Davis
Sgt M Coles
25 Sep
CALAIS
1020-1350
FO H P Peck and crew
(raid abandoned due to low
cloud)
26 Sep
CALAIS
1020-1350
FS G Myles
FS W D Aveyard
WO F G Mills RAAF
Sgt V Rielly
Sgt A F Crawford
FS H C Field RAAF
FS F J Schell RAAF
27 Sep
CALAIS
0720-1035
PO M G Baxter
Sgt L G Parsons
Sgt G C Brooker
FS R Hopkinson RAAF
Sgt E J Rossiter
Sgt J W Schuler
Sgt F Ramsey
5/6 Oct
SAARBRUCKEN
1845-0040
FO H P Peck and crew
6 Oct
DORTMUND
1650-2200

FO H P Peck and crew
7 Oct
KLEVE
1120-1540
FO H P Peck and crew
14 Oct
DUISBURG
0630-1055
FO H P Peck and crew
14/15 Oct
DUISBURG
2240-0355
FO H P Peck and crew
18 Oct
BONN
0825-1330
FO H P Peck and crew
19 Oct
STUTTGART
1725-2355
FO N G Flaxman
FO M G Stewart
FO G K Soderberg
FO J Adamson
Sgt E Turley
Sgt W J Thurman
Sgt K F Sadler
22 Oct
NEUSS
1300-1725
FO H P Peck and crew
23 Oct
ESSEN
1610-2130
FO H P Peck and crew
25 Oct
ESSEN
1230-1720
FO H P Peck and crew
28 Oct
COLOGNE
1310-1745
FO H P Peck and crew
30 Oct
COLOGNE
1745-2300
PO D C Barnett
Sgt H C Grimsay
FS J P Sullivan
FO E J I Hurditch
Sgt C G Staplehurst
FS L C Nichols
Sgt F Murrell
4 Nov
SOLINGEN
1130-1600
FO R Curling
FS J W H Murdin
WO T Hine
FS L W Middleditch
Sgt D S White
Sgt C W S Robinson
Sgt H Wallace
5 Nov
SOLINGEN
1020-1520
FO H P Peck and crew
8 Nov
HOMBERG
0805-1245
FO N G Flaxman and crew
11 Nov
CASTROP
0820-1300
FO N G Flaxman and crew
15 Nov
DORTMUND

1235-1750
FO H P Peck and crew
16 Nov
HEINSBURG
1300-1730
FO H P Peck and crew
20 Nov
HOMBURG
1240-1745
FO H P Peck and crew
21 Nov
HOMBURG
1235-1640
FO H P Peck and crew
23 Nov
GELSENKIRCHEN
1245-1755
FO H P Peck and crew
26 Nov
FULDA
0750-1330
FO H P Peck and crew
27 Nov
COLOGNE
1220-1715
FO R Curling and crew
WO T R Allan in, WO Hine out
28 Nov
NEUSS
0240-0720
FO H P Peck and crew
30 Nov
BOTTROP
1050-1510
FO W H Thorbecke
FL A Westbrook
Sgt R A Adams
FO H S Villiers
Sgt T J Hogen
FS J W Scouler
Sgt A Staele
2 Dec
DORTMUND
1255-1715
FO W H Thorbecke and crew
Sgt A J Smith RCAF in, FS
Scouler out
8 Dec
DUISBURG
0825-1255
FS L Stille
Sgt J W Morgan
Sgt S G Hewett
Sgt N Brabazon
Sgt E Eastwood
Sgt H Dewey
Sgt E Jones
11 Dec
OSTERFELD
0835-1310
FS L Still and crew
12 Dec
WITTEN
1103-1600
FO N G Flaxman and crew
15 Dec
SIEGEN
1125-1415
FO N G Flaxman and crew
(recalled and jettisoned the
'cookie' in the Channel)
16 Dec
SIEGEN
1120-1650
FO N G Flaxman and crew
24 Dec
BONN A/F

Above: A-Able, LM594 of No 576 Squadron pictured after its 100th op.

Below: ME746 of No 166 Squadron after its 100th sortie on 11 March 1945. The air and groundcrews present the aircraft with a mock DSO, held between Harold Musselmann DFC and Corporal Dennis Terry. At far right is the squadron CO, Wing Commander Vivian.

Above: Pictured here at the end of the war, ME746 sports 125 bombs, the DSO and DFC, and has 'R2' marked on the nose. With Pilot Officer S. Todd and crew, the two groundcrew men on the left are Corporals Dennis Terry and Sid Woodcock.

Below: The presentation of mock DSO and DFC to 'Nan' (ME758) of No 12 Squadron at Wickenby, in April 1945. The bomb tally shows 106 operations, plus one swastika and two searchlights shot out. In the centre is the squadron CO, Wing Commander Mike Stockdale and to his left, with arms folded, is Squadron Leader Peter Huggins, OC B Flight.

Above: ME758 now has 108 bomb symbols and markings for six 'Manna' trips. The crew depicted flew this aircraft on its last 'Exodus' trip, 11 May 1945. Standing: K. Bratby, Squadron Leader P. S. Huggins DFC, Len Laing, Geoff Robinson; front: Len Jackson, Sam Pechet and Tommy Thompson.

Below: ME801 of No 576 Squadron pictured after its 100th op, 15/16 March 1945.

Left: An almost identical shot of ME801, but a DSO ribbon has been added next to the DFC and the swastika, along with a leek forward of the bomb tally.

Right: 'Fair Fighter's Revenge' was ME812 of No 166 Squadron and is shown here having flown 33 ops.

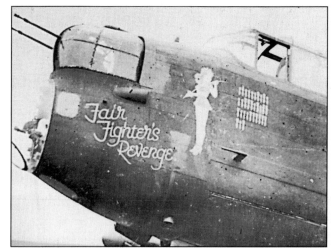

Right: Sid Coole and crew took ME812 on its first operation and completed 14 altogether in this aircraft. At rear from l to r: Ray Scargill, William Ansell, Alf Holyoak, Arthur Downs; front: Charles Birtwhistle, Sid Coole, Bob Rennie. All were awarded the DFM.

Left: ME803 of No 115 Squadron with groundcrew personnel, shows a total of 100 bombs. Unfortunately this Lanc's artwork — a 'lovely lady' — seems to be too worn and corroded to be made out in this picture.

Right: ME812 sports 104 bomb symbols in May 1945.

Opposite page, top: Jack Playford RCAF in the cockpit of 'Able Mabel', ND458, showing 129 bombs and two swastikas. This aircraft eventually completed 132 trips.

Left: 'Able Mabel's' Jack Playford shakes hands with the Lanc's groundcrew chief, Sergeant W. Hearn. From l to r: Leading Aircraftsman J. Cowls (fitter-airframes), Corporal R. Withey (fitter), Hearn and Playford, Leading Aircraftsman J. Robinson and Aircraftsman J. Hale (both fitters-engines).

Above: John Chatterton and crew took ND578 on its first op and flew a further 15 in the aircraft. From l to r: John Davidson, Bill Champion, Bill Barker, John Chatterton, D. J. Reyland, Ken Letts and John Michie. Chatterton and Reyland received DFCs, the others DFMs.

Left: 'Yorker's' groundcrew (rear row only): Jock Biggar (rigger and 'bomb' painter), Harry Prior (engines), Palmer, Dick Pinning (engines) and Alan Rubenstein (NCO i/c).

Below: ND578 KM-Y of No 44 (Rhodesia) Squadron. Aircrew from l to r: Pete Roberts, Steve Burrows, M. J. Stancer; Wing Commander F. W. Thompson DFC (9th from left), Bill Clegg (11th), Flight Lieutenant G. E. Mortimer (Gunnery Leader) far right.

Top right: 'Yorker' at Spilsby after its 107th op — it went on to complete a total of 123.

Bottom right: No 100 Squadron's ND644 HW-N showing 112 ops. From l to r: Aircraftsman C. F. Turrell (fitter-engines), Leading Aircraftsman J. Atkinson (fitter-engines), Flight Lieutenant H. G. Topliss, Sergeant H. W. Williams (NCO i/c), Leading Aircraftsman B. Gorst (fitter-airframes).

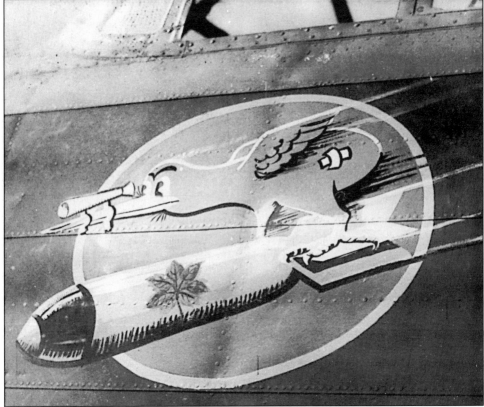

Top left: ND709 F2-J of No 635 Squadron, still showing the 'M' of No 35 Squadron, with which it flew two trips. With the Lanc's groundcrew are Derrick Coltman and Denis Linacre.

Bottom left: ND709's nose emblem, a flying kiwi, with bomb sight, and a maple leaf on the bomb, representing the Lanc's New Zealand skipper and Canadian navigator.

Above: Wing Commander D. W. S. Clark DFC (left) and crew: Harry Laskowski, Denis Linacre, D. A. Tulloch, J. Smith(?), Derrick Coltman and J. A. Rayton.

Below: Harry Laskowski and Derrick Coltman with ND709, showing 54 bomb symbols. Note how they are painted on in an upward direction.

Left: Squadron Leader P. R. Mellor and crew: Gerry Shaw and Jock Blair (in doorway), F. E. Prebble, Peter Mellor, L. Freeman, A. Rowbothom, and (in front) E. E. Freake. Note the PFF insignia beneath their brevets.

Below: The 100th op being marked on ND709, 14 February 1945. This aircraft then went to No 405 (RCAF) Squadron and flew another 11 trips.

Right: ND875 on the occasion of reaching its 100th op, flown by Squadron Leadwer P. F. Clayton (bottom right) and crew with No 156 (PFF) Squadron, 24 March 1945.

Right: 'The Captain's Fancy' was NE181 of No 75 (NZ) Squadron, and is pictured here when nearing its 100th op.

Right: Flying Officer Gordon Cuming and crew of No 75 (NZ) Squadron, flew NE181 on a dozen ops in late 1944. Rear l to r: Paddy McElligott, Jack Christie, Bill Scott; front: Jack Scott, Gordon Cuming, Syd Sewell.

Above: Jack Bailey DFC took EN181 on its 100th sortie on 29 January 1945 and its 101st (and last) on 2/3 February.

Left: NE181's groundcrew paint on the 101st bomb, 3 February 1945. On the ladder are Leading Aircraftsmen Taylor and Thompson; foreground left to right, Sergeant Grantham, Leading Aircraftsman F. Woolterton, Jack Bailey DFC, unknown.

Opposite page, top: George Blackler pictured in the cockpit of No 550 Squadron's PA995 on 6 March 1945.

Opposite page, bottom: George Blackler's crew who flew 27 ops in PA995. From l to r: Blackler, John Nicholson, W. Ross, H. P. Nicholls, E. Mozley, M. McCutcheon; Jack Bold is sitting.

Above: Everyone turned out to celebrate PA995's 100th op, but the Lanc failed to return from its 101st on 7 March 1945.

Below: Lancaster PB150 is believed to have reached 100 ops with 'Manna' and 'Exodus' sorties. The 'Manna' trips are recorded as Spam tins. The 'duck' wears a halo while standing on a bomb.

1525-2010
FS G E Darville RAAF
Sgt T Lisle
FS M J Keough
Sgt D Pearce
Sgt W S Godfrey
Sgt D F Dunn
Sgt H V Beckwith
28 Dec
COLOGNE
1228-1715
FS G E Darville and crew
29 Dec
KOBLENZ
1223-1709
FO N Jordan RCAF
FS W M McDonald
Sgt Mc J Robinson
FO R S Riley
Sgt L Eyre
Sgt T H Gregory
Sgt P C Laymore

1945
2 Jan
NURNBERG
1535-2305
FO W H Thorbecke and crew
3 Jan
DORTMUND
1300-1820
FO N Jordan and crew
5 Jan
LUDWIGSHAVEN
1135-1750
FO N Jordan and crew
6 Jan
NEUSS
1530-2040
FO N Jordan and crew
7/8 Jan
MUNICH
1840-0250
FO A S W Waigh
FO T S Briggs
WO T R Allen RAAF
FO T M R Lister
Sgt E Evans
Sgt P J Simmonds
Sgt W A Mitchell
29 Jan
KREFELD
1005-1530
FO N Jordan and crew
1 Feb
MUNCHEN
GLADBACH
1324-1638
FO B Morrison RCAF
Sgt J B Orr
Sgt L Totman RCAF
PO J Pollard RCAF
Sgt A G Chambers
Sgt T G Stock RCAF
Sgt D W S West
3/4 Feb
DORTMUND
1617-2142
FO B Morrison and crew
Sgt A J Smith RCAF in, Sgt
West out
7 Feb
WANNE EICKEL
1138-1734

FL N Jordan and crew
9 Feb
HOHENBUDBERG
0330-0823
FL N Jordan and crew
13/14 Feb
DRESDEN
2132-0705
FL N Jordan and crew
14/15 Feb
CHEMNITZ
2024-0442
FL N Jordan and crew
15 Feb
MINING/KATTEGAT
1635-2352
FO W H Thorbecke and crew
18 Feb
WESEL
1157-1708
FO B Morrison and crew
Sgt West back, Sgt Smith out
19 Feb
WESEL
1313-1812
FO B Morrison and crew
20/21 Feb
DORTMUND
2149-0412
FL N Jordan and crew
22 Feb
GELSENKIRCHEN
1246-1748
FO B Morrison and crew
23 Feb
GELSENKIRCHEN
1149-1720
FO A G Moore
Sgt R G Price
FS B G Dawson
Sgt L W Lund
Sgt C W A Forsey
Sgt J R Thompson
Sgt R H Davis
1 Mar
KAMEN
1140-1720
FO B Morrison and crew
2 Mar
COLOGNE
1300-1834
FO J B Cameron RCAF
FS D E Cameron RCAF
Sgt M W Guy RCAF
FS W M Leeming
Sgt E S J Seldon
FS K E Boone DFM
Sgt A D Falconer
4 Mar
WANNE EICKEL
0935-1431
FO B Morrison and crew
5 Mar
GELSENKIRCHEN
1032-1613
FO B Morrison and crew
6 Mar
SALZBERGEN
0827-1413
FO C B Moore RAAF
FS H D Patterson RAAF
Sgt E H G Potkins
Sgt F G Cowap
Sgt R B Sutcliffe

Sgt P F Cochrane
Sgt D C Carter
9 Mar
DATTELN
1042-1604
FS L W O'Connor RAAF
FS A O'Neill RAAF
FS K J McKenzie RAAF
FS W D Williamson
FS E Wiles
Sgt A Featherstone
Sgt G Goodall
10 Mar
GELSENKIRCHEN
1200-1717
FO B Morrison and crew
11 Mar
ESSEN
1129-1719
FS L W O'Connor and crew
12 Mar
DORTMUND
1231-1856
FO B Morrison and crew
14 Mar
EMSCHER
1325-1847
FL N Jordan and crew
(aircraft hit by flak)
18 Mar
HATTINGEN
1158-1718
FO B Morrison and crew
20 Mar
HAMM
0952-1534
FS C Malcolm
FS L J Lane
FS R G Winden
FS R A Robinson
Sgt W Webster
FS D H Shorter
Sgt K Wood
21 Mar
MUNSTER
0942-1526
FO B Morrison and crew
(aircraft hit by flak)
27 Mar
KONIGSBORN/HAMM
1031-1553
FO C B Moore and crew
29 Mar
HALLENDORF
1221-1919
FO J W Armfield
Sgt R R Last
FS J H Atkins RAAF
Sgt G C Anyan
Sgt W J Ellison
Sgt G Rivett
Sgt L R Pavey
9/10 Apr
KIEL
1929-0133
FL N Jordan and crew
13/14 Apr
KIEL
2030-0241
FL N Jordan and crew
14/15 Apr
POTSDAM
1809-0240
FO W M Scriven RAAF

Sgt R H Field
Sgt W B Wolfe
FS A McDonald
Sgt W F McLean
Sgt S Cadman
Sgt L T Hyde
22 Apr
BREMEN
1520-2041
FO W M Scriven and crew
1 May
'MANNA'/ THE HAGUE
1319-1540
FO B Morrison and crew
2 May
'MANNA'/THE HAGUE
1053-1313
FO C B Moore and crew
3 May
'MANNA'/THE HAGUE
1123-1349
FO J H Fiedler RAAF
FS D H A Elliott
WO R D Fitch
FS H J Doherty
Sgt E J Taylor
Sgt E Smith
Sgt L Hayward
4 May
'MANNA'/THE HAGUE
11221-1450
FO B Morrison and crew
Lt Col Morgan US (passenger)
7 May
'MANNA'/LEIDEN
1122-1403
SL C A Ogilvy
FL G J Speed (N)
FO F M Gloyne (WOP)
FO J W Tanner (BA)
FS E H Barton (FE)
LAC R Starlam (passenger)
Sgt L D Watkins (RG)
8 May
'MANNA'/THE HAGUE
1153-1425
FO F W Connolly
FO E Goose
FS W A Franks
Sgt W Lowe
Sgt A G Benson
Cpl G N Applegate (passenger) FS K Metcalfe
10 May
'EXODUS'/JUVINCOURT
0758-1352
FO E J Willis RAAF
FS C Walters RAAF
Sgt R L Robinson
FO H E Pam
Sgt C Campbell
Sgt R Farthing
11 May
'EXODUS'/JUVINCOURT
1629-2153
FL J B Cameron and crew
13 May
'EXODUS'/JUVINCOURT
1041-1225
FS L W O'Connor and crew

LM227
ITEM

This was a Mk I Lancaster built under licence by Messrs Armstrong Whitworth at Coventry as part of contract No 239. It came off the production line with four Merlin 24 engines in the spring of 1944 and was assigned to No 576 Squadron at Elsham Wolds, Lincolnshire, on 30 June 1944. Coded UL, the aircraft became I-Item when given the individual aircraft identification letter 'I' (later I2).

Item's nose-artwork consisted of a Saturn-like ringed planet surrounded by stars and a crescent moon, plus a falling star, leaving a trail of tiny stars behind it. Later a swastika was added for a night kill, and the aircraft's 21st trip was marked with a key on a parachute. The stars and other adornments were painted on by Norman Bryon, Item's engine fitter, and he remembers it took a few days to complete and he was anxious to finish it in case the Lanc failed to return before he had finished the task. He recalls:

'I was the engine fitter on this aircraft since the day she arrived brand new, till some time after the war ended. A very good aircraft, it was a pleasure to work on it. As for the emblem, I can say it was all my own work; it was all very different to everyone else's.

'Among her regular Skippers I recall Flying Officer Till. He and his crew were always pleased with her performance. The other two in our groundcrew team were LACs Bernard Clixman (engines) and Tommy Cookson (rigger).'

Item's first pilot was Pilot Officer J. R. 'Mike' Stedman, who took the aircraft to Orleans on 4/5 July 1944. He remembers:

'We joined No 576 Squadron as a crew on 20 April after converting from Stirlings to Lancasters at No 1 LFS, Hemswell. Being a "sprog" pilot and crew we were allocated spare aircraft for the first few ops. I2-Item was eventually given to us (I forget its number) later losing its letter I2 to LM227 on or about 1 July.

'It was a great thrill to have a brand new aircraft, Packard Merlins with "paddle" props which gave us an extra few thousand feet of altitude. We also had H2S and "Fishpond" which picked up enemy nightfighters and allowed us to play cat and mouse with them!

'The nose insignia had no particular meaning, I just thought it would be appropriate as I expected most of my ops would be at night. However, it turned out to be just 19 out of 32.

'We were posted to Special Duties Flight, Binbrook, for a month, rejoining LM227 on our return to Elsham Wolds on 13 August. Our last trip with LM227 was a daylight to Le Havre on 10 September.'

As the number of Item's raids mounted the bombs, painted on in rows of 10 on the lower half of the nose, reached 50, and then a second five rows of 10 were marked next to the first. A DFC ribbon was also painted above the first row, while ahead of the bombs was a swastika, denoting a combat – no doubt the one Item had on 28 July when the Lanc's gunners claimed a Me109 as probably destroyed.

Item missed D-Day by almost a month by the time it flew its first sortie. After Mike Stedman had broken the Lanc's duck, Flight Lieutenant H. B. Guilfoyle RCAF, began to fly Item, completing 17 trips and, like Stedman and his bomb-aimer, was to win the DFC. Harold Guilfoyle's Canadian rear-gunner, Sergeant Nick J. Hawrelencko was also to receive the DFM.

Harold – or more familiarly, Harry – Guilfoyle's navigator was Flying Officer George Tabner RCAF. He remembers:

'Guilfoyle and crew started with this aircraft on 7 July 1944 and we did 17 trips in her. On the 24th we were detailed as Windfinders, and again on the 28th, in order to increase the number of aircraft over the target at marking; I dropped the bombs on H2S and we claimed an Me109, credited to Nick Hawrelencko.

'On 9 August we laid mines in the Antioche Straits – a trip of 6 hours 15 minutes. We could see two other aircraft all the way to the target, near which some gunner on a ship opened up at us. Nick silenced him with the four guns from his rear turret. I dropped six mines, again on H2S.

'On our last op – 12 August, to Brunswick – we were again Windfinder aircraft and bombed on H2S. Winds broadcast to the Main Force were "Speed 99" although my actual winds were 125mph. We sent this back, breaking code but the attack was very scattered. We bombed from 22,000ft after circling and climbing over the target for 20 minutes.

'Our last two trips were extra for six of us as we volunteered to do two extra in order to finish off Nick, who had missed trips due to receiving two black eyes in a fight in London!'

Item's next regular skipper was Flying Officer D. E. Till. Derek Till flew the aircraft on no fewer than 28 sorties. Tour length during his period on the squadron was 35 ops, and his tour was completed on 24 March 1945, although in another Lancaster. Derek Till also flew one op in another 100-op veteran on No 576 Squadron, ME801. In fact he flew 37 ops in all, making two extra trips in order that his bomb-aimer could complete his tour, having missed a couple of sorties. Derek Till, who received the DFC, relates:

'Item had no mechanised throttle gate to tell you when you were up to 16lb of boost. Usually there was a wire which you had to break to take it up to the emergency boost of plus 18lb. I remember the first time I ran up the engines on Item one morning at dispersal. Without realising it I took the port inner up to plus 18 and the Flight Sergeant came roaring up and gave me hell; he was worried about burning out the exhaust stubs and, of course, it didn't do the engine any good.

'There were a couple of targets that were difficult, one being the oil refinery at Merseburg which was extremely heavily defended due to its importance. We were very nervous going off on that one because we had all been bombed-up to go the previous day but then it was scrubbed. Rumours were rife that the Germans would know about it and be ready for us and indeed the flak at the aiming point was incredible.

'On another trip I thought we had missed a course change but my navigator said everything was alright and we went on and on and on. Then my WOP came on the R/T to say Charlie had passed out. His oxygen tube had become disconnected and after writing down some things on his navigational log which was absolute nonsense, he'd fainted. He soon came-to but, of course, we were late to the target, and we felt very lonely being the last to bomb on that occasion.'

Apart from the probable victory over the Me109 on 28 July, Item's rear-gunner on a raid to Pforzheim on 23 February 1945 spotted a Ju88 just as it began firing into the Lanc from astern. Its fire caused damage to the port wing but the gunner, Sergeant Ranchuck RCAF, claimed he damaged the night fighter. Item had only just completed a major inspection and became Cat AC, returning to Avro's for repairs, but was back on the squadron by 3 March.

At the end of April Item had flown 93 sorties, then completed six 'Manna' and one 'Exodus' trips to make the 100. By and large the recorded sorties correlate, although there are two that could be at variance. On 5 August 1944, Item's usual pilot, Guilfoyle, is noted as flying aircraft 'I' but it is shown as ME800. It has been assumed the ME800 is incorrect (although this Lanc was with No 576 Squadron). The other is a raid on Essen on 11 March 1945, where the aircraft number is not recorded, but the pilot on that occasion had just previously flown on his second dickie trip with Derek Till, and may well have flown Item on one of his first crew trips.

Item did not last for long after the war, being wrecked and declared Cat E on 16 October 1945 before being Struck off Charge.

1944	Sgt W S Woolridge	Sgt J E Kearney	FO P J Dodwell (BA)
No 576 Squadron	Sgt D W Waldron	FS L Fielding	FO G E Tabner RCAF (N)
4/5 Jul	Sgt E W Shreeve	WO T P Rozy	FS J D Powewll (WOP)
ORLEANS	**5/6 Jul**	Sgt A Milne	Sgt N S Cassidy (MU)
2150-0400	DIJON	**7 Jul**	Sgt N J Hawrelechko RCAF
PO J R Stedman	2115-0615	CAEN	**12/13 Jul**
Sgt R S Porter	PO J Archibald RNZAF	1930-2330	REVIGNY
FO H A W Rumbelow(BA)	Sgt J R Cuthbert	FL H B Guilfoyle RCAF	2110-0620
Sgt E Swift	FO P J Bielle RCAF	Sgt S C Wilkin (FE)	FL H B Guilfoyle and crew

14/15 Jul
REVIGNY
2110-0530
FL H B Guilfoyle and crew
17 Jul
SANDERVILLE/V1 SITE
0320-0715
FL H B Guilfoyle and crew
18/19 Jul
SCHOLVEN
2305-0315
FL H B Guilfoyle and crew
20 Jul
WIZERNES/V1 SITE
1905-2230
FL H B Guilfoyle and crew
23/24 Jul
KIEL
2230-0400
FS A d'L Greig
Sgt N Mason
FO J R Jones
FS G G Henville RAAF
Sgt S A Johnson
Sgt H Walton
Sgt J Thresh
24/25 Jul
STUTTGART
2110-0550
FL I I B Guilfoyle and crew
25/26 Jul
BOIS DE JARDINES/
V1 SITE
0040-0425
FL H B Guilfoyle and crew
28/29 Jul
STUTTGART
2105-0535
FL H B Guilfoyle and crew
(gunners claimed Me109
probable)
30 Jul
CAHAGNES
0625-1015
FL V Moss
Sgt R A Robertson
FO R E Power
Sgt M G Callaghan
Sgt T E Sinton
Sgt V L Oatley
Sgt R Stenton
31 Jul
LE HAVRE
1800-2125
FL H B Guilfoyle and crew
1 Aug
BELLE CROIX/V1 SITE
1845-2140
FL H B Guilfoyle and crew
(raid aborted due to fog and
haze over target)
3 Aug
TROSSY ST MAXIM
1125-1535
FL H B Guilfoyle and crew
4 Aug
PAUILLAC
1325-2125
FO F H Watts RCAF
Sgt C Douglas
FO I G D Mills (BA)
FO R Hughes-Games (N)
Sgt M R Swallow
Sgt W S Clatworthy
Sgt D H Laing
5 Aug
BLAYE

1400-2200
FL H B Guilfoyle and crew
7/8 Aug
FONTENAYE
2055-0055
FL H B Guilfoyle and crew
9/10 Aug
MINING/LA ROCHELLE
2130-0345
FL H B Guilfoyle and crew
10 Aug
DUGNY
0920-1440
FO S F Durrant
Sgt R E Pearce
Sgt C J Brady RCAF
FO J Johnson
FS J R McIntrye RAAF
Sgt J G MacKay
Sgt D Spowart
11 Aug
DOUAI
1330-1730
FL H B Guilfoyle and crew
12/13 Aug
BRUNSWICK
2135-0240
FL H B Guilfoyle and crew
14 Aug
FONTAINE LE PIN
1315-1725
FO J J Mulrooney
Sgt B Beale
Sgt D M Redi RCAF
Sgt F Doaker
Sgt A P J Gostling
Sgt J T Smith
Sgt P Rayner
(FO Mulrooney failed to
return from his 29th op on 2
November 1944)
15 Aug
LE CULOT A/F
0950-1320
FO J R Stedman and crew
16/17 Aug
STETTIN
2125-0450
FO J R Stedman and crew
18/19 Aug
ERTVELDE RIEME
2230-0230
FO L Archer
Sgt J Hill
FO O J T Troy (BA)
FO R J Williams (N)
FS D T H Madell
Sgt E N Wood
Sgt E L Gidden
25 Aug
RUSSELSHEIM
2005-2355
FO J R Stedman and crew
26/27 Aug
KIEL
1940-0115
FO J R Stedman and crew
28 Aug
CHAPELLE NOTRE
DAME
1855-2150
FO J R Stedman and crew
29/30 Aug
STETTIN
2120-0640
PO K L Trent
Sgt A R Dunford

Sgt J N Wadsworth
Sgt H B Reynolds
Sgt R L Skelton
Sgt C Dalby
FO G G Riccomini
31 Aug
AGENVILLE
1245-1645
FO J R Stedman and crew
3 Sep
EINDHOVEN A/F
1520-1910
FO J R Stedman and crew
5 Sep
LE HAVRE
1555-1920
FO J R Stedman and crew
6 Sep
LE HAVRE
1700-2025
FO J R Stedman and crew
10 Sep
LE HAVRE
1620-2005
FO J R Stedman and crew
12/13 Sep
FRANKFURT
1810-0200
FO M T Wilson
Sgt R P Kimm
Sgt P F T Brooked
Sgt L N Hill
Sgt S Douglas
Sgt T A Russell
Sgt J H Addison
15/16 Sep
MINING
2150-0755
FO M T Wilson and crew
16/17 Sep
LEEUWARDEN A/F
0055-0420
FO R A Boggiano
Sgt E V Taylor
Sgt A W Black
FO A Conor RCAF
Sgt S R Coe
Sgt K Holdsworth
Sgt G R Logan
23 Sep
NEUSS
1835-2355
PO K L Trent and crew
24 Sep
CALAIS
1615-1935
PO K L Trent and crew
27 Sep
CALAIS
0910-1245
FO D E Till
Sgt D E Holland (FE)
FO J Shorthouse (BA)
FO C R Bray (N)
FO G W Griggs RAAF
Sgt K Oliver (MU)
Sgt R Hamilton (RG)
5/6 Oct
SAARBRUCKEN
1820-0110
PO K L Trent and crew
7 Oct
EMMERICH
1200-1625
PO K L Trent and crew
12 Oct
FT FREDERIK HENRIK

0635-1000
FO D E Till and crew
14 Oct
DUISBURG
0610-1120
PO K L Trent and crew
14 Oct
DUISBURG
2220-0425
FO D E Till and crew
19 Oct
STUTTGART
2135-0415
FO D E Till and crew
22 Oct
ESSEN
1610-2210
FO D E Till and crew
25 Oct
ESSEN
1255-1815
FO D E Till and crew
27 Oct
COLOGNE
1335-1930
FO D E Till and crew
30 Oct
COLOGNE
1745-2030
FO D E Till and crew
2 Nov
DUSSELDORF
1615-2105
FO D E Till and crew
4 Nov
BOCHUM
1735-2235
FL J W Acheson RCAF
Sgt R M Cameron
FO R A Flight
FS J R Tudhope RAAF
Sgt T R Bell
Sgt H Bukoski RCAF
Sgt I Lewis
6 Nov
GELSENKIRCHEN
1140-1620
FL J W Acheson and crew
9 Nov
WANNE EICKEL
0830-1255
FO D E Till and crew
11 Nov
DORTMUND
1550-2110
FO D E Till and crew
18 Nov
WANNE EICKEL
1605-2200
FO E L Saslove RCAF
Sgt R Hoyle
FO G Davies
FO N Chisick RCAF
Sgt R F Hood RCAF
Sgt A S B Campton
Sgt G W McClelland
29 Nov
DORTMUND
1225-1725
FO D E Till and crew
3 Dec
URFT DAM
0735-1200
FO D E Till and crew
4 Dec
KARLSRUHE
1630-2300

FO D E Till and crew
6/7 Dec
LEUNA
1635-0035
FO D E Till and crew
12 Dec
ESSEN
1610-2145
FO O R Herbert
Sgt E Heller
Sgt D Thornwell RCAF
PO R F Scott RCAF
Sgt J A Stephens RCAF
FS E J Wise
WO R J Turley
15 Dec
LUDWIGSHAVEN
1450-2055
FO D E Till and crew
17 Dec
ULM
1515-2240
FO D E Till and crew
22 Dec
COBLENZ
1510-2210
FO E J Pollard
Sgt E B May
Sgt A Preston (BA)
Sgt C V Dolan (N)
Sgt J Patterson
Sgt R M Goodfellow
Sgt Brown
28 Dec
BONN
1515-2105
FO D E Till and crew
29 Dec
SCHOLVEN
1515-2130
FO B H O'Neill RCAF
Sgt R J Durran
FS S W Haakstead RCAF
Sgt J G Ogilvie
Sgt F C Smith
Sgt A J W George RCAF
Sgt D Ranchuck RCAF
31 Dec
OSTERFELD
1450-2055
FO D E Till and crew
1945
4/5 Jan
ROYAN
0159-0907
FO H Benson
FO H Woolstenhulme
FO H G Mather
FO P C Milner
FO F Wilson
Sgt R C Griffiths
Sgt R Goldsbury

7/8 Jan
MUNICH
1825-0320
FO E J Pollard and crew
14/15 Jan
MERSEBERG
1905-0315
FO D E Till and crew
22 Jan
DUISBURG
1650-2155
FO D E Till and crew
28/29 Jan
STUTTGART
1953-0252
FO C T Dalziel
Sgt P R Montgomery
Sgt W E Bradbury
Sgt W E May
Sgt A Burns
Sgt C T Thorley
Sgt D O'Sullivan
(FO Dalziel and crew failed to
return 7 March 1945)
1 Feb
LUDWIGSHAVEN
1600-2235
FO D E Till and crew
2/3 Feb
WIESBADEN
2045-0250
FO D E Till and crew
3 Feb
BOTTROP
1631-2206
FO E J Pollard and crew
23 Feb
PFORZHEIM
1600-2355
FO B H O'Neill and crew
(attacked by night fighter and
damaged; Ju88 also claimed
as damaged)
7/8 Mar
DESSAU
1710-0200
FO D E Till and crew
FS Brookes in;
FO Shorthouse out
8/9 Mar
KASSELL
1700-0035
FO D E Till and crew
FS P F Sattler (2P)
11 Mar
ESSEN
1145-1705
FS P F Sattler
Sgt W E A Jeffrey
FO N Whiteley
Sgt P F Garner
Sgt W Walker
Sgt K C Durston

Sgt D F Wood
(FS Sattler and crew failed to
return 16 March 1945)
12 Mar
DORTMUND
1315-1830
FO D E Till and crew
15/16 Mar
MISBURG
1715-0050
FL F A Collins RCAF
Sgt K J Tamkin
FO C L Dalgetty RCAF
FO W A Smith RCAF
Sgt H C Cutler
Sgt W N Riley RCAF
Sgt W Millard RCAF
16/17 Mar
NURNBERG
1720-0155
FO D K Sullivan
Sgt G Charlton
Sgt G A Atkinson
Sgt I A Heath
Sgt D Phillips
Sgt R Erskine
Sgt G F Chatterton
18/19 Mar
HANAU
0030-0755
FO D K Sullivan and crew
21 Mar
BREMEN
0750-1205
FO D E Till and crew
22 Mar
HILDESHEIM
1120-1620
FO D E Till and crew
27 Mar
PADERBORN
1450-1944
FO B Simpson
Sgt J Ellis (FE)
Sgt S J Hurford (BA)
FO T E Meloche RCAF (N)
Sgt R Ridley (WOP)
Sgt G A Perfect (MU)
Sgt C J Rabey (RG)
31 Mar
HAMBURG
0630-1135
FO I L Scott
Sgt H W Batchelor
FO G R Cross
Sgt A F Marshall
Sgt S Hoskin
Sgt J A McDougall
Sgt C G Rayner
3 Apr
NORDHAUSEN
1321-1952
FO B Simpson and crew

10/11 Apr
PLAUEN
1815-0250
FO B Simpson and crew
14/15 Apr
CUXHAVEN
1945-0058
FO J Everitt
FS J Kitchen
FS T Royle
Sgt B Gilbert
PO A Popp
Sgt A Johnson
Sgt A Ward
18 Apr
HELIGOLAND
1006-1447
FO J C Hood
FS C F Lea
Sgt E A J Taylor
Sgt R A Lisk
Sgt R J Ashford
Sgt J W Harrow
Sgt W E Culshaw
25 Apr
BERCHTESGADEN
0522-1324
FO B Simpson and crew
29 Apr
'MANNA'/VALKENBURG A/F
1147-1432
FO B Simpson and crew
1 May
'MANNA'/ROTTERDAM
1456-1806
FO B Simpson and crew
2 May
'MANNA'/ROTTERDAM
1302-1548
FO B Simpson and crew
4 May
'MANNA'/VALKENBURG
1128-1422
FO B Simpson and crew
7 May
'MANNA'/ROTTERDAM
1217-1601
WO F J Carter
Sgt C A Charman
Sgt J R Pratt
Sgt G M Roberts
Sgt J F Rowan
Sgt G D Cook
Sgt D F Inkpen
8 May
'MANNA'/ROTTERDAM
1205-1525
FO B Simpson and crew
26 May
'EXODUS'/BRUSSELS
0829-1433
FO B Simpson and crew

134

LM550
LET'S HAVE ANOTHER

A Mk III with four Merlin 38s, LM550 was built at A. V. Roe's Woodford plant and came off the production line in early 1944 as part of order No 2010. Assigned to No 166 Squadron at Kirmington, Lincolnshire, LM550 was given the squadron codes of AS, and the individual letter 'B', thus becoming B-Beer. With this name, it is not surprising that the aircraft soon had a keg of beer painted on its nose and a beer mug being filled from its tap. Beneath this was a yellow heraldic scroll upon which was written 'Let's Have Another' in red. Below this came not rows of bomb symbols, but mugs of foaming ale – yellow ones for night sorties, white for day.

Beer's first pilot was Pilot Officer J. F. Dunlop RCAF who took the aircraft on its first seven ops, including two D-Day missions. James Dunlop recalls:

'The nose-artwork and slogan was designed by me and painted by one of the Kirmington groundcrew. The barrel was brown with black hoops, the scroll yellow, with the words in red. Yellow beer mugs were for night ops, white for daylight.

'Stan Parrish, the mid-upper, flew 21 trips with us before becoming tour-expired, having flown 122 sorties in the Middle East and with us. We were in fact in the aircraft ready for our 20th op when a van pulled up and a stranger, parachute in hand, informed me that Stan was screened. I made some joke about it being tough luck but he made a remark about betting me a pound that he'd be back on ops before we finished.

'He then went off to Gunnery School as an instructor. Then on our 30th and last trip, well wishers were there to give us a send-off and Stan came back to wish us luck. On the return journey, coffee was served and this voice said, "Here's your coffee, Skipper – you owe me a pound!" It was Stan Parrish, in best blue, no parachute or helmet. He had stowed away so he actually did 123 ops not 122 as the records show.

'We also had a photographer on a few trips, Pilot Officer "Horse" Ashley, who in the beginning was reluctant to come along. Coming back from one sortie, Ashley lamented that he could have done a good job if I had put the aircraft where he wanted it. By this time we were close to the enemy coast when Cy Straw spotted another bomber stream on their way to another target. "Why don't we take old Horse in again?" said Cy. Everyone was in favour so I did a steep turn and back we went.

'This time we were light, with no bombs, so we could really manoeuvre well, and I was able to put the aircraft right where Horse wanted, even though this was a much hotter target with the flak really flying. We were late back, of course, and reported missing, but Horse got his pictures.'

On Beer's eighth sortie, but with another skipper, it was engaged by two Ju88 night fighters but there was no damage on either side. Beer's first daylight – and the aircraft's first light-coloured mug – was not auspicious as the Lanc was hit by heavy flak over the target, but not seriously. This same sortie was the last for the mid-upper gunner, Flight Sergeant S. A. E. Parrish, who had now completed three operational tours with a total of 122 operations. Stan Parrish received a well earned DFM, for his total ops were not only a squadron record, but thought possible a Group record too.

James Dunlop continued to fly Beer and he completed his tour in the aircraft on 20 July 1944 and received the DFC. In all he had flown this Lanc on 21 trips. He retired from the Royal Canadian Air Force as a Colonel in 1974 and was later Manager of Gimli Airport.

Cy Straw, Jim Dunlop's rear-gunner, also recalls some of the raids and the artwork origins:

'The artwork, I think, came about because of the Canadians' (we had three in the crew) love of "dirty black stuff" – the name they called a pint of mild. When we were allotted B-Beer, the idea of a beer barrel with pint glasses instead of bombs came naturally. It was painted on by one of the groundcrew and one of the mugs had a swastika for a fighter kill. That happened with another crew when we were away on leave.

'On 5 June 1944 we went to Cherbourg to destroy large naval gun emplacements on the coast. We didn't know it was D-Day until returning over the Channel when we saw the huge armada of vessels launching the attack. We suffered heavy flak damage against Mimoyecques on 22 June, mostly to the tail area. The ammo ducts to the rear turret were damaged and the belts hit. Also a fragment went through the rear turret door and exited the top of the turret. I was the rear-gunner and must have been leaning forward at that instance. The first I knew about it, apart from the noise, was a knocking on the turret doors. I seem to recall that it was our bomb-aimer, Gordon Johnson. He told me the intercom was out. However, I could hear what was being said and was able to reply by using a signal light installed for just that purpose. The skipper was having trim problems and I could see damage to the tail fins.

'The official photographer flew with us to Caen on 7 July as well as before that date. I don't recall how many trips he flew with us. I remember at Caen we flew over at low level – the second time – so that he could get more pictures. I remember seeing gun emplacements, wreckage on the beaches and our soldiers waving to us.'

Pilot Officer H. S. Schwass, a New Zealander, began to fly Beer in June and when Dunlop left, became the Lanc's usual skipper. On his second mission his gunners shot down a FW190 night fighter (4/5 July). Henry Schwass went on to complete 15 sorties in Beer and received the DFC when he became tour-expired in October.

At the beginning of October 1944, No 153 Squadron was formed (or more correctly re-formed) at Kirmington, taking aircraft and crews from No 166 Squadron. LM550 (now coded P4, and its individual letter became 'C') was one of those re-assigned to the new squadron, having by that time notched up some 57 ops, and without a single recorded abort due to technical problems. That same month the new squadron moved to Scampton and into No 5 Group.

LM550 continued to keep a clean slate as far as mechanical problems were concerned, although the aircraft did become Cat AC on 14 October, returning to Avro's for a refit, but was back by the 28th. Flying Officer W. C. 'Bill' Capper RNZAF had started flying Beer in No 166 Squadron and for a while he became the aircraft's regular pilot in 153, flying 11 ops in it. Len Sparvell was the flight engineer:

'What can I tell you about an aircraft that did 118 ops? When you think that 921 aircrew lost their lives on No 166 Squadron alone in the two years 1943-45, it really shows you how lucky my crew and indeed LM550 was. Although we only did 11 trips in LM550 we completed our 30.

'However, LM550 wasn't so lucky for Len Nutt, our mid-upper, for when we returned from the Duren trip, he fell down the steps and broke his arm, so we had a different gunner for our last few ops. The nearest we came to trouble was when we came back from a fighter affiliation sortie on 22 November. One leg of the undercarriage wasn't showing down. After trying it again and diving her up and down sharply several times, we were still unsuccessful and told to take all emergency actions, which meant opening the upper escape hatches. This we did and they went sailing gracefully over the Lincolnshire countryside. The crew went to bracing positions leaving Bill Capper and myself to land at Scampton.

'Bill did a lovely landing, touching down on the good leg of the undercarriage and, hearing a click, took it the troublesome leg had come down; I am pleased to say, it had. The Engineering Officer wasn't pleased at losing the escape hatches, even suggesting we should get on our bikes and find them!'

George Luckraft was the wireless-operator, and he remembers:

'C-Charlie, LM550, was surely a charmed plane. It returned us safely from 11 raids on major German cities and in all survived 118 ops. We were twice coned by searchlights on the way to Dusseldorf and Bill had to dive the plane with a full bomb-load; what a relief when she responded on being pulled out of the dive. Especially as another Lancaster ahead of us went into a dive and continued straight down into the ground and burst into a huge orange glow.

'Of all the ops we did, including those in LM550, we never had inter-com failure caused by diaphragms of the microphones freezing-up. Thanks to a WAAF at the parachute room who told me to go to the Guard Room and buy nine "French letters" – one for each crew member's oxygen mask, one for the spare mask – and one for myself!

'She showed me how to cut out a square of rubber from each one, which was then stretched over the diaphragm and the top of the microphone screwed back on again. It then became moisture-proof and never became frozen-up.'

By April 1945 LM550 had reached 100 sorties, then came four 'Manna' and two 'Exodus' trips, to bring the aircraft's total to 107. It is reputed to have flown 118 ops but this cannot be substantiated from the Form 541s. Only 108 can be found, but as both Nos 166 and 153 Squadrons record LM550 as flying on the night of 7 October 1944, this has to be reduced to 107. Perhaps someone added the supposed 57 ops with No 166 Squadron and the equally supposed 51 ops with No 153 Squadron, and made it 118 instead of 108?

One of LM550's last regular skippers was Flying Officer H. W. Langford who took the aircraft on 15 raids (possibly including its 100th), plus one 'Exodus' trip. He finished his tour of flying in July 1945.

LM550 became Cat AC on 28 June 1945, returned to No 153 Squadron in August and then went to No 22 MU when No 153 Squadron disbanded. It was finally Struck off Charge on 15 May 1947.

1944
No 166 Squadron
21/22 May
DUISBURG
2220-0300
PO J F Dunlop RCAF
Sgt L O Stevinson (FE)
Sgt G R Johnson (BA)
Sgt R R Kerns RCAF (N)
Sgt N P Powell (WOP)
Sgt S A E Parish (MU)
Sgt C Straw (RG)
22/23 May
DORTMUND
2220-0300
PO J F Dunlop and crew
24/25 May
AACHEN
2335-0420
PO J F Dunlop and crew
27/28 May
AACHEN
2350-0435
PO J F Dunlop and crew
3/4 Jun
BOULOGNE
2327-0250
PO J F Dunlop and crew
5/6 Jun
CHERBOURG
2105-0145

PO J F Dunlop and crew
6/7 Jun
ACHERES
0002-0420
PO J F Dunlop and crew
10/11 Jun
ACHERES
2250-0410
PO J McLaren
Sgt D R Summers
FO S T Board
FO L H Ellerker
Sgt D F Paton
Sgt F J Collins
Sgt J T E Chalk RCAF
12/13 Jun
GELSENKIRCHEN
2226-0251
PO J F Dunlop and crew
14/15 Jun
LE HAVRE
2020-0005
PO J F Dunlop and crew
16/17 Jun
STERKRADE/HOLTEN
2240-0315
PO J F Dunlop and crew
PO W C Kuyser (2P)
17/18 Jun
AULNOYE
2330-0400

PO J F Dunlop and crew
(raid abandoned by Master Bomber due to cloud)
22 Jun
MIMOYECQUES/
V1 SITE
1400-1700
PO J F Dunlop and crew
(aircraft damaged by heavy flak)
24/25 Jun
FLERS/V1 SITE
0120-0450
PO J F Dunlop and crew
FS D B W Gilbert in (MU);
FS Parish out
27/28 Jun
CHATEAU BERNAPRE
0115-0530
PO J Double
Sgt J Read
Sgt P Peck
FS E G Farrow
Sgt H D Kirkham
Sgt L E O'Nions
Sgt K R Dean
29 Jun
DOMLEGER/V1 SITE
1145-1450
FO R L Graham
Sgt C Potter

Sgt R E Lakin
FO R W Bissonette RCAF
Sgt F A M Eade
Sgt J Wagstaff
Sgt W M G Falconer
30 Jun
OISEMONT/V1 SITE
0550-0935
PO H S Schwass RNZAF
Sgt P A Millett
Sgt D Angel
FO N J Grant RNZAF
Sgt D J Carter
Sgt S J Padman RNZAF
Sgt G Reynolds
2 Jul
DOMLEGER/V1 SITE
1215-1555
FO R L Graham and crew
4/5 Jul
ORLEANS
2155-0420
PO H S Schwass and crew
(FW190 night fighter
destroyed)
5/6 Jul
DIJON
2100-0525
PO J F Dunlop and crew
6 Jul
DIJON

1845-2210
PO J F Dunlop and crew
7 Jul
CAEN
1920-2315
PO J F Dunlop and crew
(slight flak damage to port fin)
12/13 Jul
REVIGNY
2115-0644
PO J F Dunlop and crew
14/15 Jul
REVIGNY
2125-0535
PO J F Dunlop and crew
(raid abandoned by Master Bomber)
18 Jul
CAEN/SANNEVILLE
0310-0745
PO J F Dunlop and crew
PO Ashley (photographer)
18/19 Jul
SCHOLVEN
2225-0310
PO J F Dunlop and crew
20 Jul
WIZERNES/V1 SITE
1905-2210
PO J F Dunlop and crew
PO Ashley (photographer)
FS S A E Parish DFM (stow-away!)
(Last trip of tour for PO Dunlop add crew)
23/24 Jul
KIEL
2240-0325
PO H S Schwass and crew
24/25 Jul
STUTTGART
2125-0605
PO H S Schwass and crew
28/29 Jul
STUTTGART
2130-0540
PO H S Schwass and crew
FO W B J Sedgwick (2P)
30 Jul
CAHAGNES
0630-1024
FO F D Elliott RCAF
Sgt J H Comley (FE)
Sgt M L OliphantRCAF
Sgt N W L Linton (N)
FO R B Melville RCAF
Sgt G M Canning
FS K G Rhodes
31 Jul
LE HAVRE
1810-2145
FO F D Elliott and crew
1 Aug
LA BELLE CROIX/V1 SITE
1845-2150
PO J G Davies
Sgt C L Caston
FO F Cameron
Sgt A Rollinson
Sgt W A Holt
Sgt L M Nutt
Sgt R Leigh
2 Aug
LE HAVRE
1650-2045
FO F D Elliott and crew

3 Aug
TROSSY ST MAXIM
1125-1615
FO F D Elliott and crew
5 Aug
PAUILLAC
1420-2255
PO T Donnelly
Sgt C F Sadler
Sgt W Higgins
Sgt L C Sims
FS A S Dickson RAAF
Sgt R Neal
Sgt C Madden
7/8 Aug
FONTENAY
2110-0115
PO W B J Sedgwick
Sgt D C Finch
FO H Rempel RCAF
Sgt G Cann
Sgt A E Irvine
Sgt C Offord
Sgt J Fletcher
10 Aug
DUGNY
0925-1400
PO H S Schwass and crew
11 Aug
DOUAI
1345-1725
PO H S Schwass and crew
12 Aug
BORDEAUX
1120-1825
FS K G Groves RAAF
Sgt E Clay
FS Selkirk RAAF
FS N D McDonnell RAAF
FS H B J Kirkby RAAF
FS W C Healy RAAF
FS W H J Foot
12/13 Aug
FALAISE
0016-0356
PO J G Davies and crew
Sgt F E Forge, FS E J Sharpe and Sgt R B Ward in;
Sgts Caston, Holt and Nutt out
15 Aug
LE CULOT
0955-1328
PO H S Schwass and crew
FS H W Cooper in;
FO Grant out
18/19 Aug
REIME
2215-0120
PO H S Schwass and crew
25/26 Aug
RUSSELSHEIM
2005-0440
PO H S Schwass and crew
26/2 Aug
KIEL
2020-0115
PO H S Schwass and crew
29/3 Aug
STETTIN
2100-0600
PO J G Davies and crew
FS P W Byers RAAF in;
FS Sharpe out
3 Sep
GILZE RIJEN A/F
1550-1850

PO H S Schwass and crew
5 Sep
LE HAVRE
1555-1930
PO H S Schwass and crew
12/13 Sep
FRANKFURT
1750-0125
PO H S Schwass and crew
16/17 Sep
STEENWIJK A/F
2145-0107
PO H S Schwass and crew
23 Sep
NEUSS
1855-2350
FO A J E Laflamme RCAF
Sgt F Etherington
FO G J Monckton
FO G McArthur
FS D H Schofield
Sgt S Pollitt
Sgt F Toogood
25 Sep
CALAIS
0720-1040
FO A J E Laflamme and crew
26 Sep
CALAIS
0955-1330
FO A J E Laflamme and crew
27 Sep
CALAIS
0905-1215
FO A J E Laflamme and crew
3 Oct
WESTKAPELLE
1250-1555
FO W C Capper RNZAF
Sgt L W Sparvell (FE)
FS G B Bale RNZAF
FS T C Morris (N)
Sgt G Luckraft(WOP)
Sgt L M Nutt (MU)
Sgt J D Ramsey (RG)
5/6 Oct
1820-0100
SAARBRUCKEN
FO W C Capper and crew

No 153 Squadron
7 Oct
EMMERICH
1140-1615
FO W C Capper and crew
11 Oct
FT FREDERIK HENDRIK
1535-1810
FO W Holmes RCAF
Sgt S Martin
FO R C Taylor
Sgt H J Burton
Sgt E S Neil
Sgt A D Kell
14 Oct
DUISBURG
0705-1135
PO J Searle
Sgt S Robinson
Sgt R P R Wavish
Sgt W Thomason
Sgt K G Hunt
Sgt W T Flavell
Sgt T Thomas
28 Oct
COLOGNE

FO L M Taylor
Sgt J F Yearsley
FS J F Howe
Sgt A Allan
Sgt F S Thornton
Sgt S R Hurst
Sgt F G Hammacott
30 Oct
COLOGNE
1745-2335
SL T W Rippingale DSO
Sgt R C Taylor
FO H L Howling
FO H B Coxon
PO C I Edwards
Sgt D Lewington
Sgt W Craig
31 Oct
COLOGNE
1800-2305
FO W C Capper and crew
PO P H Morris RAAF (2P)
2 Nov
DUSSELDORF
1610-2305
FO W C Capper and crew
4 Nov
BOCHUM
1730-2240
FO W C Capper and crew
6 Nov
GELSENKIRCHEN
1150-1615
FO W C Capper and crew
FL G E Dury (2P)
9 Nov
WANNE EICKEL
1545-2120
FO W C Capper and crew
FO E C Blackman (2P)
16 Nov
DUREN
1310-1745
FO W C Capper and crew
18 Nov
WANNE EICKEL
1545-2120
FO W C Capper and crew
21 Nov
ASCHAFFENBURG
1540-2220
FL J P Holland
Sgt C H Beauchamp
FO E G Turner
Sgt R M Nattress
Sgt G Turner
FL T Burgoyne
FO C A Groves
27 Nov
FREIBURG
1605-2315
FO H W Langford
Sgt W D Thompson
FO B F Rea-Taylor
FO D S McDonald
Sgt T W Jones
Sgt D W Hallam
Sgt K A Hawkins
29 Nov
DORTMUND
1230-1725
FO W C Capper and crew

1945
2 Jan
NURNBERG
1450-2330

FO G B Potter
Sgt G P Woolley
FO W H Thomas
FS J Boyle
FS J S Askew RAAF
Sgt D Smith
Sgt H J Hambrook
4 Jan
ROYAN
0220-0905
FO K W Winder
Sgt D George
FO M A Smith RCAF
Sgt A K Rabin
Sgt R Evans
Sgt G B Hamilton RCAF
Sgt T O'Gorman
7/8 Jan
MUNICH
1805-0225
FO G R Bishop RCAF
Sgt J S B Syme
FO Z R Cherko RCAF
FO W A Jackson RCAF
WO E W Lott RCAF
Sgt T F Bolak RCAF
FS R E Dash RCAF
22 Jan
DUISBURG
1645-2225
FL H W Langford and crew
28 Jan
STUTTGART
1925-2335
FO L A Wheeler
Sgt V P G Morandi
FO E C Durman RNZAF
FS E F Fish
WO W I A Turner RAAF
Sgt A Hodges
Sgt A Scott
3 Feb
BOTTROP
1605-2200
FL H W Langford and crew
8/9 Feb
POLITZ
1910-0330
FL W Holman RCAF
Sgt A Martin
WO V S Reynolds RCAF
FO R C Taylor RCAF
FS H J Burton
Sgt E S Neil RCAF
Sgt A D Kall RCAF
13/14 Feb
DRESDEN
2130-0740
FO K A Ayres
Sgt W C Taylor
FO R Mains
FS R J McMinn
FS D Head

Sgt R Wilson
Sgt R Cox
(Ken Ayres DFC and crew
failed to return 12 March,
Ayres being killed)
1/22 Feb
DUISBURG
1940-0125
FO J Rhodes
Sgt M F Kingdom
FO D G Webb
FO P C H Clark
FS J E Livock
Sgt T J Bicknell
FS H Cuthbertson
(Jack Rhodes and crew failed
to return 1 March, Rhodes
being killed)
23 Feb
PFORZHEIM
1545-2350
FL A F McLarty RCAF
FS D Huddlestone
FO D S Crawford
FO J M Stevenson
Sgt J Calderbank
Sgt W Brear
Sgt C Peacock
1 Mar
MANNHEIM
1145-1805
FL H W Langford and crew
2 Mar
COLOGNE
0655-1220
FL H W Langford and crew
8/9 Mar
KASSEL
1700-0035
FL H W Langford and crew
11 Mar
ESSEN
1130-1705
FL H W Langford and crew
FL L A E Trusler in;
FO Rea-Taylor out
12 Mar
DORTMUND
1255-1825
FL H W Langford and crew
FO Rea-Taylor back
13 Mar
GELSENKIRCHEN
1720-2300
FL H W Langford and crew
15/16 Mar
MISBURG
1700-0105
FO V S Martin RCAF
Sgt D N Baker
Sgt N E Fenerty RCAF
FO J Essen RCAF
Sgt H C Hauxwell

Sgt R Gray RCAF
Sgt J Western
16/17 Mar
NURNBERG
1720-0145
FL H W Langford and crew
18/19 Mar
HANNAU
0030-0815
FS D W Veale RAAF
Sgt J H Harrison
FS L J Mountcastle
Sgt N Fraen
Sgt D S Stewart
Sgt T Keegan
Sgt F H Lloyd
21 Mar
BREMEN
0736-1222
FL H W Langford and crew
22 Mar
HILDESHEIM
1114-1642
FL H W Langford and crew
3 Apr
NORDHAUSEN
1312-1944
FL J A McWilliams RCAF
FS J Howitt
Sgt H C Muddle
FO J F Arnoldi RCAF
FO E H Mulligan RCAF
FS W J Smith RCAF
FO E Ruse RCAF
FS J W Stewart RCAF
4/5 Apr
LUTZKENDORF
2103-0518
FL H W Langford and crew
PO J F Douglas (2P)
9/10 Apr
KIEL
1916-0139
FL P H S Kilner
FS L O Spinks
Sgt G H Bridger
Sgt W G Corcovan
Sgt K P Barker
Sgt R S Mepstead
Sgt W A Pinkham
10/11 Apr
PLAUEN
1829-0306
FL P H S Kilner and crew
14/15 Apr
POTSDAM
1746-0255
FL H W Langford and crew
18 Apr
HELIGOLAND
0956-1433
FO J Searle and crew
Sgt J B Mitchell in;

Sgt J Thomas out
22 Apr
BREMEN
1520-2012
FL H W Langford and crew
(raid aborted by Master
Bomber due to cloud and
smoke)
25 Apr
BERCHTESGADEN
0500-1159
WC G F Rodney DFC AFC
FS K H Dickinson
FS V J Vaughan
FS F R C Melville
Sgt D T Davies
Sgt J Brady
FL J T G Weaver
(starboard inner failed, jetti-
soned bombs over southern
Germany)
29 Apr
'MANNA'/THE HAGUE
1140-1519
WO H Eckershall
Sgt E C Hawker
FO M Downes
FO W R Proctor RCAF
Sgt R J Taylor
Sgt N A J Webb
Sgt J Duffield
2 May
'MANNA'/ROTTERDAM
1242-1546
WO H Eckershall and crew
7 May
'MANNA'/ROTTERDAM
1208-1532
FL J A McWilliams and crew
8 May
'MANNA'/THE HAGUE
1102-1436
FO J A Heaton
Sgt A Evans
Sgt N Kirkman
FS W Edmunds RAAF
Sgt A Owen
Sgt H Crossett
Sgt J Gist
11 May
'EXODUS'/BRUSSELS
FL P N Speed
PO F P Wittingstall
FO C A Meadows
FO R H Bates RNZAF
FO C D Hill
Sgt R H Fowler
Sgt J B Mitchell
26 May
'EXODUS'/BRUSSELS
FL H W Langford and crew

LM594
A ABLE

Built at Avro, Woodford, this Mk III Lancaster with Merlin 38 engines was produced in early 1944. In May it was assigned to A Flight of No 576 Squadron at Elsham Wolds, Lincolnshire, and coded UL, with the individual identification 'G2'. This was later changed to 'A2' (Able) in the summer. (The previous two Lancasters marked A2 were lost in July – LM532 on the 4th, PB253 on the 28th.)

Able had an auspicious day on which to start its operational life, Flying Officer V. Moss taking the aircraft to Vire on the night of D-Day. One of Able's early pilots was the A Flight Commander, Squadron Leader B. A. Templeman-Rooke DFC, who had been with No 100 Squadron in 1943 (he received the DSO in March 1945 and was promoted to Wing Commander).

By the summer, Able's regular skipper had become Flight Sergeant Archibald D'Largy Greig who would fly 18 trips in this Lanc, before receiving his commission and the DFC; followed by Pilot Officer Clarence Frank Phripp RCAF, who would fly 22 ops in Able before becoming tour-expired in January 1945, and also winning the DFC.

Phripp and crew had encounters with night fighters, the first being on 2 November. Nearing the target an aircraft approached the Lancaster showing a white light. As it closed with the bomber, it seemed to be emitting sparks as the range decreased. The rear-gunner ordered a corkscrew to port and both gunners opened fire. The enemy machine was lost to sight and darkness surrounded them once more.

On 11 November, another night fighter, this time a FW190, approached during the trip home, the mid-upper gunner firing at 50yds range as the pilot went into another evasive corkscrew manoeuvre. Their final – and successful – combat came on 16/17 January 1945, the last sortie of their tour. Two jet aircraft were engaged by the gunners and both were shot down, one being identified as a Me163 rocket fighter. The latter was seen to lose height before it hit the ground and exploded.

Able's last regular captain was Flying Officer R. Carter RAAF, he and his crew taking the Lanc on 15 trips. The aircraft's 100th sortie appears to have been a 'Manna' trip in May 1945 and in all one can count 104 sorties, which includes five 'Manna' and two 'Exodus' trips.

After the war Able went to No 1651 CU, then to No 16 Ferry Unit, finally being Struck off Charge on 13 February 1947. One hundred bomb symbols appeared on Able's nose with what appears to be a walking figure in a striped jumper with the letter 'A' on it, saluting and carrying a bomb.

1944
No 576 Squadron
6/7 Jun
VIRE
2220-0350
FO V Moss
Sgt R A Robertson
FO R E Power RCAF
Sgt M G Callaghan
Sgt T E Sinton
Sgt R Stanton
Sgt V L Oatley
9/10 Jun
FLERS
0010-0625
FO V Moss and crew
12/13 Jun
GELSENKIRCHEN
2230-0240
SL B A Templeman-Rooke
DFC
Sgt F N Ashton
FS S R Strand
FS W E Sollis
Sgt B G Holmes
Sgt J Boyd
Sgt F A West
14/15 Jun
LE HAVRE
2035-0055
PO W R Ireland
Sgt F Davis
Sgt J P Broadey
Sgt J Southwell
Sgt W Logan
Sgt A Hubeman
Sgt G J Brain
16/17 Jun
STERKRADE
2315-0415
PO J Archibald RNZAF
Sgt J R Cuthbert
FO P J Biollo RCAF
Sgt J T Kearneyn
Sgt L Fielding
Sgt A Milne
WO R S Pyatt RCAF
17/18 Jun
AULNOYE
2350-0420
PO A J Aldrige
Sgt A L Lewenden
Sgt A R Giles RAAF
Sgt R B Rennie
Sgt S Ormonroyd
Sgt N Costigan
Sgt M F Nelson
22 Jun
MIMOYECQUES/
V1 SITE
PO J Archibald and crew
1410-1740
23/24 Jun
SAINTES
2205-0555
PO J Archibald and crew
27 Jun
CHATEAU BENAPRE/
V1 SITE
0135-0535
PO J Archibald and crew
29 Jun
DOMLEGER/V1 SITE
1155-1500
PO W R Ireland and crew
2 Jul
DOMLEGER/V1 SITE

1209-1612
Flt Off R J Servis USAAF
Sgt A Balfour (FE)
Sgt J M Weir
Sgt R T Gordon
Sgt J Coates
Sgt E Reed
Sgt T A Clark RCAF
(Flt Off Servis failed to return
24 July; he and Sgt Balfour
were posted missing, the
others all evaded)
4/5 Jul
ORLEANS
2150-0335
SL B A Templeman-Rooke
and crew
5/6 Jul
DIJON
2055-0510
SL B A Templeman-Rooke
and crew
7 Jul
CAEN
1910-2310
SL B A Templeman-Rooke
and crew
12/13 Jul
REVIGNY
2100-0600
WC B D Sellick and SL B
A Templeman-Rooke's
crew
17 Jul
SANDERVILLE/V1 SITE
0315-0710
SL B A Templeman-Rooke
and crew
18/19 Jul
SCHOLVEN BUER
2300-0305
FL V Moss and crew
20 Jul
WIZERNES
1850-2220
SL B A Templeman-Rooke
and crew
23/24 Jul
KIEL
2235-0430
FO F H Watts RCAF
Sgt G Douglas
FO I G D Mills RCAF
FO R Hughes-Games RCAF
Sgt M R Swallow
Sgt W S Clatworthy
Sgt D H Laing
24/25 Jul
STUTTGART
2135-0545
SL B A Templeman-Rooke
and crew
25/26 Jul
BOIS DES JARDINES
0055-0420
SL B A Templeman-Rooke
and crew
28/29 Jul
STUTTGART
2130-0635
PO J B Bell
Sgt H C Gore
Sgt T E Seabrook
Sgt R Hughes
Sgt R E Badger
Sgt S G Parry
Sgt P Turton

7/8 Aug
FONTENAYE
2100-0110
FS A D'L Greig
Sgt N Mason
FO J R Jones
FS G G Henville RAAF
Sgt S A Johnson
Sgt H Walton
Sgt J Thresh
10 Aug
DUGNY
0910-1420
FS D'L Greig and crew
11 Aug
DOUAI
1400-1800
FO C D Thieme
Sgt L G Playfoot
FO H S Ravenhill
Sgt H W Vine
FS O R Davison
Sgt J L Morgan
Sgt J O O'Shea
12 Aug
BORDEAUX
1125-1815
FS A D'L Greig and crew
15 Aug
LE CULOT
0950-1320
FS A D'L Greig and crew
16/17 Aug
MINING/BALTIC
2115-0445
FS A D'L Greig and crew
18/19 Aug
RIEME
2235-0155
FS A D'L Greig and crew
25/26 Aug
RUSSELSHEIM
FO S F Durrant
2015-0520
Sgt R E Pearce
Sgt S J Brady RCAF
FO J Johnstone
FS J R McIntyre RAAF
Sgt J G Mackay RCAF
Sgt D Spowart
26/27 Aug
KIEL
2020-0155
PO K L Trent
Sgt A R Dunford
Sgt J N Wadsworth
Sgt H B Reynolds
Sgt R L Skelton RAAF
Sgt C Dalby
FP G C Riccomini
29/30 Aug
MINING/STETTIN BAY
FS A D'L Greig and crew
2145-0610
31 Aug
AGENVILLE
1305-1645
FS A D'L Greig and crew
3 Sep
EINDHOVEN
1530-1915
FS A D'L Greig and crew
5 Sep
LE HAVRE
1615-1930
FS A D'L Greig and crew

6 Sep
LE HAVRE
1700-2030
FS A D'L Greig and crew
8 Sep
LE HAVRE
0615-1000
FS A D'L Greig and crew
(raid aborted by Master
Bomber due to cloud)
10 Sep
LE HAVRE
1625-2015
FS A D'L Greig and crew
12/13 Sep
FRANKFURT
1830-0130
FS A D'L Greig and crew
15/16 Sep
MINING/PALLEUA
2145-0635
FS A D'L Greig and crew
16/17 Sep
LEEUWARDEN A/F
0050-0440
FO K F Mills
Sgt J Mills
Sgt A G Martin
FS G L McLean RAAF
FS R R Maxwell RAAF
Sgt W A Murray
Sgt D Massey
17 Sep
FLUSHING
FS A D'L Greig and crew
1620-1905
20 Sep
CALAIS
1450-1810
FS A D'L Greig and crew
3 Sep
NEUSS
1820-2325
FS A D'L Greig and crew
24 Sep
CALAIS
PO C F Phripp RCAF
1610-2005
Sgt D K Cleaver
FO L A Poxon RCAF
Sgt K M Kerns RCAF
Sgt W Wilson
Sgt R E Streatfield
Sgt R Norman
27 Sep
CALAIS
0905-1235
PO C F Phripp and crew
3 Oct
WESTKAPELLE
1305-1610
PO C F Phripp and crew
5/6 Oct
SAARBRUCKEN
1830-0125
PO C F Phripp and crew
7 Oct
EMMERICH
1205-1615
PO C F Phripp and crew
14 Oct
DUISBURG
0645-1055
PO C F Phripp and crew
14/15 Oct
DUISBURG
2155-0355

142

PO C F Phripp and crew
15 Oct
WILHELMSHAVEN
1730-2205
PO T C Dawson
Sgt W J Stockwell
Sgt E Gedling
Sgt J S M Shearing
Sgt A Stewart
Sgt G A Fergus
Sgt A G Murray
19/20 Oct
STUTTGART
2115-0415
PO C F Phripp and crew
26 Oct
MINING/FRISIANS
1720-2110
FO J J Mulrooney
Sgt B Beale
FO C W P Mortel RAAF
FS F J L Paton RAAF
WO J N Casey RAAF
FS F B McGrath RAAF
FS R S M Coventry RAAF
27 Oct
COLOGNE
1310-1820
FO C A Rhude
Sgt J C Duncan
FO D C Goughnour
Sg G F Fitzgerald
Sgt Blythe
Sgt G A Falleur
Sg A G Murray
29 Oct
DOMBERG
1130-1420
FO W A Stewart
Sgt W Boyle
Sgt A N Schuett RCAF
FO H R Stevenson
Sgt E A Dermont RCAF
Sgt L Thomas
Sgt J T Challenger
30 Oct
COLOGNE
1720-2320
FO W A Stewart and crew
2 Nov
DUSSELDORF
1630-2115
FO C F Phripp and crew
4 Nov
BOCHUM
1720-2150
FO C F Phripp and crew
6 Nov
GELSENKIRCHEN
1155-1625
FO C F Phripp and crew
9 Nov
WANNE EICKEL
0835-1245
FO C F Phripp and crew
11 Nov
DORTMUND
1550-2100
FO C F Phripp and crew
16 Nov
DUREN
1250-1810
FO H Benson
Sgt H Woolstenhulme
FO H G Mather
FO P C Milner
PO F Wilson

Sgt R C Griffiths
Sgt R Goldsbury
18 Nov
WANNE EICKEL
1540-2045
FO C F Phripp and crew
27 Nov
FREIBURG
1610-2230
FO C F Phripp and crew
29 Nov
DORTMUND
1210-1705
FO C F Phripp and crew
4 Dec
KARLSRUHE
1635-2230
FO C F Phripp and crew
6/7 Dec
MERSEBURG
1635-0040
FO C F Phripp and crew
12 Dec
ESSEN
1605-2205
FO O R Halnan
Sgt W G Young
FO G T Shepherd RCAF
FO E N Weldon RCAF
Sgt F R Lait
Sgt H G C Farey RCAF
Sgt R H Gray
28 Dec
BONN
1530-2105
FO F A Collins RCAF
Sgt K J Tamkin
FO G L Dalgetty
PO W A Smith
Sgt H C Cutler
Sgt W N Riley
Sgt W Millard
29 Dec
SCHOLVEN BUER
1515-2045
FO C F Phripp and crew
31 Dec
OSTERFELD
FO C T Dalziell
1455-2100
Sgt P R Montgomery
Sgt W E Bradbury
Sgt W R May
Sgt A Burns
Sgt C T Thorley
Sgt D O'Sullivan

1945
2 Jan
NURNBERG
1450-2300
FO C F Phripp and crew
4/5 Jan
ROYAN
0140-0845
FO C F Phripp and crew
FS P E T Brookes in, FO
Poxon out
7/8 Jan
MUNICH
1855-0245
FO C F Phripp and crew
14/15 Jan
MERSEBURG
1915-0315
FO C F Phripp and crew
FO Poxon back

16/17 Jan
ZEITZ-TROGLITZ
1730-0105
FO C F Phripp and crew
(Last trip of tour; gunners
shot down two jet fighters
over target)
2/3 Feb
WIESBADEN
2130-0310
FO R Carter RAAF
Sgt D Mayer (FE)
FS C Townsend (BA)
Sgt D R Ovenden (N)
FS H R Wilson RAAF
Sgt F J Dray (MU)
Sgt J E Davies (RG)
3 Feb
BOTTROP
1618-2204
FO C T Dalzeill and crew
8/9 Feb
POLITZ
1910-0340
FO R Carter and crew
13/14 Feb
DRESDEN
1910-0340
FO R Carter and crew
11 Mar
ESSEN
1140-1655
FL R M Crowther RCAF
FS J F Ryan RAAF(2P)
Sgt R F Wilshire
FS W Thorpe
FS H L Basson
FS B T Wilmot
FS B Rink RCAF
FS J R Scarfe RCAF
(FS Ryan and 'his' crew failed
to return 16 March)
12 Mar
DORTMUND
1305-1830
FO R Carter and crew
15/16 Mar
MISBURG
1715-0050
FO R Carter and crew
16/17 Mar
NURNBERG
1715-0140
FO R Carter and crew
18/19 Mar
HANAU
0110-0820
FS G A Abercrombie
Sgt R R Ewart
Sgt W Milburn
Sgt R J Gratton
Sgt V Serge
Sgt Weston
PO S H Braithwaite
22 Mar
HILDESHEIM
1120-1630
FO R Carter and crew
24 Mar
DORTMUND
1300-1810
FO R Carter and crew
27 Mar
PADERBORN
1442-1933
FO R Carter and crew
Sgt D U Hogg (2P)

31 Mar
HAMBURG
0614-1121
FO R Carter and crew
3 Apr
1323-2006
NORDHAUSEN
Sgt D U Hogg
Sgt J R Reed (FE)
Sgt J W Forster (BA)
Sgt S Ganige (N)
Sgt J Tyrer (WOP)
Sgt W J Monksfield
Sgt H J Allen (RG)
(Sgt Hogg and crew failed to
return 4 April)
4/5 Apr
LUTZKENDORF
2113-0511
FO R Carter and crew
9/10 Apr
KIEL
1910-0131
FO R Carter and crew
10/11 Apr
PLAUEN
1855-0225
FO R Carter and crew
18 Apr
HELIGOLAND
0948-1421
FO R Carter and crew
22 Apr
BREMEN
1510-2004
FO R Carter and crew
25 Apr
BERCHTESGADEN
0526-1338
FO B Hinderks RCAF
Sgt J M Hallsmith
FO H McCarthy RCAF
FO C Proctor RCAF
Sgt F White
Sgt J Hudson
Sgt D Currie RCAF
29 Apr
'MANNA'/VALKENBURG A/F
FO C A Titchner
Sgt W J Faulkner
Sgt A W Watson
1149-1445
Sgt F A Kesterton
Sgt P R Elkington
Sgt M V Jones
Sgt G W Somerville
1 May
'MANNA'/ROTTERDAM
1451-1742
FO W N Holmes RCAF
Sgt D J Nimmo
FO F W Christisen
FO M K Scruton RCAF
Sgt W Brown
FS E C Grife RCAF
FS W W A Parsons RCAF
3 May
'MANNA'/VALKENBURG
1020-1317
FL R W McCurdy RCAF
Sgt J Chisholm
FO W J Driscoll RCAF
FO G W Stevens RCAF
WO A G Little
Sgt J Nazarko RCAF
FS F A Parker RCAF

4 May
'MANNA'/VALKENBURG
1134-1418
FL B H O'Neill RCAF
PO R J Durren
FO S W Haakstead RCAF
FS J G Ogilvie

FO F C Smith
PO A J W George RCAF
FS D Renchuk RCAF
7 May
'MANNA'/ROTTERDAM
1231-1546
FO C A Titchner and crew

11 May
'EXODUS'/BRUSSELS
1129-1839
FO C A Titchner and crew
Sgt Jones out
16 May
'EXODUS'/BRUSSELS

0742-1320
FO C A Titchner and crew
Sgts Jones back, Sgt
Somerville out

ME746
ROGER SQUARED

Produced at Metro Vickers, Manchester, as part of aircraft order No 2221, this Mk I Lancaster fitted with four Merlin 24 engines rolled out of A. V. Roe's works on 2 April 1944. The aircraft was assigned to No 166 Squadron at Kirmington, Lincolnshire, on 14 April – just two weeks prior to the arrival of LM550, another 100-op veteran – and was given the same code (AS) and individual letter 'C', then 'R' (later 'R2'). Gordon Rodwell DFM, who was Squadron Leader A. S. Caunt's bomb-aimer, recalls:

'When ME746 arrived at Kirmington she was allocated to the Flight Commander, Stan Caunt. We air-tested it and then due to the un-serviceability of another aircraft it was "borrowed" the next night. It had the letter AS-C – Stan's initials, and was known simply as "Charlie".

'Stan finished his tour on 5 June and the headless crew was taken on by Wing Commander Garner and also lent to Wing Commander Reddick who worked at Group.'

Unlike LM550 with its beer barrel and beer-mug symbols, ME746 had nothing other than a steadily growing number of bombs on its nose, in neat rows of 10. The man who painted on all the bombs was Corporal Dennis Terry, the aircraft's fitter-airframes. He remembers painting on the DFC ribbon after H. J. Musselman was awarded this 'gong', and when the aircraft itself was 'awarded' a DSO after the 100th op this was painted on too, by which time the aircraft letter had changed to R2.

ME746's first operation was to Cologne on the night of 20/21 April, with Flight Sergeant F. A. Mander and crew. It was their only trip in this aircraft and they failed to return on 27 May, Fred Mander being killed. ME746 became a regular aircraft to two senior pilots on the squadron, firstly the Flight Commander, Squadron Leader Arthur S. Caunt, who was to win the DFC, and the Squadron CO, Wing Commander Donald A. Garner, who was to win the DSO. Another pilot who flew this aircraft early in his tour was Pilot Officer J. F. Dunlop who was later to fly LM550 B-Beer for most of the rest of his ops.

ME746 operated on D-Day, against V1 sites and railway targets, the Caen attack prior to the Allied break-out from the Normandy beach-head, and all the other summer raids that occupied Bomber Command. Flying Officer W. C.

Hutchinson became the regular pilot during the summer, flying 18 trips in the aircraft and winning the DFC when his tour ended in August. Later ME746 flew on eight ops with Flight Lieutenant A. P. Gainsford DFC, a New Zealander.

ME746 collected the odd bit of flak damage and was even hit by fire from another Lancaster during a raid on Revigny on 12/13 July, but the Lanc was almost totally free of any mechanical problems and soon its total of ops rose towards the 100 mark. The aircraft's regular skipper by then had become Flying Officer H. J. Musselman RCAF, who received an immediate DFC in January 1945. Flying in ME746 on 29 December, they lost the starboard inner engine and suffered an oil leak in the port outer but carried on, only to be hit by flak which put holes through the starboard wing-flap and mainplane. As they could no longer reach the target – the Scholven/Buer oil refinery – and were losing height, Harold Musselman bombed a last-resort target, Duisburg, from low level before heading home.

On a trip to Kassel on 8 March 1945 ME746 had an inconclusive scrap with a night fighter, and the next sortie was the Lanc's 100th op. This was flown on 11 March, a daylight to Essen with Harold Musselman in command. Again Dennis Terry remembers this event, as Musselman proceeded to beat-up the control tower at Kirmington upon his return. Terry and Corporal Sid Woodcock (ME746's fitter-engines) cycled like mad to the airfield to see it, getting there just in time to see the final low-level fly-by.

Pilot Officer S. Todd was ME746's final more regular skipper, taking the aircraft on 11 raids and five 'Manna' sorties. In all, ME746 is supposed to have flown 116 operational sorties, but it also flew six 'Manna' and three 'Exodus' trips, so a possible 124 in all.

The aircraft continued to serve with the squadron until 3 September 1945 at which time it became Cat AC and went back to Avro's for repair and overhaul. ME746 returned to 166 on 22 September, but when the squadron disbanded the Lanc moved to No 103 Squadron on 12 November, then to No 57 Squadron two weeks later. On 29 December it was on the strength of Lindholme and on 21 February 1946 the aircraft was sold to Hestons Ltd, struck off RAF charge and reduced to scrap.

1944
No 166 Squadron
20/21 Apr
COLOGNE
2340-0415
FS F A Mander
Sgt H J Evans (FE)
FS C J Overand (BA)
Sgt V L J Newall (N)
FS O P McFadden (WOP)
FO B White (MU)
WO H Shaw (RG)
22/23 Apr
DUSSELDORF
2230-0350
SL A S Caunt
Sgt D J Baveystock (FE)

FS G G H Rodwell(BA)
FO B L Thomson RAAF (N)
FS E Ridout (WOP)
Sgt A R Williamson (MU)
FS L Wayte (RG)
24/25 Apr
KARLSRUHE
2205-0510
SL A S Caunt and crew
FO C Squire in, Sgt Williamson out
27/28 Apr
FRIEDRICHSHAVEN
2125-0555
PO J F Dunlop RCAF
Sgt L D Stevinson
Sgt G R Johnson RCAF

Sgt R R Kerns RCAF
Sgt N P Powell
FS S A E Parish
Sgt C Straw
30/1 May
MAINTENON
2130-0210
PO J F Dunlop and crew
3/4 May
MAILLY-LE-CAMP
2155-0325
WC D A Garner with SL Caunt's crew
Sgt L V Robbins in, FO Squire out
9/10 May
MARDYCK

2240-0130
PO J F Dunlop and crew
11/12 May
HASSELT
2200-0120
PO J F Dunlop and crew
19/20 May
2210-0305
ORLEANS
SL A S Caunt and crew
PO H A Brown RAAF in, FS Ridout out
21/22 May
DUISBURG
2235-0255
WC D A Garner and SL Caunt's crew

FL T R Danby DFC in, Sgt Robins out
22/23 May
DORTMUND
2220-0255
SL A S Caunt and crew
Sgt A G Manuel DFM in, FL Danby out
24/25 May
AACHEN
2345-0435
FO W I Warmington
Sgt E Askey
FS G S Gissing RAAF
FO J F Clark
Sgt D Oakley
Sgt O W Sansby
Sgt H Peterson RCAF
3/4 Jun
BOULOGNE
2332-0245
WC D A Garner and crew
FS P C Robins in, Sgt Manuel out
5/6 Jun
CHERBOURG
2110-0150
SL A S Caunt and crew
Sgt G L Nordbye RCAF in, FS Robins out
6/7 Jun
ACHERES
0005-0500
FO R L Graham
Sgt G Potter
Sgt R E Lakin
FO R W Bissonette RCAF
Sgt F A M Eade
Sgt J Wagstaff
Sgt W M G Falconer
10/11 Jun
ACHERES
2255-0400
FO W I Warmington and crew
12/13 Jun
GELSENKIRCHEN
2231-0316
FO W C Hutchinson
Sgt N Harris (FE)
FS H R Easton RCAF
WO A T Mathews RCAF(N)
Sgt E A Lummis RCAF (WOP)
Sgt R B Parry (MU)
Sgt D G Prescott(RG)
14/15 Jun
LE HAVRE
2015-2355
WC A D Garner and crew
Sgt F Watson in, Sgt Nordbye out
17/18 Jun
AULNOYE
2340-0440
FO R L Graham and crew
(raid abandoned by Master Bomber; unable to identify target due to cloud)
22 Jun
MIMOYECQUES/V1 SITE
1410-1725
FO W C Hutchinson and crew
23/24 Jun
SAINTES
2210-0520
FO W C Hutchinson and crew
24 Jun
FLERS/V1 SITE

FO W C Hutchinson and crew
0130-0500
28 Jun
CHATEAU BERNAPRE/V1 SITE
0130-0500
FO W C Hutchinson and crew
29 Jun
DOMLEGER/V1 SITE
1140-1505
FO W C Hutchinson and crew
30 Jun
OISEMONT/V1 SITE
FO W C Hutchinson and crew
0600-0945
2 Jul
DOMLEGER/V1 SITE
1210-1540
WC D A Garner and crew
4/5 Jul
2150-0410
ORLEANS
FO W C Hutchinson and crew
5/6 Jul
DIJON
2105-0550
FO W C Hutchinson and crew
6 Jul
FORET DU CROC/ V1 SITE
1855-2230
PO D J Dickie RNZAF
Sgt W L McDonald
FO E R Pickering
FO J Mason
Sgt K E Corbett
Sgt R Turner
Sgt S L Giles
7 Jul
CAEN
1925-2345
FO W C Hutchinson and crew
12/13 Jul
REVIGNY
2125-0625
FO W C Hutchinson and crew
(raid abandoned over target by Master Bomber)
14/15 Jul
REVIGNY
2115-0535
FO W C Hutchinson and crew
(raid abandoned over target by Master Bomber)
18 Jul
CAEN/SANNERVILLE
0315-0715
FO W C Hutchinson and crew
Sgt D G Hutchinson in
18/19 Jul
SCHOLVEN
2235-0320
PO D J Dickie and crew
20 Jul
WIZERNES/V1 SITE
1910-2220
WC D A Garner and crew
FL J Barritt in, Sgt Watson out
23/24 Jul
KIEL
2240-0330
FS R W Miller RCAF
Sgt W A Adams
FO D W Harding
FO E A F Hall
Sgt A H MacDonald
Sgt J R Scott RCAF
Sgt C A Pike RCAF

24/25 Jul
STUTTGART
2120-0555
FS R W Miller and crew
Sgt Schafer in, Sgt Scott out
25/26 Jul
STUTTGART
2115-0540
FO W C Hutchinson and crew
26/27 Jul
MINING/HELIGOLAND
2310-0330
FO W C Hutchinson and crew
28/29 Jul
STUTTGART
2120-0545
FO W C Hutchinson and crew
30 Jul
CAHAGNES
0635-1010
FO W C Hutchinson and crew
31 Jul
LE HAVRE
1825-2120
FO W C Hutchinson and crew
1 Aug
LA BELLE CROIX/V1 SITE
1850-2145
FO F D Elliott
Sgt J H Comley
Sgt M L Oliphant
Sgt N W L Linton
FO R B Melville
Sgt G M Canning
Sgt K G Rhodes
(raid abandoned over target by Master Bomber due to fog and haze)
2 Aug
LE HAVRE
1650-2020
WC D A Garner and crew
Sgt A R Williamson in, FL Barritt out
(aircraft damaged by flak)
4 Aug
PAUILLAC
1245-2140
FL A P Gainsford DFC RNZAF
FO F G Kitson
FS J E F Rowe
FS W R Williams RAAF
FS F E Bennett
FO H G Cook RCAF
Sgt G H Davidson RCAF
5 Aug
PAUILLAC
1425-2210
PO A G S Watkins
Sgt I M Morganstein (FE)
Sgt D H Barr (BA)
Sgt H G Carson RCAF (N)
Sgt A P Maitland
Sgt F J McCahon RCAF(MU)
Sgt F W Walker (RG)
7/8 Aug
FONTENAY
2105-0125
FL A P Gainsford and crew
10 Aug
DUGNY
0930-1430
FO W L Shirley RNZAF
Sgt O O Evans
Sgt G B McFee RCAF
FO F D Hill
FS C J Mitchell RAAF

Sgt N W Sykes
PO C J Burch
11 Aug
DOUAI
1345-1719
WC Reddick and Garner's crew
(sustained flak damage to starboard outer and port tyre)
14 Aug
FONTAINE-LE-PIN
1315-1719
FS K G Graves RAAF
Sgt E Clay RAAF
FS Selkirk RAAF
FS N D McDonnell RAAF
FS H B J Kirkby RAAF
FS W C Healy RAAF
FS W H J Foot
15 Aug
LE CULOT
0950-1350
FL A P Gainsford and crew
16/17 Aug
STETTIN
2055-0455
FL A P Gainsford and crew
25/26 Aug
RUSSELSHEIM
2005-0500
FL A P Gainsford and crew
26/27 Aug
KIEL
2010-0140
FL A P Gainsford and crew
29/30 Aug
STETTIN
2100-0605
FL A P Gainsford and crew
31 Aug
AGENVILLE
1315-1645
FL A P Gainsford and crew
3 Sep
GILZE RIJEN A/F
1550-1950
FO J G Davies
Sgt C L Caston
FO F Cameron
Sgt A Rollinson
Sgt G B Bawley
Sgt R B Ward
Sgt R Leigh
20 Sep
SANGATTE
WC D A Garner and crew
1520-1850
25 Sep
CALAIS
0710-1155
FL A E Jones
Sgt J L James
Sgt C L Cullen
FS J J L McDonell RAAF
Sgt D L Williams
Sgt R V Trafford
Sgt A Simpson
(raid abandoned over target by Master Bomber due to cloud)
27 Sep
CALAIS
0905-1230
FL A E Jones and crew
5 Oct
SAARBRUCKEN

1310-1445
FL A E Jones and crew
11 Oct
FT FREDERIK HENDRIK
1445-1805
FO J A Sherry RCAF
Sgt A Martin
FO D M Bennett RCAF
FO M Bernyk RCAF
Sgt K Surman
Sgt C Young RCAF
Sgt J C Daze RCAF
(raid abandoned over target
due to smoke; this crew
failed to return 31 December
1944, James Sherry being
killed)
14 Oct
DUISBURG
0625-1110
PO N Appleton
Sgt J W Lockie
FS R J Taylor
FO D N Bain
FS W L Ball
FS L J Denney
FS B H Cook
19/20 Oct
STUTTGART
1700-0035
FO W B J Sedgwick
FO W H J Underwood (2P)
Sgt D C Finch
FO H Rempel RCAF
Sgt G Cann
Sgt A W Irving
FS J Rees
FS H I Scott RCAF
4 Nov
BOCHUM
1730-2320
FO D G Stuart
Sgt L M Christie
Sgt K M Wilding
FO W H Ford RCAF
Sgt B Robinson
FO M D Woods RCAF
Sgt J E Horner
6 Nov
GELSENKIRCHEN
1150-1625
FO E P Burke RAAF
Sgt T E Carr
FS S M Meggitt
Sgt G J E John
FS G W Kirk
Sgt G Anson
Sgt T R Wood
9 Nov
WANNE EICKEL
0750-1310
FO J C Parry RCAF
Sgt E W Ball
Sgt L O Kirton RCAF
Sgt C G J Mason
Sgt F R Cole RCAF
Sgt D G Pegge RCAF
Sgt D W Morrow
11 Nov
DORTMUND
1600-2140
FO H J Musselman RCAF
Sgt J R Cogbill (FE)
Sgt F Reid (BA)
FS H H Park RCAF (N)
Sgt R Williamson(WOP)
FS J M Donnelly (MU)

FS K Forrest (RG)
16 Nov
DUREN
1230-1755
FO H J Musselman and crew
18 Nov
WANNE EICKEL
FO H J Musselman and crew
1550-2230
21 Nov
ASCHAFFENBURG
1545-2200
FO H J Musselman and crew
27 Nov
FREIBURG
1600-2310
FO H J Musselman and crew
29 Nov
DORTMUND
1155-1735
FS R N Dickson RCAF
Sgt D Bell
FS R C Caswell RCAF
FO D S Derbecker RCAF
Sgt R Gregg RCAF
Sgt H E Barker RCAF
Sgt D M Williams RCAF
4 Dec
KARLSRUHE
1625-2305
FO J C Parry and crew
6/7 Dec
MERSEBURG/LEUNA
1635-0105
FO H J Musselman and crew
12 Dec
ESSEN
1550-2200
FO H J Musselman and crew
Sgt M D Mulcaster in, Sgt
Cogbill out
17 Dec
ULM
1515-2305
PO R D Ritchie
Sgt F O Mulcaster
FO J E Walker RCAF
FO M D Greene RCAF
Sgt S O Badcock RCAF
Sgt J E Anderson RCAF
Sgt L E Dunlop RCAF
24 Dec
COLOGNE
1450-2045
FO H J Musselman and crew
Sgt J Parry in, Sgt Mulcaster
out
27 Dec
RHEYDT
1205-1715
FO E P Burke RAAF
Sgt T E Carr (FE)
PO J S Meggitt RAAF
FS G K H John (N)
FS G W Kirk (WOP)
FS G Anson (MU)
FS T R Wood (RG)
(crew failed to return 16
January 1945)
28 Dec
MUNCHEN GLADBACH
1545-2110
FO H J Musselman and crew
Sgt Cogbill back
29 Dec
SCHOLVEN/BUER
1540-2055

FO H J Musselman and crew
(lost an engine and flak damage; bombed Duisburg)

1945
7/8 Jan
MUNICH
1830-0305
FO H J Musselman and crew
14/15 Jan
MERSEBURG/LEUNA
1850-0340
FL J Glenesk RCAF
Sgt G G Anderton
FO W H Couch RCAF
FO D E Junker RCAF
Sgt D H Fenwick
FO J N West RCAF
FO C Mizzen RCAF
16/17 Jan
ZEITZ
1730-0150
FL J Glenesk and crew
22 Jan
DUISBURG
1650-2010
FO H J Musselman and crew
(abandoned raid as aircraft
filled with smoke although
no fire was located)
1 Feb
LUDWIGSHAVEN
1540-2220
FO H J Musselman and crew
2/3 Feb
WIESBADEN
2015-0250
FO H J Musselman and crew
3 Feb
BOTTROP
1630-2135
FO H J Musselman and crew
7/8 Feb
KLEVE
1900-2400
FO H J Musselman and crew
8/9 Feb
POLITZ
1940-0325
FO H J Musselman and crew
13/14 Feb
DRESDEN
2125-2400
FL H J Musselman and crew
(sortie abandoned when all
instruments went u/s)
14/15 Feb
CHEMNITZ
1955-0445
FL H J Musselman and crew
PO S Todd RCAF (2P)
20/21 Feb
DORTMUND
2135-0345
FL H J Musselman and crew
21/22 Feb
DUISBURG
1940-0215
FO E J Jenkins
Sgt S Appleby
Sgt D S Struthers
Sgt R J Khoo
Sgt R Bradley
Sgt J W Inman RCAF
Sgt E J Hammond
23 Feb
PFORZHEIM

1540-2350
FO E J Jenkins and crew
1 Mar
MANNHEIM
1145-1805
FL H J Musselman and crew
2 Mar
COLOGNE
0700-1215
FL H J Musselman and crew
5/6 Mar
CHEMNITZ
1610-0215
FL H J Musselman and crew
7/8 Mar
DESSAU
1645-0215
FL H J Musselman and crew
8/9 Mar
KASSEL
1700-0035
FL H J Musselman and crew
11 Mar
ESSEN
1125-1655
FL H J Musselman and crew
12 Mar
DORTMUND
1250-1825
FL H J Musselman and crew
13 Mar
ERIN
1720-2255
FL H J Musselman and crew
16/17 Mar
NURNBERG
1715-0210
PO S Todd RCAF
Sgt W A Hall
Sgt J Connell
FO C N Dumbleton
Sgt J G Brame
Sgt J M Torris
Sgt R L Edwards
18/19 Mar
HANAU
0040-0805
PO S Todd and crew
21 Mar
BREMEN
0740-1230
PO S Todd and crew
Sgt S Pryce in, Sgt Torris out
22 Mar
HILDENHEIM
1120-1640
PO S Todd and crew
24 Mar
HARPENERWEG
1255-1830
FS J L Briggs
Sgt J Christie
Sgt R A Fraser
Sgt K W Fortune
Sgt G A Cearns
Sgt F Tucker
Sgt S Pryce
25 Mar
HANNOVER
0630-1225
PO S Todd and crew
27 Mar
PADERBORN
1445-1945
FS E Farrington
Sgt J Nimmo
Sgt J C Roberts

148

Sgt S F Allen
Sgt L P Carter
Sgt G W W Goodger
Sgt A W Peacock
31 Mar
HAMBURG
0625-1145
FS E Farrington and crew
3 Apr
NORDHAUSEN
1312-2009
FS T C Anderson RCAF
FS G R Cockerill
FO A T McCulloch RCAF
Sgt R A Ruttan RCAF
Sgt D C Smith
Sgt H Pearce RCAF
Sgt T W Stevenson RCAF
4/5 Apr
LUTZKENDORF
2108-0526
FS N J Fletcher RAAF
Sgt W A Hall
Sgt L W J Parsons
FS W M Wilson
Sgt R B Berry
Sgt R N Thompson

Sgt A D Kerr
9/10 Apr
KIEL
1933-0118
PO S Todd and crew
10/11 Apr
PLAUEN
1812-0253
PO S Todd and crew
4/15 Apr
POTSDAM
1750-0312
PO S Todd and crew
18 Apr
HELIGOLAND
1022-1446
PO S Todd and crew
22 Apr
BREMEN
1507-2019
PO S Todd and crew
(raid abandoned over target
by Master Bomber due to
smoke and cloud)
25 Apr
BERCHTESGADEN
0522-1328

PO S Todd and crew
29 Apr
'MANNA'
1210-1458
PO S Todd and crew
30 Apr
'MANNA'
1600-1900
FS T C Anderson and crew
1 May
'MANNA'
1345-1703
PO S Todd and crew
2 May
'MANNA'
1210-1501
PO S Todd and crew
3 May
'MANNA'
1122-1420
PO S Todd and crew
7 May
'MANNA'
1231-1606
PO S Todd and crew
10 May
'EXODUS'/BRUSSELS

0740-1245
FL R Lawson
FS H Allen
FO Brown
FO M F Vickerman
FO Small
Sgt J Campbell
Sgt W B Mandeville
11 May
'EXODUS'
FS G Barlow
1605-2320
FS A R Kischner
FO J Doyle
Sgt C Butler
FS P Sullivan
Sgt W Edge
26 May
'EXODUS'
1425-2100
FO K D Foxall
Sgt B A Skelton
WO R A C Lundie
FO B J Brook
FS D C Smith
Sgt S L Horrobin

149

ME758
NAN

Just 12 aircraft on from ME746 and another 100-plus veteran, this Lancaster came off the production line at Metro Vickers with four Merlin 28 engines, also as a Mk I in early 1944. Assigned to No 12 Squadron at Wickenby, Lincolnshire, it was given the squadron code of PH and the individual letter N-Nan.

Nan began ops with Pilot Officer N. Rollin and his crew on a raid to Lyons on 1/2 May, who then flew five missions in this aircraft before they went off to join No 156 Squadron PFF, where they were to fly some ops in another centenarian featured in this book, ND875, on a few occasions.

Like most aircraft, ME758 had the usual variety of crews in the first weeks on the squadron. With one crew skippered by Pilot Officer L. Pappas RCAF, it had a fight with a Me109 night fighter which it shot down during a raid on Tergnier on 31 May/1 June. As it happened, Len Pappas and crew had another fight exactly one month later and shot down a second night fighter, for which he received the DFC and his rear-gunner, Sergeant Berthan Swanson, the DFM – but this was in another Lancaster.

After the Tergnier trip, Sergeant T. F. Holbrook (later Flying Officer) and crew took over Nan and flew it on 25 ops. Holbrook became tour-expired in September 1944 and received the DFC.

Then there was another period during which time Nan had no regular crew, although Flying Officer E. King, Flying Officer K. T. Wallace RCAF and Flying Officer Colin H. Henry DFC, RNZAF, flew the Lanc more often than not. Wallace also flew later with No 156 Squadron PFF where he often flew ND875, too.

Nan was almost totally free of mechanical problems, although towards the end of its career the aircraft's starboard engine caught fire on 2 February 1944, causing the crew to abort a trip to Wiesbaden. This particular Lanc was also lucky to escape serious damage from the German defences; only once was damage recorded when its hydraulics were hit by light flak at 06.12 hours during the attack to support Operation 'Goodwood' – the Allied breakout from Caen on 17/18 July 1944.

Otherwise Nan did all the usual trips Lancasters flew during the last year of the war, including both D-Day ops, Caen of course, V1 sites, and to Dresden on 13/14 February 1945. Flying Officer Arthur J. D. Leach became a regular captain with Nan in 1945, flying 16 ops in the aircraft and winning the DFC. However, it was Flying Officer F. M. Baird who took Nan on its 100th raid, a mining sortie off Oslo on 22/23 March 1945, then flew the Lanc on two more raids before 'Manna' trips began at the end of that month. Nan flew six 'Manna' sorties and two 'Exodus' trips, so in all 108 bombing raids were flown plus these later eight sorties for a possible 116 total.

When Nan had reached 106 operations on 18 April someone must have realised what this Lanc and its crews had achieved, for there was a ceremony to present Nan with both the DSO and DFC, mock medals and ribbons being made and hung from the cockpit ledge. Beneath them were the 106 bomb symbols, 10 rows of 10, with another 'block' being started with the other six. Above them was the swastika, representing the Me109 destroyed before the Invasion, and two searchlight beams which represented the two occasions on which Nan had been coned over Germany but had managed to evade them.

Like some of the other 100 ops-plus Lancs there could have been a slight mix-up in this particular aircraft's bombing log. It had red bombs for night and yellow for daylight raids, and clearly the first yellow bomb is shown as the aircraft's 19th. However, the Form 541 shows Nan flew its first daylight trip on 22 June 1944, which is thus recorded as its 22nd trip. Perhaps whoever added up Nan's 'score' failed to notice that its first trip – while noted as ME758 with the letter 'M' next to it, and for two ops in early May the letter 'N' was shown but with the serial numbers of ME759 and MG758 – ME759 was with No 9 Squadron and there is no such Lancaster serial as MG758, so clearly both are wrong.

To add to the confusion, Nan's 100th bomb is also yellow, but the mining sortie to Oslo was a night op. A straightforward count might indicate its 100th was two raids later on 27 March, a daylight to Paderborn. This would at least make the last bombing sortie on 25 April Nan's 106th – the number this aircraft is usually credited with. However, these problems occur with almost all the 100 ops-plus Lancs as already mentioned.

Nan lived to see out the war but did not survive long enough to see out the first six months of peace, being Struck off Charge on 19 October 1945 and 'reduced to produce'.

1944	2143-0249	**21/22 May**	FO A R Witty
No 12 Squadron	PO E K Farfan	DUISBURG	Sgt J P S Payner
1/2 May	FS C J Gordon	2248-0353	Sgt T P Crook
LYONS	Sgt V Poole	PO J Downing and crew	Sgt I E Squires
2145-0441	FO T H Vansickle	**22/23 May**	Sgt S Swaine
PO N Rollin	Sgt G Neal	DORTMUND	Sgt T S Gibb
FO W A Chmilar (N)	Sgt J D Rollason	2213-0253	**31/1 Jun**
Sgt S Hudson (WOP)	Sgt J Bell	PO N Rollin and crew	TERGNIER
Sgt J W Pollack (BA)	**9/10 May**	**24/25 May**	2348-0443
Sgt H G Phillips(FE)	MERVILLE	AACHEN	PO L Pappas RCAF
Sgt E Murray (MU)	2157-0118	2330-0445	FO L M McLean
Sgt D W Hornesby(RG)	PO N Rollin and crew	PO J W Landon	Sgt K B Phipps
3/4 May	**19/20 May**	FS M C Griffiths	FO A L Garrick
MAILLY-LE-CAMP	ORLEANS	Sgt J Goodlett	Sgt C Anderson
2138-0333	2211-0316	Sgt J O Davis	Sgt R K Redmond
PO N Rollin and crew	PO J Downing	Sgt J D Anderson	Sgt B C Swanson RCAF
6/7 May	FO E D Figg	PO J Longbottom	**2/3 Jun**
AUBIGNE-RACEN	Sgt J Alten	Sgt J M Nicol	BERNAVAL
0034-0504	Sgt T C Clitheroe	**27/28 May**	2322-0337
PO N Rollin and crew	Sgt S V Day	AACHEN	Sgt T F Holbrook and crew
7/8 May	Sgt W C Porter	2355-0425	**4/5 Jun**
BRUZ	FS H Lavellee	Sgt T F Holbrook	SANGATTE

0139-0434
Sgt T F Holbrook and crew
5/6 Jun
ST MARTIN VARREVILLE
2130-0213
Sgt T F Holbrook and crew
Sgt J P S Reid in, Sgt Payner
out
6/7 Jun
ACHERES
0007-0455
Sgt T F Holbrook and crew
(Master Bomber ordered
crews to abandon raid due to
cloud)
9/10 Jun
FLERS A/F
0101-0545
Sgt T F Holbrook and crew
11/12 Jun
EVREUX
0045-0509
Sgt T F Holbrook and crew
12/13 Jun
GELSENKIRCHEN
2238-0302
Sgt T F Holbrook and crew
14/15 Jun
LE HAVRE
2103-0036
Sgt T F Holbrook and crew
Sgt Payner back
15/16 Jun
BOULOGNE
2114-0032
Sgt T F Holbrook and crew
17/18 Jun
AULNOYE
2316-0415
Sgt T F Holbrook and crew
(abandoned on instruction
from Master Bomber due to
cloud)
22 Jun
MARQUISE/V1 SITE
1358-1713
Sgt T F Holbrook and crew
23/24 Jun
SAINTES
2222-0529
PO K A Underwood
Sgt H J Heavener
Sgt D W O'Brien
FO L C Boyes
Sgt J F Marshall
Sgt H Ball
Sgt G H Beevers
24 Jun
FLERS/V1 SITE
0126-0448
PO J B Starr
PO J W Steuart
FS A C Kndt
FS P C Barr
Sgt A T Vass
Sgt W D North
FS F Mills
27/28 Jun
VAIRES
0035-0509
PO J B Starr and crew
29 Jun
SIRACOURT/V1 SITE
1200-1553
FS H H Turner
WO B E Vipond
Sgt H Idle

FS E Getty
Sgt W Marshall
Sgt J P Ewing
Sgt F A Forster
30/1 Jul
VIERZON
2215-0356
WO B E Vipond and crew
2 Jul
DOMLEGER/V1 SITE
1220-1544
WO T F Holbrook and crew
4/5 Jul
ORLEANS
2147-0345
WO T F Holbrook and crew
5/6 Jul
DIJON
2101-0515
WO T F Holbrook and crew
7 Jul
CAEN
1913-2316
PO R G Hancox
PO T B Booth (N)
Sgt R G W Wareham
WO M W Moore (BA)
Sgt D A Nelson (FE)
Sgt T E Hayes (MU)
Sgt H A Burt (RG)
12/13 Jul
TOURS
2106-0327
WO T F Holbrook and crew
18 Jul
CAEN
0404-0850
Sgt F B Snell
Sgt G W Allen
FS A S D Brown
PO E Standing
Sgt N J Moore
Sgt A Wilson
Sgt R J Sharp
(aircraft hit by flak)
18/19 Jul
SCHOLVEN
2311-0323
PO A E Lowry RCAF
Sgt A D McPherson
Sgt A Hetherington
FO S A Bacon
Sgt Brackenbury
Sgt P B Potts
Sgt W R Classen
20/21 Jul
COURTRAI
2338-0259
WO T F Holbrook and crew
23/24 Jul
KIEL
2238-0336
WO T F Holbrook and crew
30 Jul
CAUMONT
0602-1000
WO T F Holbrook and crew
31/1 Aug
FORET DE NIEPPE/V1 SITE
2203-0137
FO A J Thompson
Sgt V O Langdon
FO J Simpson
FS A A Wheatley
Sgt J E D Squires
Sgt D H Moyer
Sgt E Jones

2 Aug
LES CATELLIERS/V1 SITE
1547-1615
PO T F Holbrook and crew
3 Aug
TROSSY ST MAXIM/V1 SITE
1154-1615
PO T F Holbrook and crew
13/14 Aug
FALAISE
FO G S Whyte
0032-0403
WO L V Dredge
FS C S R Horne
WO J V Parkinson
Sgt R O Walters
Sgt T Gilmour
Sgt J G Kernhan
15 Aug
VOLKEL A/F
0936-1326
FO D W McLean RAAF
WO J Phillips
FS J E Kelly
FS A B Llewellyn
Sgt F W Niblett
FS K Rowley
FS I Hunter
16/17 Aug
STETTIN
2045-0517
FO D W McLean and crew
(FO McLean and crew failed
to return 24 October, Doug
McClean being killed)
25/26 Aug
RUSSELSHEIM
1958-0457
FO T F Holbrook and crew
26/27 Aug
KIEL
2004-0136
FO T F Holbrook and crew
31 Aug
ST REQUIER
1325-1639
FO T F Holbrook and crew
5 Sep
LE HAVRE
1630-1947
FO E King
Sgt L F Elms
Sgt W E Spilsbury
Sgt A R Bailey
Sgt G W Smith
PO H S Brown
Sgt C V Swann
8 Sep
LE HAVRE
0625-1019
FO E King and crew
12/13 Sep
FRANKFURT
1835-0141
FO C H Henry RNZAF
Sgt F N Hesketh
FO N L Chesson
Sgt G C Heywood
Sgt J K Penrose
FS E D Martin
Sgt F W Kendall
16/17 Sep
HOPSTEN
2158-0120
FO C H Henry and crew
17 Sep
WESTKASPELLE

1705-1943
FL T Nisbett
Sgt J K Pinder
Sgt C K Spencer
Sgt R V Holt
FS V C Kirchner
PO C E Simons
PO E L Williams
20 Sep
CALAIS
1531-1849
FL T Nisbett and crew
23 Sep
NEUSS
1851-2340
FO E King and crew
25 Sep
CALAIS
0720-1045
FS I L McIntyre
Sgt N L Woods
Sgt J F Craigen
FS J R Syratt
FS K W Tong
Sgt E Reed
Sgt C G Rutzou
26 Sep
CAP GRIS NEZ
1055-1355
FO E King and crew
5/6 Oct
SAARBRUCKEN
1847-0050
FO E King and crew
7 Oct
EMMERICH
1153-1612
FO E King and crew
FO A S Mallik (BA) in
14 Oct
DUISBURG
0648-1127
FO K T Wallace RCAF
Sg R B Glasper (FE)
FO C L Carlson RCAF
FS W D Bonter RCAF (N)
Sgt H S Hoare (WOP)
Sgt S A Bedford (MU)
FS J S Ross (RG)
15 Oct
WILHELMSHAVEN
1730-2145
FO K T Wallace and crew
19 Oct
STUTTGART
1649-2348
FO K T Wallace and crew
23 Oct
ESSEN
1632-2224
FO K T Wallace and crew
28 Oct
COLOGNE
1300-1815
FO A J D Leach
Sgt W White
FS C W Young
Sgt J W Moore
Sgt H M Powell
Sgt A W Lush
Sgt A M Hiebert
29 Oct
DOMBURG
1206-1441
FL A J Thompson
Sgt F W Mills
FL J E T Mansfield

FS A A Wheatley
FS J Hyde
Sgt M J McMillan
FS E Jones
30 Oct
COLOGNE
1748-2355
FO C H Henry and crew
31 Oct
COLOGNE
1749-2309
FO K T Wallace and crew
2 Nov
DUSSELDORF
1635-2155
FO L L Galbraith
Sgt G C Eaton
FO J C Blyth
FO W M Ogle
Sgt D C Johnston
Sgt E M Warren
Sgt B Gallagher
4 Nov
BOCHUM
1741-2225
FO K T Wallace and crew
6 Nov
GELSENKIRCHEN
1150-1626
FL A J Thompson and crew
WO D M Pugh and Sgt V C
Langdon in
FL Mansfield and FS Hyde out
9 Nov
WANNE EICKEL
0743-1310
FL A J Thompson and crew
16 Nov
DUREN
1244-1805
FO J L Walters
Sgt J M Gibb
FS W Blythe
Sgt S R Harris
FS K O Langham
Sgt C Butler
Sgt M Boulding
18 Nov
WANNE EICKEL
1602-2124
FO J L Walters and crew
21 Nov
ASCHAFFENBURG
1515-2205
FO C H Henry and crew
27 Nov
FREIBURG
1604-2320
FO P C L Bird
Sgt G W Robinson
FS S Pechet
FO W N Thompson
WO E Bratby
Sgt L W Lang
Sgt L R Jackson
29 Nov
DORTMUND
1217-1735
FO P C L Bird
4 Dec
KARLSRUHE
1644-2255
FO C H Henry and crew

1945
16/17 Jan
ZEITZ

1746-0136
FO L L Galbraith and crew
1 Feb
LUDWIGSHAVEN
1527-2248
FL A J D Leach and crew
FS Middleman in, Sgt Powell
out
2/3 Feb
WIESBADEN
2046-0032
FO W Kroeker RCAF
FS O Brooke
FO C E Modeland
FL W D Smith
FS J F Woodcherry
FO C W G Biddlecombe
FO G T Wood
(abandoned when starboard
outer engine caught fire)
13/14 Feb
DRESDEN
2131-0720
FO P C L Bird and crew
14/15 Feb
CHEMNITZ
1956-0523
PO N A Wickes
Sgt C McCabe
FS R J White
Sgt E Parker
FS T A Connolly
Sgt F S Saunders
Sgt D G Horton
20/21 Feb
DORTMUND
2145-0426
FL A J D Leach and crew
FS Powell back
21/22 Feb
DUISBURG
1844-0204
FL A J D Leach and crew
23/24 Feb
PFORZHEIM
1543-2337
FL A J D Leach and crew
28 Feb
DUSSELDORF
0823-1136
FL A J D Leach and crew
(raid abandoned, recalled
due to adverse weather;
jettisoned bombs over
sea)
1 Mar
MANNHEIM
1159-1808
FL A J D Leach and crew
2 Mar
COLOGNE
0709-1225
FL A J D Leach and crew
5/6 Mar
CHEMNITZ
1641-0220
FL A J D Leach and crew
7/8 Mar
DESSAU
1711-0249
FO K W Mabee
Sgt K W Clarke
FO A R Hovis
Sgt L E Rae
Sgt D W Debonnaire
Sgt M M Barker
Sgt T K Imperious

8/9 Mar
KASSEL
1700-0037
FL A J D Leach and crew
11 Mar
ESSEN
1145-1721
FO K W Mabee and crew
12 Mar
DORTMUND
1300-1856
FL A J D Leach and crew
13/14 Mar
DAHLEUSCH
1724-2327
PO N A Wickes and crew
Sgt E L Cullup in, Sgt
Saunders out
16/17 Mar
NURNBERG
1716-0156
FL A J D Leach and crew
18/19 Mar
HANAU
0006-0732
FL A J D Leach and crew
20/21 Mar
MINING/HELIGOLAND
0224-0645
FO F M Baird
Sgt G W G Monk
FO G B MacPherson
FO M Couse
Sgt J D'Arcy
Sgt D D Boyd
Sgt D J Duncan
21/22 Mar
BRUCHSTRASSE
0037-0646
FO C Grannum
Sgt E J Wicks
Sgt A Babayan
Sgt J J Wakeham
Sgt F Gankrodger
Sgt W McCormick
FS J Green
22/23 Mar
MINING/OSLO FIORD
2129-0400
FO F M Baird and crew
25 Mar
HANNOVER
0654-1250
FO F M Baird and crew
27 Mar
PADERBORN
1440-1946
FL W Kroeker and crew
(FL Kroeker's crew failed to
return 4 April, Kroeker being
killed)
4/5 Apr
LUTZKENDORF
2051-0510
FL A J D Leach and crew
9/10 Apr
KIEL
1944-0016
FL A J D Leach and crew
FO G F Sage (2P)
10/11 Apr
PLAUEN
1806-0236
FL A J D Leach and crew
18 Apr
HELIGOLAND
1010-1514

FL T MacPherson
Sgt J Harrison
Sgt M L White
FO S Borkowitz
FS J B Hosie
Sgt J J Picco
Sgt P B Adams
22 Apr
BREMEN
1532-2018
FL T PacPherson and crew
(abandoned over target due
to cloud and smoke)
25 Apr
BERCHTESGADEN
0507-1319
FO G F Sage
Sgt G Horridge
FO M P Shewring
FO T W Owen
FO K P C Smith
FO W Faulkner
FL R C Candy
30 Apr
'MANNA'
1500-1809
FO W H Cumming
Sgt J C Steele
FS W J Booth
FO J O J Bates
FS L J Minchin
Sgt R T Diamond
Sgt E P Bryant
1 May
'MANNA'/VALKENBURG
1243-1535
FL C Grannum and crew
2 May
'MANNA'/VALKENBURG
1030-1312
FL C Grannum and crew
3 May
'MANNA'/VALKENBURG
1220-1531
FL C Grannum and crew
4 May
'MANNA'/ROTTERDAM
1314-1559
FO J L Wallace
Sgt V Tracey
FO R Brooke
FO D M Harrison
FS G Smith
FS R W Middlemass
Sgt M Brigg
7 May
'MANNA'/ROTTERDAM
1319-1707
FL C Grannum and crew
9 May
'EXODUS'/BRUSSELS
pm
FL C Grannum and crew
11 May
'EXODUS'/BRUSSELS
pm
SL P S Huggins
Sgt G W Robinson
WO S Pechet
FO W N Thompson
WO K Bratby
Sgt R W Lang
WO L R Jackson

ME801
NAN

Built at Metro Vickers, Manchester, as part of contract No 2221, this Lancaster came off the production line in the spring of 1944, with four Merlin 24 engines, and was assigned to No 166 Squadron on 16 May. Almost immediately it was re-assigned to No 576 Squadron at Elmsham Wolds from 18 May and given the squadron code letters UL, and individual letter 'C' (actually 'C2'). ME801 carried 'C' until October 1944 when it became N2 – the previous 'N' (PD235) having been lost on 24/25 September.

Pilot Officer S. G. Hordal RCAF took ME801 on its first 15 trips from 21 May to 24 June, including both nights of D-Day. Stephen Hordal flew the aircraft just once more before he completed his tour and went to No 1666 CU, receiving the DFC.

ME801 was then taken over by Pilot Officer J. McDonald, who took the aircraft on 24 raids before he completed his tour in October, winning the DFC. He failed to bomb on one occasion, during a raid on Le Havre on 6 September, because the radio failed. He therefore did not receive orders to reduce height due to the weather, was unable to identify the target and had to bring his bombs back. A month earlier one of his crew was slightly wounded by flak over Agenville.

Although fairly trouble-free, ME801 did have to return early from a trip to Duisburg on 14/15 October with engine trouble, but that was its only reported failure. When the aircraft became 'N', the squadron had just changed bases, moving from Elmsham Wolds to Fiskerton. Its skipper had been Flying Officer D. Fletcher for six raids during this period (he and his crew were killed in a crash in another Lanc at Manston, coming back from Bonn on 28 December). ME801 was then taken over by Flight Lieutenant H. Leyton-Brown who, on his first trip in the aircraft, actually led the daylight formation against the oil refinery at Wanne Eickel. He made five sorties in the aircraft during November, then another three in January 1945. On another sortie on 20/21 February they played cat-and-mouse with a FW190 but it did not attack.

A few nights later, however, on the 23rd, Nan was attacked by a Ju88 night fighter shortly after bombing. They were at 8,000ft with a bright moon off to port when the rear gunner, Warrant Officer J. J. Hiscocks, saw the '88 when it crossed at a range of 400yds from the dark part of the sky across the fires of the target – Pforzheim – as it climbed slowly beneath the bomber stream. Hiscocks opened fire with a five-second burst and saw the starboard wing of the '88 start to break-up. He had just ordered the pilot to corkscrew to port when, in a final defiant jester, the '88 pilot fired a quick but ineffectual burst at the Lanc. Hiscocks fired again from 300yds and saw the fighter begin to disintegrate then burst into flames and was later seen to explode on the ground.

Flying Officer D. E. Till, who flew most of his tour in another centenarian – LL227, did one trip in Nan, and remembers:

'Chemnitz on 14 February 1945 was the longest trip we'd ever done up to that point and we were concerned about the route being fairly straight due to the fuel load, which was kind of bad news, but we made it.

'Boredom was always a problem on long trips and if you didn't keep your gunners alert they might doze off. After all, on night ops we had all been up some hours before taking off. Like a lot of people, we had this practise of sticking the aircraft on its sides every ten minutes or so, to ensure nothing was going on underneath us. It was probably more something to do rather than anything useful because there was still quite a time gap between doing it, but it kept everyone occupied.

'We knew nothing then about night fighters having upward-firing guns but we knew we were vulnerable from below, so we adopted this tactic as soon as we got over enemy territory, this and a bit of weaving.'

Leyton-Brown continued flying ME801, taking it on seven more ops making 18 in all. The last, on 15/16 March 1945, was his 36th and last trip of his tour. It was also ME801's 100th operation. On this raid Leyton-Brown spotted a Ju88 just 20yds on his port beam over the target but apparently its pilot did not see them, at least he did not attack. Tour-expired, Leyton-Brown received the DFC.

Flying Officer Don Graham then took Nan on ops 102 to 105, and then assorted crews brought the total up to 109 ops by the end of April. The aircraft is recorded as flying 113 sorties in all, so these must include 'Manna' ops, making in fact 114, plus three 'Exodus' trips.

With the previous centenarian, ME758, we read that it was Struck off Charge on 19 October 1945. ME801 preceeded this Lanc by three days, being struck-off after a crash that left the aircraft Cat E. ME801 had no known nose-artwork other than its rows of bomb symbols, a swastika for the Ju88 destroyed, and a DFC ribbon.

1944	Sgt D E Jones RCAF	2350-0420	VIRE
No 166 Squadron	**22/23 May**	PO S G Hordal and crew	2150-0300
21/22 May	DORTMUND	**2/3 Jun**	PO S G Hordal and crew
DUISBURG	2215-0245	CALAIS	**9/10 Jun**
2225-0315	PO S G Hordal and crew	2230-0245	FLERS A/F
PO S G Hordal RCAF	**24/25 May**	PO S G Hordal and crew	0005-0300
Sgt G B Valentine	AACHEN	**5/6 Jun**	PO S G Hordal and crew
FO H Gerus RCAF	2340-0410	VARREVILLE	**12/1 Jun**
FS A M Cambrin RCAF	PO S G Hordal and crew	2120-0130	GELSENKIRCHEN
Sgt F Sheppard	**27/28 May**	PO S G Hordal and crew	2230-0240
Sgt F B Edwards	AACHEN	**6/7 Jun**	PO S G Hordal and crew

14/15 Jun
LE HAVRE
2020-2350
PO S G Hordal and crew
Sgts N C Stafford, J Armstrong
and J S McGinn in;
Sgts Shepherd, Edwards and
Jones out
16/17 Jun
STERKRADE
2255-0320
PO S G Hordal and crew
Sgts Shepherd, Edwards and
Jones back
17/18 Jun
AULNOYE
2330-0330
PO S G Hordal and crew
22 Jun
MIMOYECQUES/
V1 SITE
1345-1655
PO S G Hordal and crew
23/24 Jun
SAINTES
2140-0505
PO S G Hordal and crew
24 Jun
FLERS/V1 SITE
0115-0440
PO S G Hordal and crew
27/28 Jun
CHATEAU BENAPRE/V1 SITE
0130-0540
FS H D Murray
Sgt C C Adams
Sgt R Lee
Sgt D A Barnes
Sgt F Thackeray
Sgt P J Taylor
Sgt W O Kenyon
29 Jun
DOMLEGER/V1 SITE
1135-1445
PO S G Hordal and crew
30 Jun
OISEMONT/V1 SITE
0605-1005
PO J McDonald
Sgt J Beeson
Sgt J S Pym (BA)
Sgt J Fall
Sgt J Thomson
Sgt J F F McDonald RCAF
Sgt K A Grant RCAF
2 Jul
DOMLEGER/V1 SITE
1220-1615
PO D F J Baxter RCAF
Sgt J Ponder
Sgt D Shoebridge
Sgt E J McLasky
Sgt A H Tredwell
Sgt R F Hillman
Sgt E W Boakhout
5/6 Jul
DIJON
2120-0620
PO J McDonald and crew
7 Jul
CAEN
1945-2355
PO J McDonald and crew
12/13 Jul
REVIGNY
2130-0645
PO J McDonald and crew

14/15 Jul
REVIGNY
2115-0515
PO J Archibald RNZAF
Sgt J R Cuthbert
FO P J Biolle
Sgt J E Kearney
FS L Fielding
WO T P Barry
Sgt A Milne
(PO Archibald and crew
failed to return 28 July)
17 Jul
SANNERVILLE/V1 SITE
0340-0740
PO J McDonald and crew
18/19 Jul
SCHOLVEN/BUER
2320-0335
PO J McDonald and crew
20 Jul
WIZERNES
1910-2300
PO J McDonald and crew
23/24 Jul
KIEL
2230-0325
PO J McDonald and crew
24/25 Jul
STUTTGART
2130-0625
FL D J Masters
Sgt A Wightman
FO W C Johnstone RCAF
FS B W Wakely
FS Donovan RCAF
Sgt R E Smith
Sgt A S Smart
25/26 Jul
STUTTGART
2130-0550
FO J McDonald and crew
28/29 Jul
STUTTGART
2105-0305
FO J McDonald and crew
30 Jul
CAHAGNES
0640-1145
FO J J Mulrooney
Sgt B Beale (FE)
FO C W P Mortel (BA)
FS F J L Paten RAAF
WO J N Casey RAAF
FS F B McGrath RAAF
FS R S M Coventry RAAF
31 Jul
LE HAVRE
1810-2140
FL D J Masters and crew
1 Aug
LA BELLE CROIX/V1 SITE
1850-2210
FS A D'L Greig
Sgt N Mason
FP J R Jones
FS F G Henville RAAF
Sgt S A Johnson
Sgt H Walton
Sgt J Thresh
3 Aug
TROSSY ST MAXIM
1140-1620
FS A D'L Greig and crew
4 Aug
PAUILLAC
1325-2125

FS A D'L Greig and crew
5 Aug
BLAYE
1445-2230
FS A D'L Greig and crew
7/8 Aug
FONTENAYE
2110-0120
FO J J Mulrooney and crew
10 Aug
DUGNY
0925-1430
PO N Layden
Sgt S B Naftel
FS F H Lowing
FS J E Wright
FS H J G Sibley
Sgt H Elliott
Sgt E Smith
11 Aug
DOUAI
1400-1730
FO S F Durrant
Sgt R E Pearce
Sgt C J Brady RCAF
FO J Johnstone
F J R McIntyre
Sgt J G MacKay
Sgt D Sprowart
14 Aug
FONTAINE-LE-PIN
1325-1730
FO S F Durrant and crew
15 Aug
LE CULOT
1005-1345
FO S F Durrant and crew
16/17 Aug
STETTIN
2100-0505
PO N Layden and crew
18/19 Aug
RIEME
2220-0235
FO J McDonald and crew
25/26 Aug
RUSSELSHEIM
2005-0455
FO J McDonald and crew
26/27 Aug
KIEL
1950-0125
FO J McDonald and creew
29/30 Aug
STETTIN
2110-0550
FO J McDonald and crew
PO L F Moore and Sgt S H
Bolton in;
Sgts Beeson and Pym out
31 Aug
AGENVILLE
1305-1700
FO J McDonald and crew
Sgts Beeson and Pym back
(Sgt Pym slightly wounded by
flak over target)
5 Sep
LE HAVRE
1630-2015
FO M T Wilson
Sgt R P Kimm
Sgt P E T Brooked
Sgt L N Hill
Sgt S Douglas
Sgt T A Russell
Sgt J H Addison

6 Sep
LE HAVRE
1702-2035
FO J McDonald and crew
(failed to bomb due to radio
failure)
8 Sep
LE HAVRE
0640-1030
FL P P Hague
Sgt T E Allen
PO K D Clements RCAF
Sgt R Low
Sgt C C Routledge
Sgt H Henson
WO J W Sargent
10 Sep
LE HAVRE
1630-2020
FO J McDonald and crew
12/13 Sep
FRANKFURT
1815-0135
FO K L Trent
Sgt A R Dunford
Sgt J N Wadsworth
Sgt H B Reynolds
Sgt R L Skelton RAAF
Sgt C Dalby
FO G C Riccomini
15/16 Sep
MINING
2220-0445
FO K L Trent and crew
17 Sep
FLUSHING
1625-1920
FO K L Trent and crew
20 Sep
CALAIS
1500-1820
FO J McDonald and crew
23 Sep
NEUSS
1830-2345
FO J McDonald and crew
25 Sep
CALAIS
0720-1020
FO J McDonald and crew
(aborted due to weather,
when half way across the
Channel)
26 Sep
CAP GRIS NEZ
1055-1400
FO J McDonald and crew
5/6 Oct
SAARBRUCKEN
1815-0100
FO J McDonald and crew
14 Oct
DUISBURG
0645-1055
FO J McDonald and crew
14/15 Oct
DUISBURG
2230-0105
FO J McDonald and crew
(returned early with engine
trouble)
22 Oct
ESSEN
1605-2200
FO H R McLelland RAAF
Sgt A Rhodes
FS J A Kennedy RAAF

156

FO W R Curtis RAAF (N)
FS N G McGill RAAF
FS M Chapman RAAF
FS J P Coe RAAF
25 Oct
ESSEN
1305-1815
FO D Fletcher
Sgt P A Lake
Sgt L Angus
Sgt C G Campbell
Sgt L J Bull
Sgt G Warren RCAF
Sgt J Norris
27 Oct
COLOGNE
1310-1815
FO D Fletcher and crew
30 Oct
COLOGNE
1750-0030
FO D Fletcher and crew
2 Nov
DUSSELDORF
1625-2200
FO D Fletcher and crew
4 Nov
BOCHUM
1725-2225
FO D Fletcher and crew
6 Nov
GELSENKIRCHEN
1200-1625
FO D Fletcher and crew
9 Nov
WANNE EICKEL
0805-1250
FL H Leyton-Brown
Sgt R A Hawkins (FE)
FS G Paterson (BA)
Sgt L Peters (N)
Sgt E Johnson (WOP)
Sgt J P McMullen (MU)
Sgt G Lester (RG)
11 Nov
DORTMUND
1555-2105
FL H Leyton-Brown and crew
16 Nov
DUREN
1245-1730
FL H Leyton-Brown and crew
18 Nov
WANNE EICKEL
1550-2050
FL H Leyton-Brown and crew
21 Nov
ASCHAFFENBURG
1530-2145
FL H Leyton-Brown and crew
27 Nov
FREIBURG
1610-2235
FO D C Smith RAAF
Sgt K Warner
FS R G Collins RAAF
FS D K E Anderson RAAF
FS D K Rowe RAAF
Sgt C G R Jones
Sgt S V Lloyd
29 Nov
DORTMUND
1210-1720
FO D C Smith and crew
(slight flak damage)
3 Dec
URFT DAM

0735-1155
FL L Arthur
Sgt J Hill
FO O J T Troy
FO R J Williams
FS D T H Madell
Sgt E N Woods
Sgt E L Gidden

1945
5 Jan
ROYAN
0209-0916
FO E J Pollard RAAF
Sgt E B May
Sgt A Preston
Sgt C V Dolan RCAF
Sgt J Patterson
Sgt R M Goodfellow
Sgt Brown
7/8 Jan
MUNICH
1825-0240
FL A Leyton-Brown and crew
FS Hill (2P)
14/15 Jan
MERSEBURG
1920-0335
FO H J Rowe RCAF
Sgt T Parkinson
FO G L Dunn RCAF
FO R F Brothers RCAF
Sgt J A F Demant
Sgt E A Lindberg RCAF
Sgt F A Chidley RCAF
16/17 Jan
ZEITZ
1745-0135
PO R R J Young
Sgt K G Greatland
FO H N Cheeseman
Sgt G R James
Sgt D P Bannister
Sgt H E Ward
Sgt W M Webb
22 Jan
DUISBURG
1650-2150
FL H Leyton-Brown and crew
28/29 Jan
STUTTGART
1950-0239
FL H Leyton-Brown and crew
1 Feb
LUDWIGSHAVEN
1600-2310
FL C D Thieme
Sgt K Wallace
WO J U Lowing RAAF
FS H W Vine
FS C B Robinson RAAF
Sgt L Hull
Sgt C H Crouch
7/8 Feb
KLEVE
1900-0010
FL C D Thieme and crew
8/9 Feb
POLITZ
1905-0350
FL C D Thieme and crew
(FL Thieme and crew were
shot down on 21 February,
but he and three of his crew
managed to evade and
return to Allied lines)

14/15 Feb
CHEMNITZ
2000-0440
FO D E Till
Sgt W E Holland (FE)
FO J Shorthouse (BA)
FO C R Bray (N)
FO G W Griggs RAAF
Sgt K Oliver (MU)
Sgt R Hamilton (RG)
20/21 Feb
DORTMUND
2135-0335
FL H Leyton-Brown and crew
WO J J Hiscocks in, Sgt Lester
out
21/22 Feb
DUISBURG
1930-0120
FL H Leyton-Brown and crew
23 Feb
PFORZHEIM
1545-2325
FL H Leyton-Brown and crew
FO D B Graham (2P)
(rear-gunner shot down a
Ju88 night fighter)
28 Feb
NEUSS
0845-1005
FL H Leyton-Brown and crew
1 Mar
MANNHEIM
1145-1755
FL H Leyton-Brown and crew
2 Mar
COLOGNE
0715-1220
FL Sleight
PO W B Hazel
PO C Hathaway
FL C Dyson
FO J Hatchard
FS B R Phillips
FS F H Taylor
5/6 Mar
CHEMNITZ
1645-0150
FL H Leyton-Brown and crew
7/8 Mar
DESSAU
1700-0215
FL H Leyton-Brown and crew
8/9 Mar
KASSEL
1700-0025
FL H Leyton-Brown and crew
11 Mar
ESSEN
1150-1655
FO W N Holmes RCAF
Sgt D J Nimmo
FO F W Christisen
FO J K Scruton RCAF
Sgt W Brown
FS E O Grife RCAF
FS W W Parsons RCAF
12 Mar
DORTMUND
1325-1825
FL H Leyton-Brown and crew
15/16 Mar
MISBURG
1720-0045
FL H Leyton-Brown and crew
21 Mar
BREMEN

0755-1250
PO D K Sullivan
Sgt G Charlton
Sgt G S Atkinson
Sgt I A Heath
Sgt D Phillips
Sgt R Erskine
Sgt G F Chatterton
24 Mar
DORTMUND
1305-1830
FO D B Graham
Sgt C R Huxtable (FE)
FS P Brookes (BA)
Sgt W Hatton (N)
Sgt S F Gascoigne
Sgt E J Boniface (MU)
Sgt E J Moore (RG)
31 Mar
HAMBURG
0625-1135
FO D B Graham and crew
4 Apr
LUTZKENDORF
2130-0520
FO D B Graham and crew
9 Apr
KIEL
1910-0135
FO D B Graham and crew
10/11 Apr
PLAUEN
1805-0300
FL J R Tile
Sgt P J Robinson
Sgt D L Howells
Sgt W Vaughan
Sgt J R Currie
Sgt S H Wragg
Sgt D I Vicary
14/15 Apr
CUXHAVEN
1935-0055
FS R C Sayers
Sgt J G Blair
Sgt P Patterson
Sgt K E Stott
Sgt L V Hayes
FS H E Bell RAAF
Sgt R J Hagger
18 Apr
HELIGOLAND
1010-1450
PO K Fry
FS H W Parkins
FS H Woodliff
Sgt D G Smith
FL H T Shewan
Sgt A G Younger
Sgt J L Watkins
25 Apr
BERCHTESGADEN
0530-1345
FS R C Sayers and crew
Sgt J D Cormack in, Sgt Blair out
29 Apr
VALKENBURG A/F
1150-1448
FO A Roberts
FS L A Piddington
FS B Rosario
Sgt J Smale
Sgt R Briggs
Sgt S Davies
Sgt B Benson
1 May
'MANNA'/ROTTERDAM

1535-1818
FO A Roberts and crew
2 May
MANNA'/ROTTERDAM
1256-1549
FO H Drew
Sgt G Booth
Sgt A W Stone
Sgt A T Turnton

FS W Guthrie
Sgt L Cousins
Sgt J P Riley
3 May
'MANNA'/VALKENBURG
1018-1315
FO H Drew and crew
8 May
'MANNA'/ROTTERDAM

1207-1528
FO D B Graham and crew
11 May
'EXODUS'/BRUSSELS
1125-1959
FO D B Graham and crew
16 May
'EXODUS'/BRUSSELS
0737-1616

FO D B Graham and crew
26 May
'EXODUS'/BRUSSELS
0830-1548
FO D B Graham and crew

ME803
L FOR LOVE

Another Lancaster Mk I from batch No 2221, built by Metro Vickers, Manchester, ME803 had four Merlin 24 engines and rolled off the production line in the spring of 1944. Assigned to C Flight of No 115 Squadron at Witchford, Cambridgeshire, on 20 May 1944, it initially carried the code letters A4-D and began flying ops on the night of 31 May/1 June in the hands of Flying Officer J. G. Sutherland RAAF.

Its second sortie was on the night of D-Day, again with Sutherland in command. In fact Sutherland was to be this Lanc's regular skipper until August, taking it on a total of 19 trips. ME803's next regular captain was Flying Officer F. A. Stechman, who flew the aircraft on a further 18 sorties. Freddie Stechman received the DFC at the end of his tour.

In October ME803 changed codes to KO-D and by November had changed its individual identification letter to 'L' (L-Love) which it would keep until the spring of 1945 when C Flight changed its codes again, ME803 then becoming IL-B.

ME803 flew regularly on operations with hardly a mechanical problem until it had to abandon a daylight trip to Nordstern on 23 November 1944 when the port outer engine went u/s, by which time the aircraft had flown more than 60 ops. On its next sortie, a daylight to Cologne, ME803 had its first reported flak hit which caused slight damage to the mid-upper turret.

It now seemed as if this particular Lanc was starting to live on borrowed time. On yet another daylight, to Dortmund on 2 December, ME803 was nearly hit by falling bombs from another bomber above, and on 21 December the aircraft lost its port inner engine but the pilot, Flying Officer J. P. R. Mason RCAF, flew on and bombed the target. On its first sortie of 1945 the Gee became u/s, the electrics failed and in consequence ME803 could not drop its bombs.

On the next raid the aircraft was hit by flak in both wings and undercarriage, and on the following trip flak put holes in the port elevator and port wing. Shaking off these misfortunes ME803 continued to press on but on 3 February it was coned by searchlights for 10 minutes during a raid on Dortmund and the crew bombed the wrong target – Botrop. Seeing Target Indicators (TI) which appeared to be dummy ones, they took a dead-reckoning course to other TIs, then found they were over the wrong target, by which time it was too late to do anything else but bomb Botrop instead.

On 22 February ME803 was hit again when flak damaged a hydraulic pipe. However, that ended the run of bad luck but with its total ops just a couple short of the century, ME803 was sent to No 54 MU on 15 March, having become Cat AC. The aircraft's last regular skipper was Flying Officer G. P. Pickering, who flew the Lanc on 21 ops during its problem period, but Godfrey Pickering survived his tour and received the DFC. He recalls this move to No 54 MU:

'Shortly after finishing our tour we had to take the "Old Girl" to Oakington for a Major Service. She was found to have a cracked main spar in the port wing. After having repairs completed she returned to active service and completed a few more raids before the war ended. Not bad for an "Old Girl".'

Cliff Cunnington, Pickering's rear-gunner, recalls the nose-artwork: 'We called the Lancaster L for Love. She had a naked lady painted on her and there was a growing number of bomb symbols, each representing a raid completed.'

Returning to the squadron on 22 March, ME803 became IL-B and quickly took its total ops to 100 during April. In total this Lanc was credited with 105 ops, which included four 'Manna' and three 'Exodus' trips.

On 21 May 1945 ME803 went to No 1659 CU then to No 39 MU on 18 September, from where it was Struck off Charge on 20 November 1946.

1944
No 115 Squadron
31/1 Jun
TRAPPES WEST
0001-0445
FO J G Sutherland RAAF
FS V Catchlove RAAF (N)
Sgt A Wright (BA)
FS H Huddlestone (WOP)
Sgt C Wright (FE)
Sgt S Cooper (MU)
FS D Day RAAF (RG)
5/6 Jun
OUISTREHAM
0327-0637
FO J G Sutherland and crew
6/7 Jun
LISIEUX
0021-0256
FO J G Sutherland and crew
10/11 Jun
DREUX
2307-0257
FL D W Martin
FO R Fisher
FS J Waple
Sgt T S Longhurst
Sgt C Bridges
FS R Champlion
FS L Johns RAAF
11/12 Jun
NANTES
2355-0450
FL D W Martin and crew
12/13 Jun
GELSENKIRCHEN
2326-0241
FO J G Sutherland and crew
14/15 Jun
LE HAVRE
2358-0218
PO P Frankland

FO H H Skinner
FS Stewart-Smith
Sgt E Martin
Sgt F Rutter
Sgt J Doering RCAF
Sgt R Bayly
15/16 Jun
VALENCIENNES
2316-0226
PO P Frankland and crew
21 Jun
DOMLEGER/V1 SITE
1803-2028
FL D W Martin and crew
(raid aborted by Master
Bomber due to cloud)
23/24 Jun
L'HEY/V1 SITE
2308-0138
FO J G Sutherland and crew
27/28 Jun
DIENNAIS
23329-0209
FL D W Martin and crew
30 Jun
VILLERS BOCAGE
1815-2105
FO J G Sutherland and crew
2 Jul
BEAUVOIR/V1 SITE
1238-1548
FO J G Sutherland and crew
FO D F O'Sullivan in, Sgt
Wright out
5/6 Jul
WATTEN/V1 SITE
2312-0132
FO J G Sutherland and crew
9 Jul
NUCOURT/V1 SITE
0425-0755
FO J G Sutherland and crew

Sgt Wright back
12 Jul
VAIRES
1810-2205
FO J G Sutherland and crew
(raid abandoned by Master
Bomber due to cloud)
15/16 Jul
CHALONS-SUR-MARNE
2200-0410
FO J G Sutherland and crew
18 Jul
EMIEVILLE
0410-0710
FO J G Sutherland and crew
18/19 Jul
AULNOYE
2240-0205
FO J G Sutherland and crew
20/21 Jul
HOMBERG
2324-0244
FO J G Sutherland and crew
23/24 Jul
KIEL
2240-0350
FS B S Wadham
FO R C Halkyard RAAF
FS H Ellis
FS H Evans RAAF
Sgt R Walker
Sgt T Cheeseman
Sgt H Minns
(also dropped leaflets)
28/29 Jul
STUTTGART
2147-0537
FS B S Wadham and crew
30 Jul
AMAYE-SUR-SHULLES
0610-0905

Sgt A Middleton
FS N Cottle
Sgt J Courley
Sgt L Rowles
Sgt T Barr
Sgt T Winters
1 Aug
COULONVILLERS
1614-1909
FS R Cooper
FO D M Curtis
Sgt T Webb
Sgt G Piper
Sgt J Bracey
Sgt J Herrod
FS T Moore RCAF
(raid abandoned over target
due to cloud)
3 Aug
BOIS DE CASSON
1150-1545
FL J G Sutherland and crew
4 Aug
BEC-D'AMBES
1346-2110
FO A J Osborne
Sgt W McNeil RCAF
Sgt P McGowen
Sgt J Simcox
Sgt H Edwards
Sgt D Cattle
Sgt J Cowan
5 Aug
BORDEAUX
1430-2200
FL J G Sutherland and crew
7/8 Aug
MARE DE MAGNE
2159-0119
FL J G Sutherland and crew
8/9 Aug
FORET DE LUCHEUX

2205-0145
FS J Perry
Sgt E Bennett
Sgt E Wood
Sgt J Lynch
Sgt A Clarke
Sgt N Atkins
Sgt N Elvish
9/10 Aug
FORT D'ENGLOS/V1 SITE
2204-0004
FL J G Sutherland and crew
11 Aug
LENS
1420-1730
FO F A Stechman and crew
15 Aug
ST TROND A/F
FO F A Stechman and crew
1012-1327
16/17 Aug
STETTIN
2120-0515
FS B S Wadham and crew
FS J Perry (2P)
18/19 Aug
BREMEN
2150-0255
FO A J Osborne and crew
25/26 Aug
RUSSELSHEIM
2025-0510
FS J Perry and crew
26/27 Aug
KIEL
2010-0130
PO B S Wadham and crew
29/30 Aug
STETTIN
2120-0655
PO B S Wadham and crew
31 Aug
PONT REMY
1620-1945
FO F A Stechman and crew
6 Sep
LE HAVRE
1625-1940
FL T W Miller RAAF
FS A Taylor RAAF
FS L Dickinson RAAF
WO D L Wood
Sgt F R Ruston
Sgt A Pavelyn
Sgt J McCue
8 Sep
LE HAVRE
0610-0945
FO F A Stechman and crew
FO S C Josling (2P)
(raid abandoned over target
by Master Bomber)
10 Sep
ALVIS IV/LE HAVRE
1550-1910
FO F A Stechman and crew
11 Sep
KAMEN
1615-2030
FO F A Stechman and crew
12/13 Sep
FRANKFURT
1855-0110
FO F A Stechman and crew
17 Sep
BOULOGNE
1015-1255

FO F A Stechman and crew
17/18 Sep
ZALTBOMMEL
1920-2210
FL T W Miller and crew
20 Sep
CALAIS
1450-1745
FO F A Stechman and crew
23 Sep
NEUSS
1935-2340
FO F A Stechman and crew
25 Sep
CALAIS
0830-1125
FO F A Stechman and crew
(raid abandoned by Master
Bomber due to cloud)
26 Sep
CAP GRIS NEZ
1135-1405
FO J C Josling
FO L A Wood RCAF
FS H G H Park
FO S J Sargent
Sgt F S Martin
Sgt W D Leitch
Sgt D G Hunt
27 Sep
CALAIS
0750-1025
FL E W Talbot
WO R O Mackley RAAF
Sgt F L Bonar
FS L B Swifte RAAF
Sgt W J Tandy
Sgt F Hawkins
Sgt A Byne
5 Oct
SAARBRUCKEN
1720-2235
FL F A Stechman and crew
FL L D Easterman (2P)
6 Oct
DORTMUND
1703-2223
FL F A Stechman and crew
7 Oct
EMMERICH
1259-1621
FL P B Brown RNZAF
FO J A Willman
FS E J Banks
WO R Critchley
Sgt P R Spittal
FS D W Pratt
FO G B McDonald RNZAF
1 Oct
DUISBURG
0650-1100
FO F A Stechman and crew
FL L W Thorne (2P)
14/15 Oct
DUISBURG
2310-0341
FO F A Stechman and crew
19/20 Oct
STUTTGART
2215-0410
FO F A Stechman and crew
PO J Palmer (2P)
21 Oct
FLUSHING
1120-1415
FO F A Stechman and crew
FO Scott (2P)

25 Oct
ESSEN
1327-1717
FO K V Gadd RCAF
FS C Hawkins
FO W T Smith RCAF
FS R Bradfield
FS D Marsh
FS G Brown RCAF
FS D Miller RCAF
31 Oct
COLOGNE
FO S M Grant RAAF
1800-2242
FS W Parlett
FO B Keating RAAF
FS R Bourne RNZAF
Sgt G Parsloe
Sgt J Duthie
Sgt G Fry
11 Nov
CASTROP-RAUXEL
0835-1255
FO E G Fillimore
FS H Charlesworth RAAF
FO W Worster RCAF
FS J Breather RAAF
Sgt N Bannister
Sgt A Dennis
Sgt C Bradshaw
15 Nov
DORTMUND
1244-1719
FO G P Pickering
Sgt P O'Reilly (N)
Sgt J Kielback RCAF
Sgt L Morgan (WOP)
Sgt T Patterson (FE)
Sgt C Clowes (MU)
Sgt C Cunnington (RG)
20 Nov
HOMBERG
1246-1702
FO G P Pickering and crew
21 Nov
HOMBERG
1227-1638
FO G P Pickering and crew
27 Nov
COLOGNE
1218-1707
FO G P Pickering and crew
(slight flak damage to
mid-upper turret)
29 Nov
NEUSS
0259-0707
FL G O Russell DFC
WO C Wilson (2P)
Sgt M C Preston
FS G Alexander RNZAF
FO J Currin
Sgt E S Shaw
Sgt A MacIntyre
Sgt D J Whitehead
30 Nov
OSTERFELD
1034-1506
FO J L Snyder RCAF
FO K G Logue RCAF
Sgt M J Kehoe
WO L Walker
Sgt A Hammond
Sgt E Keate
Sgt V Miller
2 Dec
DORTMUND

1318-1716
FO J L Snyder and crew
(aircraft nearly hit by falling
bombs from an aircraft over-
head)
4 Dec
OBERHAUSEN
1147-1622
FO J L Snyder and crew
12 Dec
WITTEN
1103-1554
FO G P Pickering and crew
16 Dec
SIEGEN
1117-1655
FO G P Pickering and crew
21 Dec
TRIER
1222-1705
FO J P R Mason RCAF
FO R I MacKay RCAF
FO L W P Barker RCAF
Sgt G L Taylor
Sgt B Baldero
Sgt M M Nauss RCAF
Sgt T M Michaik RCAF
(lost port inner engine but
continued to target)
27 Dec
TRIER
1251-1628
FO G P Pickering and crew
28 Dec
COLOGNE
1123-1652
FO G P Pickering and crew
31 Dec
VOHWINKEL
1629-2202
FO G P Pickering and crew
PO A H Dick (2P)

1945
1 Jan
VOHWINKEL
1629-2202
FO G P Pickering and crew
FO R L Burbridge RCAF (2P)
(Gee u/s, electrics failed,
unable to bomb)
2 Jan
NURNBERG
1537-2312
FO F H Graham RCAF
FO R E Valentine RCAF
FO H B Tims RCAF
Sgt J Haigh
Sgt D Squires
Sgt W H Dubois RCAF
Sgt R J Giesel RCAF
5 Jan
LUDWIGSHAVEN
1120-1721
FO G P Pickering and crew
(slight flak damage)
13 Jan
SAARBRUCKEN
1139-1745
FS D F Cameron
Sgt V Higgins
Sgt G H Scott
Sgt G Graham
FS H Somerville
Sgt R Davison
Sgt J Boyd

16/17 Jan
WANNE EICKEL
2326-0446
FO R L Burbridge RCAF
Sgt H E Harrison
FO C W Walker RCAF
Sgt P W Pollack
Sgt A W Fisher
Sgt H R Sidney
Sgt L Ireland
22 Jan
DUISBURG
1701-2148
FL G P Pickering and crew
FS L Arakiel (MU) in, Sgt
Clowes out
29 Jan
KREFELD
1015-1530
FL G P Pickering and crew
PO J Corner in, Sgt Clowes
out
1 Feb
MUNCHEN
1259-1810
FL G P Pickering and crew
GLADBACH
Sgt Clowes back
2/3 Feb
WIESBADEN
2042-0249
FL G P Pickering and crew
WO A H Gibbins (2P)
3 Feb
DORTMUND
1618-2121
FL G P Pickering and crew
(bombed Botrop)
9 Feb
HOHENBUDBERG/KREFELD
0321-0832
FL G P Pickering and crew
13/14 Feb
DRESDEN
2151-0655
FL G P Pickering and crew

WO Gibbins (2P)
16 Feb
WESEL
1236-1731
FL G P Pickering and crew
18 Feb
WESEL
1201-1649
FL G P Pickering and crew
WO M J Carberry RAAF (2P)
19 Feb
WESEL
1316-1902
WO M J Carberry RAAF
FS G G Shepphard
FO A Tracey
FS A L Johnson RAAF
Sgt K W Shepherd
Sgt J D Bollard
Sgt S J Cheetham
22 Feb
OSTERFELT
1233-1732
FL G A Sherwood RCAF
FO T D Lees RCAF
FO L Ing
FS E Thompson
FS B S Jones
FS A D Lennox RCAF
FS W H Hole RCAF
25 Feb
KAMEN
0930-1500
FL O F Hill
FS R A Hurnell
FS A W Whitehead
FS E W Oakden
Sgt D N Clarke
Sgt B R Smith
Sgt A G Walker
26 Feb
DORTMUND
1042-1634
FL R E Roberts
FS A K Olliffe
Sgt C E Searle

Sgt B W Forst (BA)
PO S Liston (Under Gnr)
Sgt R Strange (FE)
Sgt B D Smith (MU)
WO R S Groves (RG)
27 Feb
GELSENKIRCHEN
1126-1625
FO C A Knapp
FS C L Denison RCAF
FS K Edginton RCAF
FS F C Vines
Sgt E Spencer
Sgt E Whitley
Sgt B C Treagus
9/10 Apr
KIEL
1934-0131
FL O F Hill and crew
14/15 Apr
POTSDAM
1803-0304
FL E W Talbot and crew
18 Apr
HELIGOLAND
1021-1512
FO J Bettle RNZAF
FS I W James
FS C MacDonald
FS H H Judge
Sgt G Oliver
Sgt J Milby
Sgt J Galley
22 Apr
BREMEN
1540-2040
FO H A Hernan
Sgt W Pollock
FS J Fowler RCAF
FO R H Leonard
FS J E Milner
Sgt P Mitchell
Sgt A Bevan
30 Apr
'MANNA'/ROTTERDAM
1653-1938

FO H A Hernan and crew
1 May
'MANNA'/THE HAGUE
1407-1649
FO N J Byers
FS H E Spencer RAAF
FO J L Dunbar
Sgt F Tomlinson
FS H Burbridge
Sgt C H Wood
Sgt T Oram
4 May
'MANNA'/THE HAGUE
1205-1453
FO B J Doucette RCAF
FS R G Chatfield
FS D Sutter RCAF
Sgt B B Ashford
FS L Craven
no M/U
Sgt C A Copp (RG)
7 May
'MANNA'/THE HAGUE
1215-1447
FO H A Hernan and crew
10 May
'EXODUS'/JUVINCOURT
FO H A Hernan and crew
1053-2004
11 May
'EXODUS'/JUVINCOURT
1648-2208
FL A C Hulme
FS J S Selwyn
FS W J Lawson
Sgt B R Williams
no BA
Sgt W G Pounder
Sgt G Brown
12 May
'EXODUS'/JUVINCOURT
0754-1335
FO N J Byers and crew

162

ME812
FAIR FIGHTER'S
REVENGE

Built at Metro Vickers, Manchester, as part of contract No 2221, this Lancaster came off the production line in the spring of 1944 and was assigned to No 166 Squadron at Kirmington, Lincolnshire in May. Given the squadron code letters AS and individual letter 'F', it flew its first op on the night of 3/4 June in the hands of Flight Sergeant S. G. Coole, who had begun his tour of ops at the end of March.

This bomber had a female figure painted on its nose, holding a duelling sword, with the words 'Fair Fighter's Revenge' forward of her and beneath the front gun turret. The bomb log which began beneath the cockpit soon increased until there were five 10-bomb rows followed by another similar block next to the first. After seven night raids, which included both nights of D-Day, ME812 flew its first daylight op on 22 June. This and the Lanc's other daylight sorties were identified by having stars (or perhaps tiny suns) marked above the bomb symbol.

Sidney Coole (who had won an immediate DFM in May, together with his mid-upper gunner, Sergeant Ray Scargill, for bringing their badly damaged bomber home for a crash-landing, after a fight in which Scargill shot down their attacker) took ME812 on 16 trips over the first months of its time on No 166 Squadron, completing his tour in August. Other pilots who flew the aircraft were men who had also flown other 166 veteran Lancs, namely W. C. Hutchinson, and Henry Schwass and Douglas Dickie – the latter two both New Zealanders who would complete tours and receive the DFC. In fact, Schwass and his crew actually shot down a FW190 during a sortie on 4 July flying in aircraft 'B'.

Arthur Downs was Sid Coole's flight engineer and recalls how ME812 got its name:

'When we arrived at No 166 Squadron we were given a new Lancaster LM521 and because its letter was "F" it was agreed that we name her "Fair Fighter", hence the female holding the rapier. This aircraft served us well until our 11th operation when, after bombing Aachen on 27/28 May, we were shot-up pretty badly and set on fire over Belgium by a Me110. The fighter was shot down by the mid-upper gunner and after putting the fire out we managed to get back to England but crashed in a wood at Woodbridge. The aircraft broke in half and was a write-off.

'We were then given another new Lancaster, code letter "F" (ME812). The same artwork was painted on the nose with the addition of one word – Revenge – and this is why it was called Fair Fighter's Revenge.'

Fair Fighter's Revenge flew steadily on during the summer and autumn of 1944, against all the usual targets, such as V1 sites, communication centres and against German positions in Normandy, only having one recorded problem due to intercom trouble on 23 September which caused the crew to abort. Arthur Downs also remembers the trip to Revigny on 12/13 July which lasted 9 hours 30 minutes, of which they flew the last 3 1/2 hours on three engines.

Then, at the beginning of October, having completed some 53 ops, Fair Fighter's Revenge was re-assigned to No 153 Squadron when this unit was formed from crews and aircraft of No 166 Squadron at Kirmington, and then it moved to Scampton.

With No 153 Squadron the Lanc became P4-F, but within a month it became Cat AC following night fighter damage to the starboard aileron and wing, but the aircraft was back on duty before the month was out. In December Flight Lieutenant L. K. Firth RCAF took over Fair Fighter's Revenge and completed 24 ops in the aircraft by April 1945, winning the DFC.

The aircraft's 100th sortie was flown in early April 1945 and in total it was credited with 105 operational sorties, which included four 'Manna' missions and at least one 'Exodus' flight. The squadron was disbanded in September and ME812 went to No 20 MU at Aston Down. The aircraft was Struck off Charge in October 1946.

1944
No 166 Squadron
3/4 Jun
2324-0315
BOULOGNE
FS S G Coole DFM
Sgt A W Downs (FE)
Sgt R S Rennie (BA)
FS C L Birtwhisle (N)
Sgt A G Holyoak (WOP)
Sgt R Scargill DFM
FS A G Manuel DFM
5/6 Jun
2105-0140
CHERBOURG
FS S G Coole and crew
6/7 Jun
ACHERES
0005-0525
FO W C Hutchinson
Sgt N Harris
FS H R Eaton RCAF
WO A T Mathews RCAF
Sgt E A Lummis
Sgt R B Parry
Sgt D G Prescott
7/8 Jun
VERSAILLES
0005-0430
FO W C Hutchinson and crew
10/11 Jun
ACHERES
2250-0355
FO W C Hutchinson and crew
12/13 Jun
GELSENKIRCHEN
2229-0304
PO J McLaren

Sgt D Summers
FO S T Broad
FO L H Ellerker
Sgt D F Paton
Sgt F J Collins
Sgt J T E Chalk RCAF
14/15 Jun
LE HAVRE
2015-0010
FS S G Coole and crew
17/18 Jun
AULNOYE
2340-0355
PO J McLaren and crew
(raid abandoned over target
by Master Bomber due to
cloud)
22 Jun
MIMOYECQUES
PO H S Schwass RNZAF
1405-1730
Sgt P A Millett (FE)
Sgt D Angel (BA)
FO N J Grant RNZAF (N)
Sgt D J Carter(WOP)
FS S J Padham RNSAF (MU)
Sgt G Reynolds (RG)
23/24 Jun
SAINTES
2205-0540
PO S G Coole and crew
23/24 Jun
FLERS/V1 SITE
0140-0530
PO S G Coole and crew
27/28 Jun
CHATEAU BERNAPRE/V1 SITE
0130-0510

PO S G Coole and crew
29 Jun
DOMLEGER/V1 SITE
1155-1455
PO S G Coole and crew
30 Jun
0555-1000
OISEMONT/V1 SITE
PO S G Coole and crew
2 Jul
DOMLEGER/V1 SITE
1210-1545
PO S G Coole and crew
5/6 Jul
DIJON
2105-0535
PO S G Coole and crew
7 Jul
CAEN
1925-2320
PO S G Coole and crew
PO F D Elliott RCAF (2P)
(aircraft hit by flak; hole in
bomb-aimer's perspex)
12/13 Jul
REVIGNY
2120-0646
PO S G Coole and crew
FS W H H Ansell in, FS
Manuel out
(returned on three engines)
18 Jul
SANNERVILLE/CAEN
1925-2245
PO S G Coole and crew
20 Jul
WIZERNES
1910-2245

FS R W Miller RCAF
Sgt W A Adams
FO L W Harding
FO E A F Hall
Sgt A H MacDonald
Sgt J Schafer RCAF
Sgt C A Pike
23/24 Jul
KIEL
2235-0340
PO S G Coole and crew
PO W L Shirley RNZAF (2P)
24/25 Jul
STUTTGART
2115-0610
PO D J Dickie RNZAF
Sgt W L McDonald
FO E R Pickering
FO T Mason
Sgt K E Corbett
Sgt R Turner
Sgt S K Giles
25/26 Jul
STUTTGART
2120-0600
FO F D Elliott RCAF
Sgt J H Comley
Sgt M L Oliphant RCAF
Sgt N W L Linton
FO R B Melville RCAF
Sgt G M Canning
Sgt K G Rhodes
28/29 Jul
STUTTGART
2120-0550
PO D J Dickie and crew
30 Jul
LE HAVRE

1805-2130
PO D J Dickie and crew
31 Jul
LE HAVRE
1805-2130
PO D J Dickie and crew
1 Aug
LA BELLE CROIX/V1 SITE
1815-2145
PO S G Coole and crew
(raid abandoned over target
by Master Bomber due to fog)
2 Aug
LE HAVRE
1655-2025
PO S G Coole and crew
3 Aug
TROISSY ST MAXIM
1130-1550
PO S G Coole and crew
4 Aug
PAUILLAC
1335-2205
PO D J Dickie and crew
5 Aug
PAUILLAC
1415-2210
PO D J Dickie and crew
7/8 Aug
FONTENAY
2105-0105
FO D J Dickie and crew
10 Aug
DUGNY
0925-1410
FO D J Dickie and crew
11 Aug
DOUAI
1340-1740
FO D J Dickie and crew
12 Aug
BORDEAUX
1120-1825
PO D M C Jones
Sgt P R Jenkinson
Sgt E W Fletcher
Sgt J F Dormer
Sgt Milburn
Sgt J Coles
Sgt H Ferguson
14 Aug
FONTAINE
1315-1715
PO D M C Jones and crew
15 Aug
LE CULOT A/F
0955-1340
PO D M C Jones and crew
16/17 Aug
MINING/STETTIN BAY
2115-0450
FO D J Dickie and crew
18/19 Aug
REIME
2215-0425
FO D J Dickie and crew
25/26 Aug
RUSSELSHEIM
2000-0425
FL R L Graham
Sgt G Potter
FS R E Lakin
PO R H W Bissonette
Sgt F A M Eade
Sgt W M G Falconer
Sgt J Wagstaff

26/27 Aug
KIEL
2015-0120
FO D J Dickie and crew
29/30 Aug
STETTIN
2100-0540
FL R L Graham and crew
3 Sep
GILZE RIJEN A/F
1545-1955
FO W C Capper RNZAF
Sgt L W Spervall (FE)
FS G B Bale RNZAF
FS T C Morris (N) (WOP)
Sgt G Luckraft
Sgt L M Nutt (MU)
Sgt J D Ramsey (RG)
5 Sep
LE HAVRE
1615-2000
FO W C Capper and crew
6 Sep
LE HAVRE
FO R H Williams
1745-2050
Sgt J Dickson
Sgt A H Taylor
Sgt A Duignan
Sgt A R Crookes
Sgt A R Heath
Sgt J S Carter
8 Sep
LE HAVRE
0655-1045
FO A J E Laflamme RCAF
Sgt F Etherington
FO G J Monckton
FO G MacArthur RCAF
Sgt D H Schofield
Sgt S Pollitt
Sgt F Toogood
(raid abandoned by Master
Bomber due to low cloud)
10 Sep
LE HAVRE
1645-2020
FO D J Dickie and crew
FS D M Pugh RAAF in, FO
Pickering out
12/13 Sep
FRANKFURT
1750-0130
FO D J Dickie and crew
PO N Appleton (2P)
16/17 Sep
STEENWIJK A/F
2140-0109
FO D J Dickie and crew
20 Sep
SANGATTE
1520-1835
FO D J Dickie and crew
23 Sep
NEUSS
1840-2040
FO D J Dickie and crew
(sortie abandoned when
intercom failed)
26 Sep
CALAIS
0950-1310
FO D J Dickie and crew
28 Sep
CALAIS
0805-1145
FO D J Dickie and crew

(raid abandoned by Master
Bomber over target due to
cloud)
5/6 Oct
SAARBRUCKEN
1820-0040
No 153 Squadron
7 Oct
EMMERICH
1130-1615
FO A J E Laflamme and crew
PO G B Potter (2P)
14 Oct
DUISBURG
0650-1120
FO W Holman RCAF
Sgt A Martin
FS V S Reynolds RCAF
FO P C Taylor
Sgt W J Burton
Sgt E S Neil
Sgt A D Kall
19/20 Oct
STUTTGART
2125-0425
FO A J E Laflamme and crew
23 Oct
ESSEN
1630-2205
FL A E Jones
Sgt S S James
Sgt C J Cullen
FS J J L McDonnell RCAF
Sgt D L Williams
Sgt R V Trafford
Sgt A Simpson
30 Oct
COLOGNE
1740-2330
FO J Rhodes
Sgt M F Kingdom
FO D G Webb
FO P C H Clark
Sgt J E Livock
Sgt T J Bicknell
FS W L Baulk
31 Oct
COLOGNE
1755-2315
FO J Rhodes and crew
(Jack Rhodes and crew failed
to return 1 March 1945,
Rhodes being killed)
2 Nov
DUSSELDORF
1615-2145
FO L A Wheeler
Sgt V P G Morandi
FO E C Durman RNZAF
Sgt F F Fish
FS W I H Turner
Sgt A Hodges
Sgt A Scott
4 Nov
BOCHUM
1730-2230
FO L A Wheeler and crew
(aircraft damaged by German
night fighter)
29 Nov
DORTMUND
1220-1740
FO A J E Laflamme and crew
3 Dec
URFT DAM
0745-1150

FL M B French
Sgt A Boyd
FS G F A Tiger
FO L P Ogle RCAF
FS R C G Fisher
Sgt J R Dulso
Sgt R V Cox
4 Dec
KARLSRUHE
1625-2300
FO A J E Laflamme and crew
6/7 Dec
LEUNA
1635-0020
FO A J E Laflamme and crew
15 Dec
LUDWIGSHAVEN
1435-2055
FO A J E Laflamme and crew
17 Dec
ULM
1520-2310
FO A J E Laflamme and crew
22 Dec
COBLENZ
1515-2020
FO A J E Laflamme and crew
28 Dec
BONN
1525-2110
FL L K Firth RCAF
Sgt P Clowes
Sgt R W White RCAF
FO G Denbeigh RCAF
Sgt W A Jones
Sgt L E Laur RCAF
Sgt G L Lawrence RCAF
29 Dec
SCHOLVEN
1510-2105
FL L K Firth and crew
31 Dec
OSTERFELD
FL L K Firth and crew
1525-2050

1945
2 Jan
NURNBERG
1520-2315
FL L K Firth and crew
4 Jan
ROYAN
0200-9010
FL L K Firth and crew
7/8 Jan
MUNICH
1800-0325
FO G P Potter
Sgt G P Woolley
FO W H Thomas
FS J Boyle
FS J S Askew RAAF
Sgt D Smith
Sgt H J Hanbrook
22 Jan
DUISBURG
1700-2155
FL W Holman and crew
FL J T G Weaver in, Sgt Kall
out
(FL Holman and crew failed
to return 20 February 1945)
28/29 Jan
STUTTGART
1940-0300
FL L K Firth and crew

1 Feb
LUDWIGSHAVEN
1545-2245
FL L K Firth and crew
3 Feb
BOTTROP
1600-2205
FL L K Firth and crew
7/8 Feb
KLEVE
1855-0020
FL L K Firth and crew
13/14 Feb
DRESDEN
2115-0710
FL L K Firth and crew
14/15 Feb
CHEMNITZ
2000-0500
FL H W Langford
Sgt W D Thompson
FO B F Rea-Taylor
FO D S McDonald
Sgt T W Jones
Sgt D W Hallam
Sgt K A Hawkins
21/22 Feb
DUISBURG
1940-0130
FL L K Firth and crew
23 Feb
PFORZHEIM
1555-2345
FL L K Firth and crew
1 Mar
MANNHEIM
1140-1805
FO E J Parker
Sgt J J Nevens
FO H J Lodge
FO G H Small RCAF
WO R Taylor RCAF
Sgt A W Preston RCAF
Sgt L C Williams RCAF

2 Mar
COLOGNE
0645-1225
FO E J Parker and crew

5/6 Mar
CHEMNITZ
1635-0215
FL A F McLarty RCAF
FS D Huddlestone
FO D S Crawford RCAF
FO J M Stevenson RCAF
Sgt J Calderbank
Sgt C Brear
Sgt W Peacock
7/8 Mar
DESSAU
1650-0230
FL L K Firth and crew
8 Mar
ESSEN
1130-1700
FL L K Firth and crew
12 Mar
DORTMUND
1300-1835
FL L K Firth and crew
16/17 Mar
NURNBERG
1720-0145
FL L K Firth and crew
18/19 Mar
HANAU
0030-0230
FL L K Firth and crew
(sortie aborted when star-
board inner engine failed)
21 Mar
BREMEN
0757-1214
FL L K Firth and crew
22 Mar
HILDERSHEIM
1109-1634
FL L K Firth and crew
24 Mar
HARPENERWEG
1306-1823
FO J M Sharp
Sgt D S Broughton
FO B Andrews
FS J A Butler RCAF
WO L P Youle RCAF
Sgt S Evans

FS W G McKnight
27 Mar
PADERBORN
1433-1950
FL L K Firth and crew
3 Apr
NORDHAUSEN
1313-1946
FO V S Martin RCAF
FO J A Heaton
Sgt D N Baker
FS N E Fenerty RCAF
FO J Eisen RCAF
Sgt H L Hauxwell
FS R Gray RCAF
Sgt J Weston
4/5 Apr
LUTZKENDORF
2109-0528
FL P N Speed
PO F P Wittingstall
FO C A Meadows
FO R H Bates RNZAF
FO C D Hill
Sgt R H Fowler
Sgt J B Mitchell
9/10 Apr
KIEL
1907-0109
FL K L Firth and crew
FS H Eckershall (2P)
22 Apr
BREMEN
1544-2044
FL P H S Kilner
FS L O Spinks
Sgt G H Bridger
Sgt W G Corcoran
Sgt K P Barker
Sgt R S Mepstead
Sgt W A Pinkham
(raid abandoned by Master
Bomber due to cloud and
smoke)
29 Apr
'MANNA'/THE HAGUE
1123-1503
WO D W Veale
Sgt J H Harrison

FS L J Mountcastle
Sgt B Farren
Sgt D S Stewart
Sgt T Keegan
Sgt F H Lloyd
3 May
'MANNA'/ROTTERDAM
1201-1455
FO L Purvis
Sgt A Hardiman
WO Vollane
FS A Storey
FO Burke
FS A Storey
Sgt J Crowther
FS A Woolmer
4 May
'MANNA'/ROTTERDAM
1304-1551
FO R E Norris
WO G W Sutton
Sgt L R Pearson
Sgt E G Learoyd
Sgt W Sullivan
Sgt P A Cox
Sgt J Davies
7 May
'MANNA'/ROTTERDAM
1212-1546
PO J F Douglas
FS R W R Short
Sgt D H E Watson
FS S F Ward
Sgt J S Smith
Sgt J J Randall
Sgt V Simmonds
11 May
'EXODUS'
FO J A Heaton
Sgt A Evans
Sgt N Kirkman
FS W Edmunds RAAF
Sgt A Owen
Sgt H Crossett
Sgt J Gist

ND458
ABLE MABEL

Part of an order for 600 Lancasters from A. V. Roe at Chadderton, ND458 was a Mk III with four Merlin 38 engines, produced at the end of 1943. On 10 January 1944 this bomber was assigned to No 100 Squadron at Waltham near Grimsby, Lincolnshire, and given the squadron code letters of HW. With the individual identification letter A-Able, the aircraft quickly became known as 'Able Mabel', and as its bomb symbols grew, painted in 10-bomb rows beneath the cockpit, it was soon apparent how apt this Lanc's name was. After eight rows had been achieved, a second block was started next to the first. The name 'Able Mabel' was painted in front of the bombs, and between the two blocks of symbols two swastikas indicated two victories over German night fighters.

When ND458 started ops, the Battle of Berlin was in full-swing and its first trip was to the Big City on 20/21 January in the hands of an Australian Sergeant Pilot, K. W. Evans. Able Mabel went to Berlin four times but was forced to abort a fifth trip when the H2S failed not long after take-off. On a trip to Stuttgart on 20/21 February the aircraft was hit by flak and had its bomb-doors damaged but otherwise was fairly free of any problems.

Able Mabel had the usual variety of captains in its early days, although Warrant Officer P. R. M. Neal and Sergeant E. R. Belbin both flew a few missions in the aircraft. Then in March Warrant Officer (later Flying Officer) J. Littlewood began flying ND458, having done his 'second dickie' trip in the aircraft on 24/25 February, eventually taking it on 22 ops to complete his tour and receive the DFC.

Jack Littlewood took Able Mabel on the D-Day sorties and after this event the Lanc raided V1 sites and communication centres during the summer. Its bomb-doors were again damaged on 2 July whilst bombing a V1 site, and five days later there was a fault in the bomb-release mechanism and the bombs dropped onto the unopened doors, then fell through, but the doors sustained damage once again. This was obviously part of a dangerous period for Able Mabel, for on the last sortie for July the starboard wing was holed by flak, part of the trailing edge having later to be refitted.

One of the aircraft's combats with enemy fighters came over Russelsheim on 25/26 August. It was just on 1.30am and the Lanc was flying at 31,000ft not far off Luxembourg. The bomb-aimer spotted a twin-engined Me410 above on the starboard bow. The rear-gunner saw it too and ordered the skipper to corkscrew to starboard. The Messerschmitt and both Able Mabel's gunners opened fire at the same time, the rear-gunner seeing cannon and machine-gun blasts from the fighter. Able Mabel was hit and both turrets were put out of action but the fighter broke away to starboard and dived; it was later claimed as damaged.

Able Mabel had been severely damaged in the starboard elevator and rudder fin, hydraulics and both turrets, but the aircraft and its pilot, Pilot Officer C. D. Edge, brought them home. Able Mabel was out of action for almost a month, Cat AC.

The aircraft survived any further mishaps until it was slightly damaged by flak on 4/5 November, by which time its ops total was rising fairly well. These topped the 100 mark on 1 February 1945 with a raid on Ludwigshaven. Its last raid on 25 April, to barracks at Berchtesgaden, was Able Mabel's 127th trip, and the aircraft followed these with the first ever 'Exodus' trip on 27 April, then six 'Manna' sorties, making a possible total of 134.

Among Able Mabel's later pilots, Pilot Officer G. K. Veitch RAAF took the aircraft on 10 ops, while Flight Lieutenant J. D. Playford RCAF took it on 26, including the 100th. John (Jack) Playford had arrived on the squadron on 4 November 1944 although he went to No 582 Squadron for a few days in mid-February 1945. He received the DFC in December.

Able Mabel's rival on the squadron was Lancaster ND644 N-Nan, who would also achieve 100-plus ops shortly before Able Mabel, but who was destined to be lost in March. By early 1945 Able Mabel had enjoyed over 800 hours of virtually trouble-free flying. This was due in no small part to the efforts of its groundcrew, who had kept the aircraft almost totally free of major mechanical problems. These men were Sergeant W. Hearne, Corporal R. T. Withey, and LACs J. E. Robinson, J. Hale and J. Cowis.

1944	Sgt H B Peachey	Wookey out	**26/27 Mar**
No 100 Squadron	Sgt K Burchell	**1/2 Mar**	ESSEN
20/21 Jan	FO H G D Pawsey	STUTTGART	2005-0600
BERLIN	**28/29 Jan**	2345-0800	Sgt E R Belbin and crew
1640-2344	BERLIN	Sgt E R Belbin	Sgt F S Gaywood in, Sgt
Sgt K W Evans RAAF	0025-0813	Sgt R A H Cassell (FE)	Cassell out
Sgt J J Lapes	FS T F Cook and crew	Sgt H A Merchant	**30/31 Mar**
Sgt P Atha	**19/20 Feb**	FS K G Wilde RCAF (N)	NURNBERG
Sgt D Francis	LEIPZIG	Sgt F T Baldwin	2200-0100
Sgt A J Armstrong	2330-0650	Sgt E T Duckett	FS J Littlewood
Sgt C Brookes	WO P R M Neal and crew	Sgt J R Trueman	Sgt T McCartney (FE)
Sgt F Whitehouse	Sgt J S Ross in, Sgt Dixon out	**15/16 Mar**	FS B A Tovell
21/22 Jan	**20/21 Feb**	STUTTGART	Sgt J G Hughes
MAGDEBURG	STUTTGART	1905-0320	Sgt R W Gilbey
2003-0220	2335-0705	Sgt E R Belbin and crew	Sgt A G Girton
WO P R M Neal	WO P R M Neal and crew	**18/19 Mar**	Sgt J Taylor
Sgt R Cull	(slight flak damage)	FRANKFURT	**9/10 Apr**
Sgt C Starr	**24/25 Feb**	1910-0055	MINING/GDYNIA
Sgt R G Evans	SCHWEINFURT	Sgt E R Belbin and crew	2130-0545
Sgt H A T Warner	1810-0220	**24/15 Mar**	WO E R Belbin and crew
Sgt J Mason	WO P R M Neal and crew	BERLIN	Sgt Cassell back
Sgt G R Dixon	FS J Littlewood (2P);	1900-0205	**10/11 Apr**
27/28 Jan	Sgt J G Wookey in, Sgt Mason	Sgt E Walton	AULNOYE
BERLIN	out	Sgt R T Rutter (FE)	2345-0445
1758-0200	**25/26 Feb**	FS J O'Loughlin	WO E R Belbin and crew
FS T F Cook	AUGSBURG	Sgt T E Sanders	**18/19 Apr**
Sgt H Widdup	2120-0430	Sgt J R Taylor	MINING/BALTIC
FO E W Norman (N)	WO P R M Neal and crew	Sgt L Whitewood	2100-0420
Sgt J C Stewart (BA)	Sgt E J Duckett in, Sgt	Sgt J R Logan	WO J Littlewood and crew

20/21 Apr
COLOGNE
0006-0405
WO J Littlewood and crew
Sgt R P Anderson in, Sgt
McCartney out
22/23 Apr
DUSSELDORF
2245-0314
WO J Littlewood
PO M C Hennessey (2P)
Sgt J V Drew (FE)
FS T A Tovell (N)
Sgt W J Massey (BA)
Sgt H J Woodcraft (WOP)
Sgt S C Smith (MU)
FS M Biggs (RG)
24/25 Apr
KARLSRUHE
2217-0407
WO J Littlewood and own
crew
Sgt J V Dew in, Sgt Anderson
out
26/27 Apr
ESSEN
2300-0315
WO J Littlewood and crew
FL C N Waite in, Sgt Dew out
27/28 Apr
FRIEDRICHSHAVEN
2140-0550
WO J Littlewood and crew
Sgt McCartney back
30/1 May
MAILLY-LE-CAMP
2150-0315
PO E D King
Sgt W F Bloomfield
FO L H Salt
Sgt F W Cheetham
Sgt S Beardsell
Sgt D Ralston
Sgt D W Young
7/8 May
BRUZ
2140-0240
PO E D King and crew
Sgt R G Crabb in, Sgt
Bloomfield out
9/10 May
MERVILLE
2155-0110
PO E A Wainwright
Sgt R L Thomson
WO J M Rosborough
FS R A Brown RCAF
Sgt E H Wallington
Sgt L Cohen
Sgt J J O'Mara
10/11 May
DIEPPE
2240-0140
PO J Littlewood and crew
21/22 May
DUISBURG
2235-0250
PO J Littlewood and crew
22/23 May
DORTMUND
2245-0255
PO J Littlewood and crew
24/25 May
LE CLIPTON
2240-0105
PO J Littlewood and crew

27/28 May
MERVILLE
2347-0322
PO J Littlewood and crew
28/29 May
EU
2233-0143
PO R G Page
Sgt D Henderson
Sgt L C Roots
Sgt D M Jones
Sgt E W Hayes
Sgt J Todd
Sgt R Watson
31/1 Jun
TERGNIER
0006-0415
PO J M Shaw
Sgt W S Johnson
Sgt J C Locke
Sgt B W Young
Sgt W J Jones
Sgt J G Gorman
Sgt W Everitt
2/3 Jun
BERNAVAL
2335-0305
PO J Littlewood and crew
FL C N Waite in, Sgt
McCartney out
5/6 Jun
VARREVILLE
2120-0130
PO J Littlewood and crew
PO R P Anderson in, FL Waite
out
6/7 Jun
VIRE
2200-0300
PO J Littlewood and crew
FL C N Waite in, PO Anderson
out
7/8 Jun
FORET DE CERISY
2335-0400
PO J Littlewood and crew
PO R P Anderson and FO A B
Good in;
FL Waite and WO Tovell out
10/11 Jun
ACHERES
2320-0410
FO O S Milne
Sgt W H Lanning
FO R B Newman
FO R B Hutchinson
Sgt B Nundy
Sgt H Taylor
Sgt K Yeulett
11/12 Jun
EVREUX
0115-0515
FO O S Milne and crew
12/13 Jun
GELSENKIRCHEN
2300-0245
PO J Littlewood and crew
Sgt W Widdup in, PO
Anderson out, WO Tovell back
16 Jun
DOMLEGER/V1 SITE
1155-1500
PO T G Page and crew
22/23 Jun
REIMS
2250-0320

FO O S Milne and crew
24 Jun
LES HAYONS/V1 SITE
1610-1915
FO O S Milne and crew
27/28 Jun
VAIRES
0055-0515
FO O S Milne and crew
29 Jun
DOMLEGER/V1 SITE
1155-1500
PO J Littlewood and crew
30/1 Jul
VIERZON
2205-0315
PO J Littlewood and crew
FS J G Wookey in, Sgt Taylor
out
2 Jul
OISEMONT/V1 SITE
1215-1540
FO O S Milne and crew
(slight flak damage)
4/5 Jul
ORLEANS
2215-0400
PO J D Rees
Sgt M J Dunphy
FS J K Martin
FO J M Wilder
Sgt A Palmer
Sgt P L Daly
Sgt V E Locke
6 Jul
FORET DU CROC/V1 SITE
1855-2225
PO J Littlewood and crew
12/13 Jul
TOURS
2135-0320
PO J Littlewood and crew
Sgt Taylor back
14/15 Jul
REVIGNY
2125-0554
FO J G Evans RAAF
Sgt J Leitch
FO D G Heath
Sgt G P Clark
FS W H Preece
Sgt C C McKenna
PO J C Costigan
(raid abandoned over target
by Master Bomber, target not
identified)
28/29 Jul
STUTTGART
2125-0540
PO G K Veitch RAAF
Sgt P Pearce
FS H R Martin
PO H L Hamblin
FS K McIntyre
Sgt J H Smith
FS T E Hall
30 Jul
VILLERS-BOCAGE AREA
0640-1045
PO G K Veitch and crew
(starboard wing holed by
flak)
3 Aug
TROSSY ST MAXIM
1145-1605
PO G K Veitch and crew
Sgt L Cohen in, Sgt Smith out

4 Aug
PAUILLAC
1335-2135
PO G K Veitch and crew
5 Aug
PAUILLAC
1435-2250
FL C Holland
Sgt C A Taylor
PO J H Boyle RCAF
PO K Balster RCAF
FS E J Clarke RAAF
FS C Hill RAAF
Sgt J Thornton
7/8 Aug
FONTENOY-LE-MARMION
2120-0130
FO H J Healy RAAF
Sgt E Kitchen
Sgt S Owen
FS A S G Wilson
Sgt R Morgan
Sgt C J Webb
Sgt J R Buchanan
10 Aug
VINCLY/V1 SITE
1015-1140
FO C M Stuart RCAF
Sgt H Prince
FO R H Rix
Sgt P Burnett RCAF
Sgt P W T Dunn
FO J F Insell RCAF
Sgt S Kowal RCAF
(raid abandoned over target
due to cloud)
14 Aug
FALAISE
1215-1610
WO W T Ramsden
Sgt E G Stubbings
Sgt S T Howard
FO R P Simpson
Sgt R M Chestnutt
Sgt R J Williams
Sgt F Crompton
15 Aug
VOLKEL A/F
1020-1340
PO C D Edge
Sgt N Thorn
Sgt A F Lacey
FO R P Steel
FS J M Muller
Sgt F Twist
Sgt A Taylor
16/17 Aug
STETTIN
2115-0450
FO G K Veitch and crew
FO S Ross and Sgt I J Duffett
in;
PO Hamblin and Sgt Cohen
out
18/19 Aug
RIEME
2230-0145
PO C D Edge and crew
25/26 Aug
RUSSELSHEIM
2030-0515
PO C D Edge and crew
(attacked by night fighter
which was claimed as dam-
aged; Lancaster badly dam-
aged too)

169

23 Sep
NEUSS
1840-2340
FO T Batley
Sgt W J Williams
Sgt R J Perry
Sgt D Clay
WO D L Edgar
Sgt J E Fellows
Sgt H J Faulkner
25 Sep
CALAIS
0730-1125
FO G K Veitch and crew
(raid abandoned by Master
Bomber due to cloud)
26 Sep
CAP GRIS NEZ
1035-1315
FO G K Veitch and crew
28 Sep
CALAIS
0825-1140
PO P J McVerry RNZAF
Sgt J Fallon
FO W J Thorby
Sgt J H Denton
FS J M Carroll
Sgt T Myatt
Sgt A McNamara
(raid abandoned by Master
Bomber due to cloud)
5/6 Oct
SAARBRUCKEN
1855-0035
FO G K Veitch and crew
7 Oct
EMMERICH
1220-1610
FL D M D Brown
PO R P Anderson (FE)
FO S A Harvey
FO W Bathgate
Sgt A Halstead
FS W Bell
Sgt A G Tipple
14 Oct
DUISBURG
0645-1115
PO R Barker
Sgt A S Gordon
Sgt F S Elliott
Sgt A A Law
WO J M C Wilson
Sgt G Gillen
Sgt B G Aldred
14/15 Oct
DUISBURG
0030-0600
PO R Barker and crew
23 Oct
ESSEN
1615-2125
FO G K Veitch and crew
PO D Shrimpton (2P)
Sgt R Livingston(2AB)
PO P R Bond (AB) in
(crew of 9)
25 Oct
ESSEN
1255-1730
FO R T Hoyle
Sgt P T Bickley
PO G S Charles
Sgt R A Ward
Sgt A E Law
PO D R Hoptroff

Sgt L Marshall
28 Oct
COLOGNE
1320-1800
FO G K Veitch and crew
30 Oct
COLOGNE
1750-2325
FO R T Hoyle and crew
31 Oct
COLOGNE
1755-2305
FO R T Hoyle and crew
2 Nov
DUSSELDORF
1620-2135
FO F L Conn RAAF
Sgt A H Wilson
Sgt J R Hartshorne
FO J R Hughesdon
FS J Brady RAAF
Sgt F Hemmant
Sgt W F Hart
4 Nov
BOCHUM
1745-2215
FO D M Ward RAAF
Sgt W Hunter
FO J R L Linn
RCAF
PO F C Squires
WO J Safaruk RCAF
Sgt J I Griffiths RCAF
Sgt W Humphries RCAF
(aircraft slightly damaged by
flak)
18 Nov
WANNE EICKEL
1600-2120
FL J D Playford RCAF
Sgt H M Hadley
FO J Menagh RCAF
FO W J Elrick RCAF
Sgt R Kemp
Sgt H C Rasmussen RCAF
Sgt R L McKay RCAF
16 Nov
DUREN
1305-1730
FO G K Veitch and crew
27 Nov
FREIBURG
1615-2300
FL J D Playford and crew
29 Nov
DORTMUND
1215-1725
FL J D Playford and crew
3 Dec
URFT DAM
0800-1225
FO F L Conn and crew
4 Dec
KARLSRUHE
1645-2310
FL J D Playford and crew
6/7 Dec
MERSEBURG
1640-0050
FL J D Playford and crew
12/13 Dec
ESSEN
1630-0050
FL J D Playford and crew
15 Dec
LUDWIGSHAVEN
1450-2115

FO D H Shrimpton
Sgt G W Overton
FS J E Beath
Sgt R Livingstonwe
FS J H Gleghorn
Sgt R A Tilly
Sgt C Reilly
17 Dec
ULM
1515-2300
PO R Barker and crew
26 Dec
ST VITH
1255-1725
FL J D Playford and crew
28 Dec
MUNCHEN GLADBACH
1530-2110
FL J D Playford and crew
29 Dec
GELSENKIRCHEN
1515-2130
PO J A Seagroatt
Sgt M S A Wildman
FO H O Berger
FO S M Pleskett
FO D Costello
Sgt R Wadge
Sgt L Robinson
Sgt J Whittaker

1945
2 Jan
NURNBERG
1505-2325
FO R Barker and crew
(FO Barker and crew failed to
return 5/6 January in JB603
on its 112th op)
5/6 Jan
HANNOVER
1410-0025
WO W H Evans
Sgt J J Paxton
Sgt J L Pearson
Sgt A W Dack
Sgy T G Sutherland
Sgt K W J Hodges
Sgt B Burdett
6 Jan
HANAU
1600-2200
FS D W McKenzie RNZAF
Sgt R McLelland
Sgt F R Ford
FS W T D Allen RNZAF
FS J F Malvern
WO H E Thornby
Sgt C W Anderson RCAF
7/8 Jan
MUNICH
1835-0330
WO W H Evans and crew
14/15 Jan
MERSEBURG
1945-0355
FO F L Conn and crew
PO J O'Riordon in, Sgt
Hemmant out
16/17 Jan
ZEITZ TROGLITE
1735-0120
FO P J Whyler
Sgt H Doughty
Sgt F C Warin
Sgt A Robinson
Sgt L G Cox

Sgt H G Jones
Sgt A R E Witt
1 Feb
LUDWIGSHAVEN
1555-2235
FL J D Playford and crew
7/8 Feb
CLEVE
1855-2355
FO K V Fraser RCAF
Sgt W H Cooke
FS D J Sawer
FS J Wright
FS J Riddell
FS W V Gosleigh RCAF
FS H S Withers RCAF
13/14 Feb
DRESDEN
2150-0740
FL J D Playford and crew
FS C Daventry-Bull in, Sgt
Rasmussen out
14/15 Feb
CHEMNITZ
2010-0505
FL J D Playford and crew
FS K S Mitchell in, FS
Daventry-Bull out
20/21 Feb
DORTMUND
2155-0415
FL A R V Butler RCAF
Sgt W Daymont
FO T R Pryde
FS L Cox
FL R P Thompson
FA A R M Hart
Sgt J B Roadhouse
1 Mar
MANNHEIM
1150-1800
FL J D Playfrd and crew
5/6 Mar
CHEMNITZ
1705-0220
FO L Morrison RCAF
Sgt E Alvarez
FO A A Munro RCAF
FO D L Paterson RCAF
Sgt W B Stewart
Sgt C F Webb RCAF
Sgt H J Baker RCAF
7/8 Mar
DESSAU
1700-0250
FO L Morrison and crew
8/9 Mar
KASSEL
1730-0050
FO H Brown
Sgt J Wadsworth
WO J S Grubb
FS J S Metcalfe
Sgt R S Hall
SGt W H McGough
Sgt A Lloyd
11 Mar
ESSEN
1140-1705
WC T B Morton
Sgt A McDougall
FS W S Merrall
FO S McMichael
FS J Smith
FS S E Greenley
FS P Kenny

12 Mar
DORTMUND
1330-1845
WO W H Evans and crew
Sgt A Hopley in, Sgt Burdett
out
13 Mar
BRIN/BENZOL
1705-2315
FL J D Playford and crew
15/16 Mar
MISBURG
1705-0045
FL J D Playford and crew
16/17 Mar
NURNBERG
1730-0150
FO L Morrison and crew
18/19 Mar
HANAU
0035-0750
FO L Morrison and crew
21/22 Mar
BRUCHSTRASSE
0045-0650
FL J D Playford and crew
23 Mar
BREMEN

0720-1150
FL J D Playford and crew
25 Mar
HANNOVER
0645-1220
FO L Morrison and crew
27 Mar
PADERBORN
1500-1935
FL J D Playford and crew
31 Mar
HAMBURG
0625-1140
FO L Morrison and crew
4/5 Apr
LUTZKENDORF
2101-0531
FL J D Playford and crew
9/10 Apr
KIEL
1936-0107
FL J D Playford and crew
10/11 Apr
PLAUEN
1834-0315
FL J D Playford and crew
14/15 Apr
POTSDAM

1749-0258
FL J D Playford and crew
18 Apr
HELIGOLAND
0954-1444
FL J D Playford and crew
22 Apr
BREMEN
1512-2006
FL J D Playford and crew
(raid abandoned over target
due to cloud)
25 Apr
BERCHTESGADEN
0509-1330
FL J D Playford and crew
27 Apr
'EXODUS'/BRUSSELS
0644-0957
FL J D Playford and crew
30 Apr
'MANNA'/LEIDEN
1547-1842
FL J D Playford and crew
1 May
'MANNA'
1357-1730
WO P S Terry

Sgt H T Sharp
Sgt A Carr
FS T O Marsh
FS E E Boot
Sgt K H Eland
Sgt P F Fellows
2 May
'MANNA'
1155-1503
WO P S Terry and crew
3 May
'MANNA'
1146-1502
WO P S Terry and crew
5 May
'MANNA'
0631-0933
PO R E M Chaplin
FS V Quinn
Sgt D H Lewis
Sgt S T Reeves
Sgt T Cook
Sgt E Jones
Sgt J Taylor
7 May
'MANNA'
1249-1631
PO R E M Chaplin and crew

171

ND578
YORKER

Another A. V. Roe-built machine from order No 1807, this was also a Mk III with four Merlin 38 engines that came off the production line at the end of 1943-44. On 5 February 1944, ND578 was assigned to No 44 (Rhodesia) Squadron at Dunholme Lodge, Lincolnshire, and marked with the squadron code letters KM. It then received its individual identification letter 'Y' and soon became known on the squadron as Y-Yorker.

Yorker's first operation was a big one – to Berlin – and in the hands of a pilot who had joined the squadron the previous October. He was Pilot Officer John Chatterton and he would eventually fly Yorker on 15 ops, complete his tour and receive the DFC. He recalls:

'I shall always think of "Y" as my aeroplane. In those days a crew usually had their own – perhaps shared with one other crew – and the two of us pilots would "create" considerably if the Flight Commander let "odd-bods" fly it. Things were very different later when lots of crews were competing for aircraft.

'I well remember when joining No 44 Squadron, Squadron Leader Jack Shorthouse, Flight Commander B Flight, said, "You'll be sharing Yorker with Knight but he is senior (he had done about six trips in her predecessor – W4933) so he'll take precedence if you are both on ops the same night." There was the slight matter of a taxying accident with the chimney of a tar-boiler left dangerously close to the peri-track for which Knight was sent on a disciplinary course at Sheffield. During the fortnight he was away I was the uncontested driver of Y, so much so that when he came back he only did a couple more in Y before going missing in Z.

'We soldiered on in W4933 until the new year when the squadron was re-equipped with H2S Lancasters. My log-book shows "acceptance test" for ND578 as 12 February: rather a grand name for the first air-test but I wasn't likely to say NO to a new kite. I was a bit worried that I might have to share it with Sergeant Frank Levy who took her on his first op to Stuttgart on 20 February but fortunately for me he was posted to No 617 Squadron, but was lost with them on the way back from the Tirpitz raid.

'Halfway through my tour and maybe getting a bit big-headed, I was considerably affronted when the new Squadron Commander, Wing Commander F. W. Thompson called me in to his office and said he intended to share Y with me. A week or so later I realised what a tremendous favour he had done me – by keeping "my" aircraft out of the hands of a sprog crew, after all, the OC wasn't allowed to operate more than once in three or four weeks, so I would still be virtually the "sole owner".'

On completion of their tour, John Chatterton and his crew were decorated. Apart from John's DFC, Jack Reyland also received the DFC while Ken Letts, Jock Michie, Bill Champion, Jock Davidson and Mansel Scott each received the DFM. It was unusual for all seven men in a bomber crew to be decorated, but as John Chatterton explains:

'There was a short period in our squadron when the entire crew were decorated – I think as a sort of encouragement to others in the bad old days. I recently met two old pals from No 44 Squadron and the same thing happened to their crews.

'Occasionally we took new men on their first trips and, because second pilots usually got in the way of the smooth running of the flightdeck, I generally tried to wangle second navigators. There was plenty of room and he could actually be useful with the H2S set.

'We had a photo taken at the end of our tour and it is a great pity it doesn't show Scotty who was our faithful bomb-aimer for most of the tour, but he had finished his 30 trips so we had Barker for the last two or three. Our original bomb-aimer, Pete Lees, was lost with another crew early in our tour. We were stand-by crew one night when another captain's bomb–aimer (Scotty) went sick, so Lees went with them and we inherited Scotty.'

Wing Commander F. W. Thompson DFC, AFC, had already done a tour on Whitley bombers with No 10 Squadron in 1941, and came from No 1658 HCU to command No 44 (Rhodesia) Squadron. As John Chatterton says, Thompson also flew Yorker and he too came to regard Yorker as 'his' aircraft, which was perfectly natural. Thompson remembers:

'I always regarded Y-Yorker as my own lucky aeroplane. We always tried to fly with the same crew and that applied to the Squadron Commander along with all the others, although, of course, sickness, leave, etc, did at times dictate otherwise.

'I got so attached to Y-Yorker that I took great pains to see that she was detailed for operations with my most reliable crews. One sortie I recall, on 7 May 1944, was against a munitions factory at Salbris. On this one I was controlling and Wing Commander Leonard Cheshire was marking. The factory consisted of those large sheds with gaps of some 12ft between. We decided to mark the first gap.

'Cheshire called "Mark!" on time but I could not see any mark so I reported "No Mark". Cheshire went round again with the same results, and then for a third time. By now he and I were both getting rather cross. He said he put all three down the gap – and he had – but they were not visible at the height I was flying so I said to put one on the top of the second shed which he did and I was able to proceed.

'On 4 July we attacked flying-bomb stocks and on 15 July laid mines in the Kiel Canal. No 44 Squadron had a reputation for accurate mine-laying but only the most experienced crews were used. I also see I took Yorker on a daylight raid to Caen at the opening of the Caen operation. Apart from operations I did quite a lot of flying, always choosing to use Y-Yorker if possible. We regarded Yorker as a very special aircraft."

Wing Commander Tommy Thompson received the DSO during the summer. Thompson's usual navigator on No (Rhodesia) 44 Squadron was Flight Lieutenant – later Squadron Leader DFC – Steve Burrows, and he recalls:

'Whilst Master Bombing Brunswick, 22/23 May, we were coned for a long period together with the usual flak. Attacks from fighters were numerous and damage to the port inner resulted in its loss.

'We often took additional crew members who had not flown on ops before to let them see the pretty lights! We also operated on D-Day, but as far as I can remember were not told about it, but the radar screen soon showed us different.'

A whole variety of crews in fact flew Yorker, only a few becoming more or less regular for short periods, such as Flying Officer L. W. Hayler, who went on to fly 31 ops in the aircraft; Flying Officer R. Thomson RCAF, six ops; and Flying Officer H. V. Parkin, 11 ops.

Yorker had few operational problems, although it did lose its port outer 25 minutes after bombing Duren on 16 November, while the port inner engine had a runaway prop on 22/23 November, and then the engine seized. Yorker had a couple of skirmishes with night fighters, the first when Frank Levy took the Lanc on its third sortie to Stuttgart on 20/21 February. Flying at 23,000ft the rear-gunner spotted a Me210 slightly above on the port quarter coming in on a curving attack. Levy took evasive action as his rear-gunner opened fire at 600yds. The '210 was hit on the tail and dived away and was not seen again. The mid-upper could not fire as his turret had frozen up.

Then on 4 November, over Dusseldorf at 19,000ft, the rear-gunner saw fighter-flares to port and then two FW190s in formation crossed astern. The enemy fighters turned in on the port quarter flying a parallel course, closing in to 800yds. The pilot was ordered to corkscrew to port as the rear-gunner opened fire. One Focke Wulf fired a long burst but it was then lost in the manoeuvre. While this was going on, the mid-upper was keeping an eye on a twin-engined night fighter astern, but it did not approach.

Yorker completed its 100th sortie on 2/3 February 1945, with Hayler as skipper, and went on to accomplish 123 ops by 17/18 April, although the actual total of ops may only have been 121. The aircraft had no personal nose–marking other than the usual rows of bombs, 10 rows of 10, with another block starting behind the first, by which time Yorker was operating from Spilsby. However, someone had obviously been to London Zoo and had 'liberated' a sign which was placed inside the cockpit windows on the starboard side, which stated 'These Animals are Dangerous'.

Yorker became Cat AC on 18 May 1945 and went of to Avro's, Lincoln, for overhaul, returning on 24 May. It then went to No 75 (New Zealand) Squadron on 2 July, which was also at Spilsby, but was finally Struck off Charge on 27 October.

At the time of writing, John Chatterton's son pilots the the Battle of Britain Memorial Flight's famous Lancaster survivor, PA474, and has flown his father and a couple of his old crew in it.

1944
No 44 Squadron
15/16 Feb
BERLIN
1713-2355
PO J Chatterton
Sgt K F Letts (FE)
FO D J Reyland (N)
FS M M Scott (BA)
Sgt J Michie (WOP)
Sgt W H R Champion (MU)
Sgt J H Davidson(RG)
19/20 Feb
LEIPZIG
2355-0645
PO J Chatterton and crew
20/21 Feb
STUTTGART
2347-0730
Sgt F Levy
Sgt P W Groom
FO C L Cox
Sgt S S Peck
Sgt G M Maguire
Sgt A G McNally
Sgt D G Thomas
25/26 Feb
AUGSBURG
1831-2010
PO J Chatterton and crew
(aborted sortie due to R/T u/s; jettisoned bombs over North Sea)
1/2 Mar
STUTTGART
2259-0755
WC F W Thompson DFC AFC
FL S Burrows (FE)
FO P F Young
Sgt E Craven
PO P F Roberts
FS E V Burden
FS J Hall
10/11 Mar
OSSUN
2013-0409
WC F W Thompson
Sgt H W Carter
FO J L Gourlay
Sgt R B Taylor
PO R H Bennett
Sgt P Curtis
WO W Bowling

15/16 Mar
STUTTGART
1937-0310
PO J Chatterton and crew
FS W T Freeman (2N)
Sgt W T Freeman in, Sgt J H Davidson out
18/19 Mar
FRANKFURT
1941-0109
PO J Chatterton and crew
Sgt J Shaw in, Sgt Freeman out
22/23 Mar
FRANKFURT
1849-0034
PO J Chatterton and crew
Sgt Davidson back
24/25 Mar
BERLIN
1859-0153
PO J Chatterton and crew
FO T S Calder (2BA)
26/27 Mar
ESSEN
1942-0057
PO J Chatterton and crew
FS A O Kennedy (2N)
5/6 Apr
TOULOUSE
2029-0330
PO J Chatterton and crew
9/10 Apr
MINING/HELPOINT
2121-0556
PO J Chatterton and crew
FL S L J Mitchell (2N)
10/11 Apr
TOURS
2309-0439
PO J Chatterton and crew
11/12 Apr
AACHEN
2033-0035
SL S L Cockbain
FS S J Bristow
FL P Waterboys
FO C H McKenzie
FS A Dickson
FS D E Bracegirdle
FS J S Dean
18/19 Apr
JUVISY
2148-0339

2025-0123
PO J Chatterton and crew
FS R Riddoch (2N)
20/21 Apr
PARIS/LA CHAPELLE
2146-0233
PO J Chatterton and crew
FS E H Greatz (2N)
22/23 Apr
BRUNSWICK
2250-0511
PO J Chatterton and crew
24/25 Apr
MUNICH
2044-0632
PO J Chatterton and crew
FS W H Barker in, Sgt Scott out
26/27 Apr
SCHWEINFURT
2123-0645
FS G Baxter
Sgt D E Betterton
FS B A Rutherford
FS S Young
FS K Scholes
Sgt D A Taylor
Sgt D R Whitfield
28/29 Apr
OSLO KJELLER
2122-0514
WC F W Thompson
FL S Burrows (FE)
FS M J Stancer (N)
FO W Clegg (BA)
FS A Dicken (WOP)
FS J Hall
FO F D R Hildrew
FL G E Mortimer (Sqdn Gunnery Ldr)
1/2 May
TOULOUSE
2124-0550
PO W J Hough
Sgt H D Hoar
Sgt L Priestley
Sgt J J R Singer
Sgt H W Nichols
Sgt P Anderson
Sgt P S Hanna
7/8 May
SALBRIS
2148-0339

WC F W Thompson and crew
FO P F Roberts and FS E P Burden in;
FS Dicken, FO Hilbrew and FL Mortimer out
11/12 May
BOURG-LEOPOLD
2232-0206
SL S L Cockbain
Sgt R P Haly
FL N Woodhouse DFM
FO C H McKenzie
FL I Radmeyer DFC
FS D Bracegirdle
FS J S Dean
FL J White
(raid abandoned, target not properly marked; jettisoned 'cookie')
19/20 May
AMIENS
2304-0208
WC F W Thompson and crew
FL J Lowry in, FO Clegg out
21/22 May
DUISBURG
2252-0331
PO W J Hough and crew
22/23 May
BRUNSWICK
2234-0502
WC F W Thompson and crew
FS A J Cole (2P)
27/28 May
MORSALINES
2250-0300
PO W J Hough and crew
31/1 Jun
MAISY
2252-0312
WC F W Thompson and crew
PO Dicken back, FO Roberts out;
FO H W Mills DFM in (MU),
FS Burden out
(raid aborted due to wind and cloud)
5/6 Jun
LA PERNELLE
0129-0503
WC F W Thompson and crew
FS Burden back, FO Hall out

6/7 Jun
CAEN
0029-0301
WC F W Thompson and crew
8/9 Jun
PONTAUBAULT
2223-0301
FL J E White
Sgt A J Richeard
FS R C M Jones
Sgt T E Jenkins
PO L F A Marston
Sgt W M H Burnett
Sgt G W Nicholson
9/10 Jun
ETAMPES
2146-0220
WC F W Thompson and crew
12/13 Jun
CAEN
0009-0445
FL J E White and crew
Sgt D H Watts in, Sgt
Nicholson out
14/15 Jun
AUNAY
2222-0255
WC F W Thompson and crew
PO J E Oxborrow (2P)
PO S J Bristow in
16/17 Jun
BEAUVOIR/V1 SITE
2300-0251
WC F W Thompson and crew
21/22 Jun
WESSELING
2323-0353
FL J E White and crew
PO K J Gowing (2P)
24/25 Jun
POMMEREVAL/V1 SITE
2211-0138
PO D J Ibbotson
Sgt J R W Worrall
FS E N Greatz
FS I R Murray
Sgt K G Andrews
Sgt T W Whitehand
Sgt F A Wells
27/28 Jun
MARQUISE/V1 SITE
2314-0237
FO W J Hough and crew
(FO Hough and crew failed to
return 15 July)
7/8 Jul
ST LEU D'ESSERENT
2225-0315
WC F W Thompson and crew
12/13 Jul
CHALINDREY
2158-0630
FO R G Boswell
Sgt R E Lewis
Sgt R L Meakin
FO W Creed
FS H J Hunt
Sgt B Merry
Sgt V E Leslie
14/15 Jul
VILLENEUVE-ST-GEORGES
2201-0512
FO R G Boswell and crew
15/16 Jul
MINING/KIEL
2235-0428
WC F W Thompson and crew

FL H R Clarke in, FS Hall out
18 Jul
CAEN
0355-0721
WC F W Thompson and crew
19 Jul
THIVERNY/V1 SITE
1925-2335
SL G A Hildred
Sgt V G Bender
PO N O W Turner
FS L C Treloar
FS J M Parker
FO I P Hourigan
FO P Bradshaw
20/21 Jul
COURTRAI
2253-0233
WC F W Thompson and crew
24/25 Jul
STUTTGART
2150-0605
FO K M Davey
Sgt J H Rawcliffe
FL T H Grant
PO D J Roddle
Sgt J H Oliphant
FS S J Arnold
Sgt J Kerley
25/26 Jul
STUTTGART
2121-0629
FO S F Gale
Sgt A S Buchanan
Sgt L E Patterson
Sgt K H Irwin
WO J Brosnahan
Sgt R Marshall
Sgt T E Fardoe
26/27 Jul
GIVORS
2101-0630
FS C E Binion
Sgt D M Pearce
FS A O Kennedy
WO G M Gebhard
Sgt F W Stroud
FS A Micalchuk
Sgt H M Knox
28/29 Jul
STUTTGART
2211-0602
FL D J Dobson
Sgt A T McKenzie
Sgt J B Knight
FS A K Johnstone
FS R J Edge
Sgt D M Wilton
Sgt W J Dry
30 Jul
CAHAGNES
0611-1124
FO R G Boswell and crew
Sgt C E Loynd and Sgt R S
Routledge in;
Sgts Lewis and Merry out
31 Jul
JOIGNY
1725-2302
FO W C Freestone
Sgt G J Post
FS D A Gage
FO F R Woollen
Sgt R H Taylor
Sgt D M Wilton
Sgt E E Herders

29/30 Aug
KONIGSBERG
2027-0654
FO L W Hayler
Sgt R B Tink
Sgt A E Hearn (N)
FO E G Winterburn
Sgt B A Nash
Sgt B E James
Sgt W S H Knight
31 Aug
AUCHY-LES-HESDIN
1551-2023
FO L W Hayler and crew
1 Sep
BREST
1106-1655
FO L W Hayler and crew
3 Sep
DEELEN A/F
1518-1947
FO D S Lade
FS T Starkey
WO T Ancock RAAF
FO J A W McCullen
WO M Danahar
FS H J Conquest
Sgt M Benjamin
(sortie abandoned due to
mechanical defects; jetti-
soned bombs in North Sea)
(FO Lade failed to return 11
September)
9/10 Sep
MUNCHEN GLADBACH
0236-0729
FO R K Hart
FO E Yaxley (2P)
Sgt E S Smith
Sgt R P Green
FS W G Bell
Sgt A W Codrai
Sgt J A Spiers
FS B R Lillywhite
10 Sep
LE HAVRE
1526-1914
FO E Yaxley
Sgt J A M Davies
FO J F Woolcott
FS J H Alabaster
FO G H Evans
Sgt H Wlkinson
Sgt R J Wilder
11/12 Sep
DARMSTADT
2104-0248
FO E Yaxley and crew
12/13 Sep
STUTTGART
1858-0213
FO L W Hayler and crew
PO J M Parker in, Sgt Nash
out
17 Sep
BOULOGNE
0725-1045
FO P W Kennedy
Sgt E P P Olsen
FS J P Kelly
FO W J Jones
Sgt J E Short
Sgt C McBurney
Sgt G Cohen
18/19 Sep
BREMERHAVEN
1814-0012

FO L W Hayler and crew
Sgt Nash back;
FO R S Biggs in, Sgt A E
Hearn in
19/20 Sep
MUNCHEN GLADBACH
1849-2352
FO F E Wilson
Sgt H V Brian
Sgt F J J Dawe
Sgt E R Culley
Sgt R F Jenkins
Sgt S H Knight
Sgt W E Lansdowne
23/24 Sep
HANDORF A/F
1848-0023
FO L W Hayler and crew
(raid abandoned; ordered
not to bomb by Master
Bomber)
26/27 Sep
KARLSRUHE
0043-0732
FO M G Peel
Sgt G S W Saxby
Sgt J J Hall
Sgt S E Martinez
Sgt J A Mitchell
Sgt J A F Knowles
Sgt R Jackson
27/28 Sep
KAISERSLAUTERN
2109-0429
FO L W Hayler and crew
FO P F Young in, FO Biggs
out
5 Oct
WILHELMSHAVEN
0812-1309
FO L W Hayler and crew
FS J P Kelly in, FO Young out
6 Oct
BREMEN
1721-2219
FO W D Barlow
Sgt A H Thornalley
Sgt A E Simmonds
FS M R Fox
Sgt B S Clements
Sgt T H White
Sgt A J Wilkes
7 Oct
WALCHEREN
1216-1447
FO L W Hayler and crew
FO P F Young in, FS Kelly out
11 Oct
VEERE FLUSHING
1321-1553
FO R G Boswell and crew
14/15 Oct
BRUNSWICK
2301-0623
FO L W Hayler and crew
Sgt J Swaffield in, FO Young
out
19/20 Oct
NURNBERG
1710-0124
FO P W Plenderith
Sgt L Burton
FO C E Digham
FO A F Nicholls
FS J A Rodda
Sgt J W Syms
Sgt P W Dearman

176

28/29 Oct
BERGAN
2222-0545
FO L W Hayler and crew
Sgt Hearn back
(abandoned sortie when
unable to identify target)
30 Oct
WESTKAPELLE
1039-1318
FO L W Hayler and crew
Sgt J Swaffield in, Sgt Hearn
out
1 Nov
HOMBURG
1340-1814
FO R Thomson RCAF
Sgt D G Thorn
Sgt A W S Humber
Sgt J Smith
Sgt P R Wicks
Sgt A Lee
Sgt J F Padgett
2 Nov
DUSSELDORF
1631-2155
FO L W Hayler and crew
FO Smith (2P)
4 Nov
LADBERGEN
1753-2302
FO L W Hayler and crew
6/7 Nov
GRAVENHORST
1623-2151
FO R Thomson and crew
(raid abandoned by Master
Bomber)
16 Nov
DUREN
1249-1840
FL L W Hayler and crew
(port outer engine failed 25
minutes after leaving target)
21/22 Nov
GRAVENHORST
1245-0004
Capt G W Hirschfeld
Sgt A D Lorreith
FO P Yorke
FO D E Murphy
FS H S Jones
Sgt J Storr
Sgt J Mitchell
(Capt Hirschfeld and crew
failed to return 4 December)
22/23 Nov
TRONDHEIM
1554-2340
FL L W Hayler and crew
(port inner had runaway prop
and seized; raid later aban-
doned)
26/27 Nov
MUNICH
2351-0904

FL L W Hayler and crew
4 Dec
HEILBRONN
1615-2320
FL L W Hayler and crew
6/7 Dec
GIESSEN
1703-0004
FO H V Parkin
Sgt C A Green
Sgt E W Rowbottom
FS S H Henry
FS D W Kelman
Sgt L G Barker
Sgt G J Bredenkamp
11 Dec
URFT DAM
1217-1732
FO K A Smith
Sgt J Dent
FO R S Winters
FO A D Long
FS J R Pugh
Sgt D R Hall
Sgt R D Jones
(sortie called-off)
17/18 Dec
MUNICH
1614-0154
FO R Thomson and crew
18/29 Dec
GDYNIA
1652-0259
FL L W Hayler and crew
21/22 Dec
POLITZ
1705-0222
FL L W Hayler and crew
31 Dec
HOUFFALIZE
0224-0734
FO E V Parkin and crew

1945
1/2 Jan
GRAVENHORST
1701-0009
FO E V Parkin and crew
4/5 Jan
ROYAN
0110-0751
FL L W Hayler and crew
6 Jan
MINING/SPINACE
1614-2131
FO B T F Coventry
Sgt A Shuttleworth
Sgt K W Ayre
FS S Gibson
Sgt J O Wood
Sgt F T Perkins
Sgt G Lewis
(sortie aborted due to H2S
failure)

13/14 Jan

POLITZ
1639-0304
FL L W Hayler and crew
FS Hearn back
14/15 Jan
MERSEBURG
1620-0209
FO E V Parkin and crew
16/17 Jan
BRUX
1753-0301
FO E V Parkin and crew
1 Feb
SIEGEN
1607-2314
FL L W Hayler and crew
2/3 Feb
KARLSRUHE
1951-0258
FL L W Hayler and crew
FO H L Maltas (2P)
7/8 Feb
LADBERGEN
2106-0335
FO B T F Coventry and crew
FS A E Hearn and FO F R
Woodlan in;
Sgt Ayre and FS Gibson out
8/9 Feb
POLITZ
1658-0252
FO E V Parkin and crew
13/14 Feb
DRESDEN
1803-0404
FL L W Hayler and crew
FL H A B Symons (2P)
19/20 Feb
BOHLEN
2345-0802
FL L W Hayler and crew
FL D M Allen (2P)
FL L H Edwards in, FS Hearn
out
20/21 Feb
GRAVENHORST
2147-0414
FO L T Gardiner
Sgt J A C Ludlow
FS E J MacDonald
WO G C Beaton RCAF
Sgt T E Burroughs
FS H A Walsh
FS L S van Niekerk
21 Feb
GRAVENHORST
1659-2335
FL L W Hayler and crew
FS H Symonds in, FL Edwards
out
24 Feb
LADBERGEN
1406-1830
FL L W Hayler and crew
FO R S Biggs in, FS Symonds out
3/4 Mar

LADBERGEN
1841-0145
FL L W Hayler and crew
FS Hearn back
5/6 Mar
BOHLEN
1708-0253
FL L W Hayler and crew
FL A J MacKay (2P)
6/7 Mar
MINING/SASSNITZ
1830-0319
FL L W Hayler and crew
7/8 Mar
HARBURG
1822-0059
FO D M R Piggott
Sgt J C Cousins
FS J P Madley
FO T A V Russell
FO V G Carpenter
FS I G Evans
WO D Pinckard
11 Mar
ESSEN
1212-1744
FO E V Parkin and crew
12 Mar
DORTMUND
1339-1907
FO E V Parkin and crew
14/15 Mar
LUTZKENDORF
1700-0221
FO E V Parkin and crew
16/17 Mar
WURZEBURG
1738-0212
FO R Thomson and crew
20/21 Mar
BOHLEN
2344-0832
FO R Thomson and crew
23/24 Mar
WESEL
1929-0140
FO R Thomson and crew
7/8 Apr
MOLBIS
1834-0315
FO E V Parkin and crew
8/9 Apr
LUTZKENDORF
1758-0228
FO E V Parkin and crew
10/11 Apr
LEIPZIG
1826-0246
FO E V Parkin and crew
17/18 Apr
CHAM
2352-0829
FO E V Parkin and crew

ND644
NAN

Built at A. V. Roe's Chadderton factory as part of order No 1807, ND644 was a Mk III with four Merlin 38 engines, which came off the production line early in 1944. On 20 February it was assigned to No 100 Squadron at Waltham, near Grimsby, Lincolnshire, and given the squadron code letters HW. With an individual identification letter 'N' painted on its fuselage, the aircraft became N-Nan.

Nan began operations with a trip to Augsberg of 25/26 February but it was not an auspicious start for it had to abort over France when the starboard outer engine failed. Sorting this out Nan set off on its next op on 1/2 March, and although it reached the target, the wireless became u/s and Nan landed at Ford on the south coast of England rather than its home base.

Nan then began an almost trouble-free tour of duty, although it did lose another engine – the port outer – on 3/4 May, but saw action on D-Day and over Caen, plus all the other summer targets, which included V1 sites, German transport centres and battle area troops and positions in Normandy.

The first regular captain was Flight Lieutenant Peter Sherriff. He took Nan on 23 ops and received the DFC at the end of his tour, which he completed in mid-June, before moving on to instruct at No 28 OTU. Nan's next regular skipper was Pilot Officer W. Castle who took over the captaincy in June. He had flown his 'second dickie' trip with Sherriff on 9/10 May and went on to fly 20 missions in Nan before being tour-expired in August, following Sherriff to No 28 OTU.

Nan then had the usual variety of pilots during August/September, but then Flying Officer P. C. Eliff began flying the aircraft regularly, recording 19 ops. Unfortunately on their last together, to Stuttgart on 28/29 January 1945, Nan was hit by flak in the starboard wing and engine, the latter having to be feathered on the way back after the coolant pipe had been damaged. Philip Eliff had made two runs over the target and perhaps he had been in the flak too long, but nevertheless, he brought Nan home safely. Eliff became tour-expired in February, received the DFC and also went off to No 28 OTU as an instructor.

Originally Nan had what was described as 'a large painting of a lovely lady' on its front fuselage, which along with a steadily mounting 'score' of bomb symbols in neat rows of 10, was the aircraft's identity. However, when it went off for a major inspection on 5 August, returning on 2 September, the Lanc was repainted and although the bomb log remained, the 'lady' was irreverently painted out and not replaced.

As can be seen from the lists below, No 156 Squadron had a different policy from other units in that rather than taking a new pilot on his first trip with an experienced crew, the complete embryo crew and pilot went with just the experienced captain. It may have had some advantages but a pilot possibly nearing the end of his tour had the strain of flying a trip with a wholly untried crew – which must have had its moments.

As far as is known, Nan was the only one of these 100-op veterans to land in France during the war. Following a raid on Karlsruhe on 4 December 1944, the aircraft ran short of fuel and its pilot, Flying Officer R. G. Topliss had to land at Juvincourt to refuel before returning to base. Topliss flew Nan three times but had the well-known photograph of him with her ground crew taken following the Lanc's 112th op. As, supposedly, the 112th op came in March, and he had not flown Nan since 5/6 January, it does not help to confirm ND644's total number of sorties.

If the final score of 115 ops is correct, then Nan reached its 100th op in January 1945, although it could have been in December following a straight count from the Form 541. If the photograph is supposed to follow the last time Topliss flew the aircraft, then a further count puts Nan's total nearer 128.

Whatever the total, Nan was the pride of its groundcrew, who consisted of Sergeant H. W. Williams, LACs J. Atkinson (fitter-engines), B. Gorst (fitter-airframes) and AC F. Turrell (fitter-engines). The aircraft had flown over 800 hours by March but failed to return from a raid on Nurnberg on the night of the 16/17th. The mostly Canadian crew consisted of Flying Officer George A. O. Dauphanee, Flight Sergeant Mervyn R. Jeffrey, Flying Officers D. B. Douglas RCAF and William R. Vale RCAF, Pilot Officer R. S. Bailey and gunners Flight Sergeants W. H. Johnson RCAF and L. E. Bedell RCAF. Dauphanee, Jeffrey and Vale, as well as both gunners are known to have been killed and buried at Bad Tolz, Durnbach, Germany.

1944	Sgt K Pearton	1850-0130	Sgt S Beardsall
No 100 Squadron	FO F J Blute	FL P Sherriff and crew	Sgt D Ralston
25/26 Feb	Sgt H Flint	**26/27 Mar**	Sgt D W Young
AUGSBERG	**18/19 Mar**	ESSEN	**20/21 Apr**
1825-2335	FRANKFURT	1955-0040	COLOGNE
Sgt A R Oxenham	1925-0105	FL P Sherriff and crew	2338-0412
Sgt D J Fuller (FE)	PO A J T Armon	**30/31 Mar**	FL P Sherriff and crew
Sgt E Blackburn (N)	Sgt D B Cox	NURNBERG	**22/23 Apr**
Sgt A Goodall (BA)	FO R F Weedon	2200-0535	DUSSELDORF
Sgt R J Willis (WOP)	Sgt G R Boxall	FL P Sherriff and crew	2244-0308
Sgt J Barber (MU)	Sgt D Jones	**9/10 Apr**	FL P Sherriff and crew
Sgt D A Goggin (RG)	Sgt L D Bowden	MINING/BALTIC	**24/25 Apr**
(sortie aborted due to engine	FS M Robertson	2125-0535	KARLSRUHE
trouble)	**22/23 Mar**	FL P Sherriff and crew	2212-0404
1/2 Mar	FRANKFURT	**10/11 Apr**	FL P Sherriff and crew
STUTTGART	1900-0020	AULNOYE	**26/27 Apr**
2335-0640	FL P Sherriff	2330-0420	ESSEN
Sgt A R Oxenham and crew	Sgt J D Gray (FE)	FL P Sherriff and crew	2310-0320
15/16 Mar	FO J Galloway (N)	**18/19 Apr**	FL P Sherriff and crew
STUTTGART	FO R H L Girvan RCAF (BA)	ROUEN	GC I B Newbigging (2P)
1915-0325	Sgt T A Hammill (WOP)	2225-0230	**27/28 Apr**
FL J H Inns	Sgt G R Dixon (MU)	PO E D King	FRIEDRICHSHAVEN
Sgt C C Davies	Sgt G H Warren (RG)	Sgt R W Bland	2155-0545
Sgt E Usher	**24/25 Mar**	FO L H Salt	FL P Sherriff and crew
FO G C Tincler	BERLIN	Sgt F W Cheetham	

30/1 May
MAINTENON
2137-0205
PO E D King and crew
3/4 May
MAILLY-LE-CAMP
2155-0100
PO E Wainwright
Sgt R L Thomsen
FS J M Rosborough
FO R A Brown
Sgt R H Wallington
Sgt L Cohen
Sgt J J O'Mare
(sortie abandoned; port
outer engine u/s)
7/8 May
BRUZ
2150-0255
FO J C Kennedy RCAF
Sgt T T Bligh
FO A T Sparks
FO A D Hennessy
Sgt A Aveyard
Sgt F Shuttleworth
Sgt K Owst
9/10 May
MERVILLE
2150-0105
FL P Sherriff and crew
PO W Castle (2P)
Sgt E Bruce (FE)
FO F Tovey (N)
FO R H L Girvan (BA)
FO E Grundy (2AB)
Sgt R L Onions (WOP)
Sgt G R Dixon (MU)
Sgt H A G Knellor (RG)
10/11 May
DIEPPE
2235-0125
FL P Sherriff and crew
21/22 May
DUISBURG
2235-0245
FL P Sherriff and crew
22/23 May
DORTMUND
2250-0245
FL P Sherriff and crew
24/25 May
AACHEN
2235-0055
FL P Sherriff and crew
27/28 May
MERVILLE
2340-0315
FL P Sherriff and crew
28/29 May
EU
2235-0139
FL P Sherriff and crew
31/1 Jun
TERGNIER
2345-0423
PO W L Smith
Sgt J E Connolly
FO R V Barnett
FO H D Muir
Sgt J W Whiteside
Sgt D Stott
Sgt F J Graham
5/6 Jun
CAISBECQ
2120-0120
FL P Sherriff and crew
FL C N Waite and FS L F

Gladman in;
Sgt J D Gray and FO J
Galloway out
6/7 Jun
ACHERES
2220-0310
FL P Sherriff and crew
FL O Towers (2BA) in
7/8 Jun
FORET DE CERISY
2330-0345
FL P Sherriff and crew
FS R W Pye in, Sgt Dixon out
10/11 Jun
ACHERES
2255-0340
FL P Sherriff and crew
11/12 Jun
EVREUX
0115-0510
FL P Sherriff and crew
14/15 Jun
LE HAVRE
2050-0005
PO W Castle
Sgt E Bruce
Sgt E Palfreyman
FO E Grundy
Sgt R L Onions
Sgt A Melrose
Sgt H A G Kneller
16/17 Jun
DOMLEGER/V1 SITE
0020-0405
PO W Castle and crew
22/23 Jun
REIMS
2245-0310
FO H H Reid RCAF
Sgt D Judson
FO W J Smith RCAF
Sgt W A MacDonald RCAF
Sgt K E Nottage
Sgt S Krawchuck RCAF
Sgt C W Martens RCAF
24 Jun
LES HAYONS/V1 SITE
1600-1920
FO H H Reid and crew
25 Jun
LIGESCOURT/V1 SITE
0740-1045
FO H H Reid and crew
27/28 Jun
VAIRES
0050-0235
FO H H Reid and crew
29 Jun
DOMLEGER/V1 SITE
1145-1510
PO W Castle and crew
30/1 Jul
VIERZON
2205-0335
PO W Castle and crew
2 Jul
OISEMONT/V1 SITE
1220-1530
PO W Castle and crew
4/5 Jul
ORLEANS
2220-0410
PO W Castle and crew
6 Jul
FORET DU CROC/V1 SITE
1855-2225
PO W Castle and crew

7 Jul
CAEN
1940-2320
PO W Castle and crew
12/13 Jul
TOURS
2120-0340
PO W Castle
PO J R Jones (2P)
Sgt T I Scott (FE)
FS A Roxby
FO E Grundy (BA)
FS A H Lawry (WOP)
Sgt A Melrose (MU)
Sgt J M McCarthy (RG)
18 Jul
SANNEVILLE/CAEN
0335-0725
PO W Castle and crew
18/19 Jul
SCHOLVEN
2315-0305
PO W Castle and crew
20/21 Jul
COURTRAI
2350-0305
PO W Castle and crew
25 Jul
COQUEREAUX/V1 SITE
0700-1035
PO W Castle and crew
25/26 Jul
STUTTGART
2135-0600
PO W Castle and crew
28/29 Jul
STUTTGART
2125-0540
PO W Castle and crew
31/1 Aug
FORET-DE-NIEPPE/V1 SITE
2205-0124
PO W Castle
FO D M D Brown (2P)
Sgt J K Wood (FE)
Sgt E Palfreyman (N)
FO W Bathgate (BA)
Sgt R L Onions (WOP)
Sgt A Melrose (MU)
Sgt H A G Kneller
Sgt W Tipple (2MU)
3 Aug
TROSSY-ST-MAXIM
1130-1605
PO W Castle and crew
4 Aug
PAUILLAC
1330-2130
PO W Castle and crew
5 Aug
PAUILLAC
1430-2230
FO H Hassler RAAF
Sgt W F Dorman
FO H W Craig
Sgt J H Challis
Sgt A Christopher
Sgt A Mills
Sgt D Nathan
7/8 Aug
FONTENOY
2105-0027
PO W Castle and crew
10 Aug
VINCLY
1015-1420
WO W T Ramsden

Sgt E G Stebbings
Sgt S T Howard
FO R S Simpson
Sgt R A Chestnutt
Sgt R J Williams
Sgt F Crompton
(raid aborted by Master
Bomber due to cloud)
11 Aug
DOUAI
1335-1745
WO W T Ramsden and crew
12 Aug
LA PALLICE
1125-1745
FO I S Bell RCAF
Sgt T A Kewley
PO T N Shewring RCAF
FO R Watson RCAF
Sgt E A Pocock RCAF
Sgt T Brannon
Sgt C Barker
15 Aug
VOLKEL A/F
1000-1325
FO H Hassler and crew
5 Sep
GILZE RIJEN A/F
1655-2030
FO J R Copland RNZAF
Sgt D G Bayliss
Sgt K Martin
FS H Taylor
Sgt G A Lavings
Sgt K Sawkins
Sgt E Sapsed
6 Sep
LE HAVRE
1730-2110
FO P C Elliff
Sgt F W Tompkins
FO W B Hartnett
FO N P G Wallace RCAF
PO R P Hughes
Sgt V C Kite
Sgt K F Adams
8 Sep
LE HAVRE
0635-1015
FO P C Elliff and crew
(raid abandoned by Master
Bomber due to low cloud)
10 Sep
LE HAVRE
1700-2030
FO F O Parkinson
Sgt S Goodfellow
FO W A Cotton
FS C W Knight
Sgt E T P Reardon
Sgt W Robertson
Sgt C L Wood
12/13 Sep
FRANKFURT
1840-0145
FO J R Copeland and crew
16/17 Sep
HOPSTEN A/F
2145-0125
FO H Hassler and crew
17 Sep
FLUSHING
1625-1915
FO H Hassler and crew
20 Sep
CALAIS
1505-1850

180

FO C D Edge
Sgt N Thorn
FS P Lacey
FS R P Steel
FS J M Muller
Sgt F Twist
Sgt A Taylor
23 Sep
NEUSS
1840-2350
FO J R Copeland
25 Sep
CALAIS
0735-1105
FO P C Elliff and crew
(raid abandoned by Master
Bomber due to low cloud)
26 Sep
CAP GRIS NEZ
1025-1345
FO P C Elliff and crew
27 Sep
CALAIS
0850-1210
FO P C Elliff and crew
28 Sep
CALAIS
0820-1130
FO P C Elliff and crew
(raid abandoned by Master
Bomber due to cloud)
5/6 Oct
SAARBRUCKEN
1840-0045
FO P C Elliff and crew
7 Oct
EMMERICH
1210-1610
FO P C Elliff and crew
14 Oct
DUISBURG
0625-1100
FO P C Elliff and crew
14/15 Oct
DUISBURG
0020-0535
FO P C Elliff and crew
15 Oct
WILHELMSHAVEN
1745-2140
FO P C Elliff and crew
19/20 Oct
STUTTGART
1710-2315
FO E A Jackson
Sgt H W Cooke
FS C J Collier
Sgt J Irwin
Sgt P A Hopkins
Sgt J C Haggerty
Sgt J Hulsman
23 Oct
ESSEN
1635-2205
FO E A Jackson and crew
25 Oct
ESSEN
1250-1710
FO P C Elliff and crew
28 Oct
COLOGNE
1320-1825
FO P C Elliff and crew
30 Oct
COLOGNE
1755-2350
FO P C Elliff and crew

1 Oct
COLOGNE
1755-2255
FO P C Elliff and crew
2 Nov
DUSSELDORF
1625-2145
FO O Lloyd-Davies
Sgt K P Doughty
FO R Robinson
FS B Sandberg
WO R S Porter
Sgt E Sutherland
Sgt S R Vickers
4/5 Nov
BOCHUM
1725-2210
PO E Smith
Sgt E R Aldridge
Sgt E Barnett
FS P Evans
Sgt J Moores
Sgt E C Turner
Sgt J Hodgkiss
6 Nov
GELSENKIRCHEN
1200-1630
FO F I Truman
Sgt E J Motley
PO S L Caddey
FS T A Patterson
Sgt J E Hedley
Sgt W F Taylor
Sgt J Smart
11/12 Nov
DORTMUND
1550-2130
FO F I Truman and crew
18/19 Nov
WANNE EICKEL
1550-2115
FO E A Jackson and crew
Sgt J G A Isaac in, Sgt Cooke out
21 Nov
ASCHAFFENBURG
1535-2210
FS H Brown
Sgt J Wadsworth
FS J S Grubb
FS B S Metcalfe
Sgt R S Hall
Sgt H McGough
Sgt A Lloyd
27 Nov
FREIBURG
1605-2305
FO F T Quigley RCAF
Sgt F Stovell
Sgt R G Roller RCAF
FO W M Chapman RCAF
Sgt J Guy
Sgt M McMaster RCAF
Sgt J B Gibbons RCAF
29 Nov
DORTMUND
1220-1725
FO H G Topliss RCAF
Sgt H G Crowley (FE)
FL P V Knight
Sgt D L Fahey RCAF (N)
Sgt A L Pheiffer
FO J T Rock RCAF
Sgt V R Mason RCAF
Sgt G W McIntosh RCAF
3 Dec
URFT DAM
0740-1210

FO P C Elliff and crew
(raid abandoned by Master
Bomber)
4 Dec
KARLSRUHE
1640-2205
FO H G Topliss and crew
(landed at Juvincourt, France,
short of petrol)
6/7 Dec
MERSEBURG
1645-0045
FO P C Elliff and crew
12 Dec
ESSEN
1655-2225
FO L A M Fludder RAAF
Sgt L T A Williams
Sgt F B Barnes
FS G V Armstrong
FS C G Mazlin
FS I J Duffett
Sgt J Eveleigh
17 Dec
ULM
1535-2310
PO H C Smith RCAF
FO H O Berger
FS W S Merrall RCAF
FO S McMichael RCAF
Sgt J Smith
FS S E Greenley RCAF
FS P Kenny RCAF
21 Dec
BONN
1500-2040
FO P C Elliff and crew
FS F F Wright (2P)
Sgt A F Smith (2AB)
24 Dec
COLOGNE
1435-2010
FO F T Quigley and crew
FO H O Berger in, Sgt Stovell
out
(attacked by night fighter, no
damage)
26 Dec
ST VITH
1310-1725
FL C S Johnson
Sgt C Albutt
FO W Hancock
FO T G Campion
Sgt R C Vicker
Sgt R J Barnham
Sgt R W Cousins
28 Dec
MUNCHEN GLADBACH
1540-2140
PO H C Smith and crew
Sgt A McDougall in, FO
Berger out
29 Dec
GELSENKIRCHEN
1500-2125
FO P M Bunn
Sgt D J J Timm
FO R J Holford
FO R E Marsh
Sgt J E Benton
Sgt R Poulson
Sgt W Muir

1945
5/6 Jan
HANNOVER

1900-0010
FO H G Topliss and crew
16/17 Jan
ZEITZ TROGLITZ
1745-0120
FS F F Wright
Sgt K N Whitney
Sgt A Gibbons
Sgt A F Smith
Sgt R D Davies
Sgt J C Wallace RCAF
Sgt G R Youngs
28/29 Jan
STUTTGART
1945-0250
FL P C Elliff and crew
(damaged by flak; lost an
engine on homeward leg)
1 Feb
LUDWIGSHAVEN
1610-2255
FO M A Austin RNZAF
Sgt K R Noble
Sgt C Firth
FO T E Girling RCAF
Sgt K O Whittingham
Sgt A Hopley
Sgt L W Bruce
2/3 Feb
WIESBADEN
2045-0335
FL C S Johnson and crew
13/14 Feb
DRESDEN
2125-0700
FL W O Nobes RCAF
Sgt J D Kerr
FO J Kimpton RCAF
FS N L Warner RCAF
FS R A Doherty RCAF
FS C D Taylor RCAF
FS J Mitchell
14/15 Feb
CHEMNITZ
2000-0510
FL C S Johnson and crew
20/21 Feb
DORTMUND
2130-0410
FL C S Johnson and crew
21/22 Feb
DUISBURG
1925-0150
FL C S Johnson and crew
23/24 Feb
PFORZHEIM
1600-0005
FO H C Smith and crew
1 Mar
MANNHEIM
1145-1815
FL C S Johnson and crew
5/6 Mar
CHEMNITZ
1650-0150
FL C S Johnson and crew
8/9 Mar
KASSEL
1730-0105
FO G A O Dauphanee RCAF
FS M R Jeffrey
FO D B Douglas RCAF
FO W R Vale RCAF
PO R S Bailey
FS W H Johnson RCAF
FS L E Bedell RCAF

11 Mar
ESSEN
1145-1655
FO F F Wright and crew
12 Mar
DORTMUND
1300-1742

FO F F Wright and crew
(sortie abandoned due to
port inner engine going u/s
over French coast)
15/16 Mar
MISBURG
1720-0115

FO G McTavish RCAF
Sgt F Hurt
FO S N Lipsey RCAF
FO B Y Sisson RCAF
WO P N Leveille RCAF
Sgt A R Brooks RCAF
Sgt A H Boswell RCAF

16/17 Mar
NURNBERG
1755-
FO G A O Dauphanee and
crew
(failed to return)

ND709
FLYING KIWI

Another Chadderton-built Mk III, part of A. V. Roe's order No 1807, ND709 came off the production line in early 1944, equipped with four Merlin 38 engines. It moved to No 32 MU at Wyton on 2 March and from there it was assigned to No 35 PFF Squadron at Graveley, Huntingdonshire, on the 7th. Coded TL-M it flew just two operations as a supporter aircraft then, along with eight other Lancasters, ND709 was re-assigned to No 635 Squadron which was formed on 20 March from B Flight of No 35 Squadron, going to Downham Market, Norfolk. Here the squadron code was changed to F2 and the individual identification letter to J (although a faded M remained visible just ahead of the port elevator).

Initially ND709's pilot on No 635 Squadron, which was also part of No 8 Group's Pathfinder Force, was Squadron Leader J. R. Wood, but his first trip in the aircraft had to be aborted when the starboard outer failed. On the next sortie ND709 succeeded in reaching the target, but Warrant Officer J. M. Bourrassa had to fly the aircraft back on three engines again.

As PFF squadrons generally had crews whose ranks were increased by one level above main force crews, ND709 could count senior ranks among its pilots, so that when eventually it became the regular mount of the A Flight Commander, D. W. S. Clark DFC, he was of wing commander rank. Although David Clark was born in Surbiton, Surrey, he was a New Zealander, and had initially joined the RAF but was destined to transfer to the RNZAF in August 1944. He had already completed a tour of ops with No 419 (RCAF) Squadron, flying Halifaxes, with whom he had won the DFC.

It was Clark and his crew who thought up the emblem which was painted on the nose of this Lancaster, consisting of a Kiwi with an Aldis sight attached to its beak (and a plaster on its behind!) astride a falling bomb. On the bomb was a maple leaf, which referred to Clark's Canadian bomb-aimer. In fact his crew formerly had been that of Squadron Leader Wood, but Clark took them over when Wood completed his tour. Unlike most other Lancs, ND709's bomb-tally seems to have been painted on from the bottom upwards. The first three rows, each of 15 bombs, were lower than normal and as the space was used up, so the subsequent rows were mounted on top. Therefore, when the aircraft eventually topped the 100 mark, the bomb-tally had risen to the front gun turret.

Wing Commander Clark had a memorable first trip in his 'Kiwi', a sortie to Duisburg on 21/22 May 1944. En-route to the target, they were fired on by a 'friendly' four-engined aircraft and both Clark and his Canadian navigator, Pilot Officer Harry P. Laskowski were wounded, although not so seriously that they could not carry on. Clark flew to the target, bombed it on time then headed back only to have a tyre burst on landing, but they got away with it. Both men were taken to Ely Hospital, Clark with a bullet in the shoulder, Laskowski with bullet splinters to his back.

Both men were back on strength a month later, and Clark was later to receive a bar to his DFC, while Laskowski received the DFC. Clark went on to fly 16 ops in ND709, including at least one acting as Master Bomber.

By this time, the Kiwi's bomb tally was rising. The aircraft operated on D-Day night against V1 sites as well as on the Caen breakout. During the summer it was hit by flak on a few occasions, once with Clark in command during a sortie to Stettin on 16/17 August. Over the target they were coned by at least 20 searchlights and heavy flak hit the port outer engine. Clark had to order the marker flares jettisoned safe, but carried on and bombed on H2S. Only a few raids later Kiwi was hit again over Soesterburg, a fairly big hole being blasted through the starboard side of the fuselage.

There then followed a period of relative calm despite numerous sorties, but then on 28 October, Kiwi was hit again over the target – Cologne this time – damaging both starboard engines. With the port engines beginning to overheat in consequence, Flying Officer R. W. Toothill decided to put down in Belgium (the only 100-op veteran to land in Belgium during the war). Despite the aircraft suffering from a frozen-up ASI both to and from the target, he put down safely on the disused airfield at Moorsele. After repairs he flew back to base on the 30th.

By the new year, ND709's sorties were fast approaching the 100 mark. Squadron Leader P. R. Mellor took it to Oberfeld on 4 February, which was the aircraft's 96th op, and the Squadron Commander, Wing Commander S. Baker DSO, DFC and Bar, flew as Master Bomber in Kiwi on the 7th to Cleve – the 97th op. Two ops later, one being flown by Flying Officer J. D. F. Cowden, who took ND709 on a total of 13 trips, brought the aircraft up to 99. Peter Mellor and crew flew it to Chemnitz on 14/15 February, Kiwi's 100th and last trip with No 635 Squadron.

By one of those strange twists of fate, Cowden and crew flew in another Lanc that night (PB287) and failed to return. Cowden may even have been down to fly ND709 that night, for in the Form 541 he and his crew are listed as flying this aircraft, but obviously they did not. However, in the Form 540 Cowden is show as being lost in aircraft T, while Mellor did the 100th op of J.

Peter Mellor, who was to complete his tour and win the DFC (all of his crew were also decorated), remembers:

'The [100th] operation itself was uneventful. It was a clear night until we approached the target where there was a thin overcast covering the aiming point and, as we could not identify it, we retained the markers and dropped the 4,000-pounder into the glow in the clouds, then returned home. The only opposition was some scattered AA fire.

'My main feeling on returning was one of relief that I had brought back the aircraft undamaged, as I seemed to act as a magnet for any pieces of scrap metal in the sky.

'J was not really my aircraft; it had belonged to Wing Commander Clark, a New Zealander, who had a Canadian bomb-aimer, hence the insignia. After he was tour-expired the aircraft became something of an orphan but it was treated with respect by those who flew it, which included myself. It was ironic that I had it then, because my aircraft – LM524-G – was also in the nineties, so there was naturally some rivalry. As G was damaged beyond repair by fighters over Dessau on 7 March, this could be considered a recompense.

'As far as celebrations were concerned, the only thing I recall was a toast in rum to J at debriefing but I don't doubt that the groundcrew did some celebrating.'

Wing Commander 'Tubby' Baker, who was about to receive a bar to his DSO in early 1945 (and he himself completed 100 operational sorties during the war), remembers the Cleve sortie:

'The trip to Cleve with me as Master Bomber on 7 February 1945 was a very successful raid in close support of the Army, and Major Johnny Mullock flew with me to observe the flak. He was attached to HQ No 8 (PFF) Group as Flak Liaison Officer, which may or may not have been his correct appointment title.

'There were no press-boys around when ND709 returned from her 100th sortie, and the photos of her were taken by the Station Photographic Section, probably the next day, with her groundcrew.'

The Liaison Officer, Major John B. Mullock of the Royal Artillery had won the MC earlier in the war, and had only recently been awarded the RAF's DFC for flying on ops while doing his Headquarters job.

Several references to ND709 virtually end its story with the 100th sortie, but following a brief respite, it was sent to No 405 (Vancouver) Squadron RCAF, which operated out of Gransden Lodge, Bedfordshire. No 405 Squadron was also part of No 8 (PFF) Group, so ND709 continued to be a Pathfinder aeroplane.

Re-coded LQ-G, it arrived on 19 March and flew its first operation with the new unit on the 22nd – exactly a year to the day the aircraft had flown its first sortie with No 635 Squadron. In all it flew eight sorties with 405, then added two 'Manna' and one 'Exodus' operations, bringing the total ops to 111.

On 11 June 1945 the aircraft returned to its original unit, No 35 Squadron, but on 27 July went to No 1667 CU. Just before Christmas ND709 moved again, this time to No 1660 CU where it served for nearly a year, before moving again to No 1653 CU from 9 November 1946, being coded A3-U. On 9 May 1947 the aircraft was delivered to No 15 MU where it was finally Struck of Charge on 28 August.

1944
No 35 Squadron
15/16 Mar
STUTTGART
1932-0250
WO F G Tropman
FS R W Bullen
Sgt J L G Marshall
FS N W Curtis
FS J L Stevens
FS M E Ladyman
Sgt I K McGregor
18/19 Mar
FRANKFURT
1928-0036
WO J M Bourassa
FS N Rowell
FO G M Lockie
Sgt D Beaumont
Sgt R A Edie RCAF
Sgt R H Chapman
Sgt J B Fletcher

No 635 Squadron
22 Mar
FRANKFURT
1855-2245
SL R P Wood
PO H P Laskowski RCAF(N)
FL G D Linacre (BA)
PO D G Coltman
FS J Smith
WO D R Tulloch
Sgt J A Rayton
(returned early with starboard outer u/s)
26/27 Mar
ESSEN
1955-0045
WO J M Bourassa and crew
(came home on three engines)
30/31 Mar
NURNBERG
2200-0435
SL R P Wood and crew
10/11 Apr
LAON
0135-0545
FL C J K Ash
FS R J Birtles
FO C S Purkiss
Sgt W C Vessey
Sgt R H Chapman
Sgt J J Leisham
PO T W Hone
(FL Ash and crew failed to return 11 June)
11/12 Apr
AACHEN
2045-0015
PO R W Beveridge
FO M I Massey
FO J G Irwin
Sgt J J Mather
Sgt J D Smith
FO J Allinson
Sgt A R Hall
18/19 Apr
ROUEN

2205-0220
SL R P Wood and crew
20/21 Apr
OTTIGNIES
2135-0115
SL R P Wood and crew
22/23 Apr
LAON
2120-0120
SL R P Wood and crew
26/27 Apr
ESSEN
2315-0300
SL R P Wood and crew
27/28 Apr
FRIEDRICHSHAVEN
2210-0535
FL H P Connolly
FS H S Troy
FO H J Morley
Sgt A Hambley
Sgt K A Harder
Sgt J McLaughlin
Sgt S L Conley
1/2 May
MALINES
2225-0105
SL R P Wood and crew
3/4 May
MONTDIDIER A/F
2240-0210
FL H P Connolly and crew
6/7 May
MANTES/GASSICOURT
0050-0400
SL R B Roache
FL G A Stocks
FO J C Wells
FS C Chadwick (WOP)
FS D H P Womar
FO C G Whittaker
FS E H Barry
8/9 May
HAINE-ST-PIERRE
0140-0440
SL R P Wood and crew
FS H A Rudd in, FS Smith out
19/20 May
ORLEANS
2300-0050
FS D Griffiths
FS E J Waspe
FS E L C Howell
FS H P Whitehead
Sgt R Glass
FS W H Iball
Sgt A C Harding
21/22 May
DUISBURG
2250-0300
WC D W S Clark DFC
PO H P Leskowski
FL G D Lincacre RCAF
FS H A Rudd
FL J Highet
WO D R Tulloch
Sgt J A Rayton
(pilot and navigator wounded by fire from another bomber, but carried on and

bombed the target)
22/23 May
DORTMUND
2245-0221
FL J Billing
PO J E Moriarty
FS J Campbell
WO R D Curtis
PO J B Findlay
Sgt S L Edwards
PO T W Hope
27/28 May
RENNES
2315-0408
FO J Caterer
Sgt M F Haberlin
PO W C Shepherd
Sgt H Scutt
Sgt W J Beeson
Sgt D Farrell
Sgt L Benson
3/4 Jun
CALAIS
0030-0225
FO A L Johnson RAAF
WO J S Williams
FS W O Paice
PO J E L Liddle
PO E L Perman
FS R E Harrop
Sgt J Silburn
5/6 Jun
LONGUES
0300-0635
FO A L Johnson and crew
7/8 Jun
FORET DE CERISY
2340-0335
FO A L Johnson and crew
8/9 Jun
ALENCON
2200-0235
FO A L Johnson and crew
9/10 Jun
RENNES
0005-0500
FL H P Connolly and crew
11/12 Jun
TOURS
2146-0307
FO A L Johnson and crew
16/17 Jun
RENNESCURE/V1 SITE
2355-0245
FO A L Johnson and crew
24/25 Jun
MIDDLE STRAETS/V1 SITE
0025-0230
FS L J Melling
FO R F Watkins
FS L Bell
WO H R S Sullivan
Sgt E G Ostime
Sgt W H Hitchcock
Sgt J E Blyth
4 Jul
DOMLEGER/V1 SITE
1230-1550
WC D W S Clark and crew
WO C W Newman in, FL

Highet out
5/6 Jul
WIZERNES/V1 SITE
2300-0125
WC D W S Clark and crew
6 Jul
COQUEREAUX/V1 SITE
1925-2245
PO F L Boyd
Sgt H A Sygrove
FO E G Thomas
FS C L Ransford
FO D W Taylor
Sgt J MacLean
Sgt C H A Younger
7/8 Jul
CAEN
2015-0005
WC D W S Clark and crew
SL B Moorcroft
12 Jul
VAIRES/V1 SITE
1805-2155
WC D W S Clark and crew
14 Jul
REVIGNY
2155-0535
SL R B Roache and crew
FS A D Morrison
FS D R Pattersen
FL J C Wells
PO C Chadwick
FS D H P Womer
FL C G Whittaker
PO E H Barry
15/16 Jul
NUCOURT/V1 SITE
2345-0345
WC D W S Clark and crew
18 Jul
SANNERVILLE/CAEN
0405-0715
WC D W S Clark and crew
18/19 Jul
WESSELING
2310-0253
PO R L Vines
FO J D Hogg
SL R G Goodwin
FO J G Ramsden
FS J F Palen
Sgt G Herrick
Sgt J H Egan
20 Jul
FERMES DE GRAND BOIS /V1 SITE
1350-1635
FO A L Johnson and crew
24/25 Jul
FERFAY/V1 SITE
2150-0545
FO A L Johnson and crew
(raid abandoned by Master Bomber over target; Johnson was Deputy Master Bomber)
25 Jul
ARDOUVAL/V1 SITE
0700-1000
FO G S Henderson
WO R Pedrazzini

FS A T Till
Sgt J H C Ross
Sgt J H Morgan
FS F R Holledge
Sgt A V Urquhart
27 Jul
LES HAUT/V1 SITE
1720-2050
PO R M Clarke
FS W T Pethard
FS G K Hendy
Sgt C D Mountain
Sgt J H Watson
Sgt R E Catt
Sgt T Robertson
28/29 Jul
HAMBURG
2250-0335
WC D W S Clark and crew
1 Aug
LE HEY/V1 SITE
1930-2150
WC D W S Clark and crew
(raid abandoned by Master
Bomber due to fog and
cloud)
4 Aug
TROSSY ST MAXIM
1100-1455
WC D W S Clark and crew
(aircraft hit by flak in fuselage
and starboard elevator)
6/7 Aug
SPECIAL OBS FLT
2155-0021
SL F Smith RAAF
FL C G Whitehead
SL J R Dow
FL C J Fry
FS A H Mullard
FS C A Bradshaw
PO J A Wilson
7/8 Aug
SE CAEN
WC D W S Clark and crew
2140-0055
(did not bomb due to cloud
and smoke)
8 Aug
FORET DE CHANTILLY
1915-2250
SL M M Henderson
FO A H Emmott
FL J F Craik
FL J Highet
FO W J Hanks
FO A A Joseph
WO R Gardner
10 Aug
DUGNY
0945-1350
FL I B Hayes
FL W M Douglas
PO J N Steel
FS R B Warner
FS J W Emms
FS A D Clayton
Sgt B R McMaster
11 Aug
LENS
1415-1735
WC D W S Clark and crew
12/13 Aug
RUSSELSHEIM
2145-0250
SL D T Witt
SL P W G Lester

FL R W Coutts
WO S R J Harper
FS R Stuart
FS C Shaw
Sgt P Cronin
14 Aug
FALAISE
1240-1545
WC D W S Clark and crew
16/17 Aug
STETTIN
2115-0510
WC D W S Clark and crew
FS G T King in, FL Linacre out
(aircraft hit by flak on bomb
run)
25/26 Aug
RUSSELSHEIM
2055-0413
FO F L Boyd and crew
FL J C Wells
26/27 Aug
KIEL
2015-0115
WC D W S Clark and crew
Sgt D L A Twaddle in, FS King
out
31 Aug
LAMBRES
1325-1550
WC D W S Clark and crew
Sgt G Swindle in, WO
Newman out
3 Sep
SOESTERBURG
1605-1902
FL D B Williams
FS K O Handcock
FS E F Casbolt
FO F E Prebble
FS R Perry
Sgt J E Graham
Sgt J L S Robinson
(aircraft hit and damaged by
flak)
6 Sep
EMDEN
1640-2000
FL H M Johnston
PO D L Venning
PO K Usher
PO R T Padden
FO C H Brown
FS J K Ledgerwood
FS G Williams
10 Sep
LE HAVRE
1700-2004
FL A L Johnson and crew
PO J V Watson and WO R
Vere in;
POs Liddle and E L Perman
out
12 Sep
GELSENKIRCHEN
1137-1445
FO J A Rowland
FS C D McKenzie
FO H R Sindall
FO J R Donald
Sgt D A Jefferson
Sgt R Whybrow
Sgt W R Hill
15/16 Sep
KIEL
2222-0212
FL J J Lowry

FO R F Watkins
FL L Bell
WO H R S Sullivan
FS E G Ostime
FS W H Hitchcock
Sgt J E Blyth
26 Sep
CAP GRIS NEZ
1130-1338
FL I B Hayes and crew
27 Sep
CALAIS
0750-1240
SL D T Witt and crew
6 Oct
DORTMUND
FL L Henson
1715-2211
PO E J D Bill RCAF
FO H S Davis
FL C J Fry
FL C A Harding
FS A H Mullard
PO J A Wilson
12 Oct
FT FREDERIK HENDRIK
0705-0920
FO J D F Cowden
PO J R S Donohue
FL J F Craik
FL G R Hawes
FS B Botterhill
FS J T McQuillan
FL C G Whittaker
FO W Gabbott
14 Oct
DUISBURG
0718-1115
FO P E Cawthorne
Sgt G Wilson
Sgt B G Roberts
FS T Reid
Sgt R V Moore
Sgt I J Kinney
Sgt J Goulbourn
15 Oct
WILHELMSHAVEN
1725-2145
FO R K Westhope
FO M H A Hawrylak
FO H E Odlam
Sgt A Newby
Sgt D W Beach
Sgt J Bromley
Sgt D E Darvell
19 Oct
STUTTGART
1741-2329
FL G W Johnson
PO V Murphy
Sgt H M Smith
WO H Whittaker
FS L A W Howes
FS R B Benton
FS W Telford
23 Oct
ESSEN
1655-2210
FL G W Johnson and crew
FS G Smith
25 Oct
ESSEN
1325-1710
PO C L Ottaway
FS A W Brown
FS S J Booth
FS W G Blackburn

FS J R Pierce
FS G M Cornett
Sgt R S Grist
28 Oct
COLOGNE
1405-
FO R W Toothill
Sgt J A Davies
FO J B Luard
Sgt W W Colvin
Sgt F W Stone
Sgt F W Coombs
Sgt S H Fortune
(landed at Moorsele,
Belgium, with damaged
engines; returned on the
30th)
4 Nov
BOCHUM
1757-2229
FO G A Thorne
FO G N Rose
WO R M Keary
FO B Bresshof
FO G M Suttie
Sgt N M Scott
Sgt T J A Raymont
Sgt J H Parker
16 Nov
DUREN
1335-1720
FO J D F Cowden and crew
WO H A Rudd in, FL
Whittaker DFC out;
FL R A Boddington
21 Nov
WORMS
1615-2140
FO R W Toothill and crew
27 Nov
NEUSS
1755-2205
FO J D F Cowden and crew
FL J S Davison in, WO Rudd
out
30 Nov
DUISBURG
1713-2143
FO J D F Cowden and crew
2/3 Dec
HAGEN
1815-0015
FO J D F Cowden and crew
4 Dec
KARLSRUHE
1705-2235
FL H T Paddison
PO V Murphy
PO E J D Bill
FO G R Godfrey
FS C Shaw
FS W Telford
FO G H Jones
6/7 Dec
MERSEBURG
1725-0015
FL J A Rowland and crew
FO C F Jelley in, Sgt Jefferson
out
12 Dec
ESSEN
1636-2122
FL H T Paddison
Sgt G E Muchmore
PO V Murphy
WO K W Prior
FO G R Godfrey

FS R B Benton
FS W Telford
FO G H Jones
15 Dec
LUDWIGSHAVEN
1530-2130
FL H T Paddison and crew
SL F J Kelsh in, FS Benton out
17 Dec
DUISBURG
0320-0845
FO G A Thorne and crew
PO A H Mullard in, Sgt Scott
out
21 Dec
COLOGNE
1556-2026
FO G A Thorne and crew
28 Dec
MUNCHEN-GLADBACH
1610-2110
FO J D F Cowden and crew
FS C Duncan in, WO Rudd
out
29 Dec
TROISDORF
1606-2108
FO J D F Cowden and crew
WO G W Buttrick in, FS
Duncan out

1945
1 Jan
DORTMUND
1700-2120
FO J D F Cowden and crew
FS J S Davidson in, WO
Buttrick out
2 Jan
NURNBERG
1541-2245
FO J D F Cowden and crew
5 Jan
HANNOVER
1700-2120
FO J D F Cowden and crew
6 Jan
HANAU
1600-2130
FO G A Thorne and crew
Sgt Scott back
7/8 Jan
MUNICH
1900-0300
FO G A Thorne and crew
13 Jan
SAARBRUCKEN
1555-2125

FO J D F Cowden and crew
14 Jan
MERSEBURG
1940-0410
FO J D F Cowden and crew
16/17 Jan
MAGDEBURG
1900-0050
FL G C Hitchcock
FS A R Chandler
FO T A King
FO C W Spencerley
FS J Parkinson
FS A Purvis
FS D C Dunkley
Sgt W T N Trowhill
22/23 Jan
GELSENKIRCHEN
2000-0030
FL G C Hichcock and crew
PO V G Marks in, FS Dunkley
out
1 Feb
MAINZ
1641-2159
FO J A Thrasher
Sgt S E Sturgeon
FS D C Gray
WO F A Brunetta
FS H S Porter
FS A J Smith
FL L E Flatt
2/3 Feb
WIESBADEN
2105-0236
FL D B Jarvis
FO C G Hale
FS A T Jones
Sgt J Steele
FS R G Noakes
Sgt A Greeley
Sgt W S Bradford
Sgt F Murray
4 Feb
OBERFELD
1815-2242
SL P R Mellor
FL G Shaw (N)
FO H L Coulter (2N)
FL F E Prebble (BA)
FS S Blair (WOP)
FS A Rowbothom (FE)
FS E E Freake (MU)
FS L Freemam (RG)
7 Feb
CLEVE
1933-2343
WC S Baker DSO DFC*

FL R K Hawkins
FS K E Glover
FS A B R Teranzani
SL G R Hawes
Sgt W O Jones
FO J A Wilson
Maj J B Mullock MC DFC RA
8 Feb
WANNE EICKEL
0335-0820
FL C G Hitchcock and crew
FL D Swaffield, PO V G Marks
and FL L E Flatt in;
FSs Chandler, Dunkley and
Sgt Trowhill out
13/14 Feb
BOHLEN
1840-0215
FO J D F Cowden and crew
14/15 Feb
CHEMNITZ
S2043-0423
L P R Mellor and crew
FS D Kirk in, FO Coulter out

No 405 (RCAF) Squadron
22 Mar
HILDESHEIM
1150-1615
FO V R Norman RCAF
FO J M Davis RCAF (N)
FO P M Goffin RCAF
FO A L Jones RCAF (WOP)
Sgt F O'Hanlon (FE)
FS A C Byers RCAF
FS R H Baker RCAF
27 Mar
PADERBORN
1516-1925
FL M S Fyte RCAF
FL W C Irwin (N)
FO J A Welch (WOP)
FO F C Fallon (BA)
Sgt W A King (FE)
FO B F Hobley
FO W Campbell
31 Mar
HAMBURG
0642-1110
SL L G Neilly DFC
FL R M Ferguson DFC (N)
FL H G Windt (BA)
WO D D Finlay DFM (WOP)
PO A Warne (FE)
PO K A MacNair
FO G B Legros
4/5 Apr
NURNBERG

1912-0222
FL E Dereby
FO J B Miller
FO C W McClean
FS M F Botts
Sgt G M Peters
FS J R McCombe
FS S Young
8/9 Apr
HAMBURG
1955-0046
FO R B Maxwell CGM
FO N S Laidlaw
FO H W Cathron
PO J C Feasby
FS H Jones
PO D Allan
PO E D Chisholm
10 Apr
LEIPZIG
1442-2056
FL J M Hall
FO C A Weir (N)
FO J Pauley (BA)
FL E G Mader
Sgt H A Slater
FS F Koroll
PO V Smith
14/15 Apr
POTSDAM
1848-0222
SL L G Neilly and crew
25 Apr
BERCHTESGADEN
0611-1301
FL J M Hall and crew
FO G H Gray in, FO Pauley
out
30 Apr
'MANNA'/THE HAGUE
1550-1810
FL J M Hall and crew
1 May
'MANNA'/THE HAGUE
1317-1529
SL C H Hunnelle
PO E L Tenpost
PO P Young
WO J L Larrimore
PO D T Dale
FO R Littlejohn DFC
PO H L Eaton
PO C Ryan
9 May
'EXODUS'
1055-1928
FL J M Hall and crew
FO Gray out

ND875
NUTS

Part of aircraft order No 1807, this Avro-built machine came off the production line in early 1944 as a Mk III, having four Merlin 38 engines. After going to No 32 MU on 9 April it was assigned to No 7 Squadron PPF on the 14th, but a week later was re-assigned to No 156 Squadron PFF at Upwood, Huntingdonshire, and coded GT-N.

ND875 began ops on 24/25 April with a raid to Karlsruhe in the hands of Squadron Leader H. F. Slade DFC, who had with him that night the Squadron Navigation Leader, Squadron Leader A. J. Mulligan DFC (Mulligan was to receive the DSO that summer). Herbert Frank Slade flew the aircraft on 12 sorties before becoming tour-expired at the end of July, having completed a total of 58 ops with No 156 Squadron, for which he received an immediate DSO. ND875 was also flown by the CO, Wing Commander T. L. Bingham-Hall DFC, who had come to the squadron in the same month as this aircraft. Flight Lieutenant Hayden Jones was in Slade's crew, and recalls:

'On the Karlsruhe operation of 24/25 April I was the Radar Navigator or Nav/B, so-called, operating the H2S for navigation and bomb dropping. I expect we were PBM as usual (Primary Blind Markers) if there was cloud. We bombed on H2S from about 20,000ft, on the majority of our trips.

'Slade died in Australia some years ago and Tony Mulligan was lost on the ill-fated Avro Tudor [which was taking Sir Arthur Coningham to Bermuda on 30 January 1948] flying through the "Bermuda Triangle"; he was a wine sales rep by then.'

Among other notable pilots to fly ND875 were Squadron Leader A. W. G. Cochrane, who would end the war with the DSO, DFC and two Bars, Squadron Leader T. E. Ison DSO, DFC, Squadron Leader P. F. Clayton DFC, who arrived from No 582 Squadron in August 1944, and Squadron Leader Reg F. Griffin DSO, DFC. Cochrane had already flown a tour on Wellingtons and by February 1945 had himself completed 80 operations. He flew at least 14 of these as either Master or Deputy Master Bomber.

ND875 flew on D-Day, was shot-up by a night fighter on 23 June which knocked-out the port outer engine (becoming Cat AC), then flew over Caen when the Allied breakout started. On 7 October, Thomas Ison flew a 'Long-Stop' mission to Cleve, his duty being to ensure bombers did not overshoot the bomb-line and drop their loads onto Allied troops in the Nijmegen Salient.

In total, Ison took ND875 on 17 trips, and Cochrane four – including one during which, acting as Master Bomber over Goch on 7/8 February 1945, they collided with another aircraft. Although the Lanc lost a chunk of its port wing, Cochrane continued to direct the bombing and stayed over the target until all had bombed.

Its 100th op was recorded as being flown on 24 March 1945 with Squadron Leader Clayton in command, on another 'Long-Stop' sortie for a raid against a Benzol plant near Dortmund. ND875 was credited with a total of 108 raids, although not all can be verified, but the aircraft certainly flew over 100 ops.

ND875 went to No 1660 CU on 26 July, then to No 1668 CU on 20 October, where the aircraft remained until 14 March 1946 when it returned to No 1660 CU. Its final service was with No 1653 CU with effect from 9 November 1946 and then the Lanc went to No 15 MU on 9 May 1947 where it was Struck off Charge on 28 August.

1944
No 156 Squadron
24/25 Apr
KARLSRUHE
2209-0407
FL H F Slade DFC RAAF
SL A J Mulligan DFC (N)
FS T C Bower (BA)
FS B L Johnson (FE)
FL H E Jones (N2)
FS B H Andrews (MU)
PO H Boon (RG)
26/27 Apr
ESSEN
2308-0301
FL D O Blamey
FS G H Clements
FS J Dillon
Sgt K Gilbert
FL J Booth
FS G W Gracey
WO D Pedder
1/2 May
GHISLAIN
2240-0140
FL H F Slade and crew
FS E A Jackson in, FS Bower out
7/8 May
NANTES A/F
0028-0521
FL C G Hopton RCAF
FL H W Gillis
PO P J Moyes
Sgt L E Gibbs
FL R B Leigh
WO A R P Larkins
Sgt I Campbell
(SL Cecil Hopton DFC and crew failed to return 7 June, Hopton being killed)
1/12 May
HASSELT
2215-0216
PO R H Samson
PO T W Kennedy

FS R J Andrew
Sgt R G Burton
WO A A Gilchrist
FS A G Bryant
FO S F Delongree
19/20 May
MONT COUPLE
2303-0057
SL H F Slade DFC RAAF
PO C W Reeves
PO A E Egan
WO B L Johnson
FL G A R Undrell
WO B H Andrews
SL J E Blair
(raid abandoned)
21/22 May
DUISBURG
2252-0254
SL H F Slade DFC RAAF
SL A J Mulligan DFC
PO A E Egan
WO B L Johnson
FL G A R Undrell
WO B H Andrews
PO J H Detores
22/23 May
DORTMUND
2250-0236
FL R F Griffin DFC
FS L Proud
FS M W Finney
Sgt C F Pretlove
WO K E Bartleman
FS J J Corkery
FS F I Ide
27/28 May
RENNES
2342-0325
SL H F Slade and crew
Sgt E Edmund in (RG)
28/29 May
MARDYCK
2329-0111
SL H F Slade and crew
PO C W Reeves in, SL

Mulligan out
31/1 Jun
TERGNIER
0019-0402
WC T L Bingham-Hall DFC
Sgt R V Davidson (N)
FO H C Cavenagh
Sgt H R Walker
FO J E Scrivener
FS C R Alcock
FS R G Green
5/6 Jun
LONGUES
0258-0605
SL H F Slade and crew
SL Mulligan and FS E D Riley in;
PO Reeves and Sgt Edmund out
7/8 Jun
FORET DE CERISY
0058-0333
SL H F Slade and crew
8/9 Jun
FOUGERES
2206-0215
SL H F Slade and crew
PO Reeves and FS S Freeden in;
SL Mulligan and Sgt Edmund out
9/10 Jun
LE MANS A/F
2216-0235
SL H F Slade and crew
Sgt E C Bangs in, FS Freeden out
11/12 Jun
TOURS
2147-0304
SL H F Slade and crew
FS E D Riley in, Sgt Bangs out
14/15 Jun
ST POL
0210-0422
SL H F Slade and crew

FL J R Chislett and FO A McVitee in;
FL Jones and FS Riley out
15/16 Jun
LENS
2328-0210
WC T L Bingham-Hall DFC
FO H Coker
FL G H M Robinson
FL R E Maxwell
FO F Holbrook
FO F J Lockwood
FO D Plantana
16/17 Jun
RENESCURE/V1 SITE
2358-0205
PO K P C Doyle
FS D Winlow
Sgt D K Green
Sgt J T Gedney
FS A Astle
FS D J Hughes
PO J H Detores
23/24 Jun
COUBRONNES/V1 SITE
2355-0221
PO H T Griffin
FL J A Turk
FS K J Negus
FS W J Norrys
FO E Dyson
FS P M Muir
FS J A McGregor
(damaged by night fighter over French coast)
17 Jul
MON CANDON/V1 SITE
1938-2205
FL A W G Cochrane DFC
RNZAF
FO W J Evans
FO E E Court
FS J B Elder
PO H H Jenkins
FS L A Rootes
FS T E Drew

190

WO M Fleming
18 Jul
SANNERVILLE/CAEN
0413-0703
FL A W G Cochrane and crew
19 Jul
ROLLEZ/V1 SITE
1412-1720
PO V D Temple
WO H Graf
FS A E C Mundy
Sgt M J Waltham
FS A P Arnott
Sgt L E Reynolds
Sgt W V Cooper
20 Jul
FORET DE CROC/V1 SITE
1412-1731
SL T E Ison DFC
FL H F Morrish
Sgt L T Walton
Sgt C E Moss
FL F E Keighley
PO A J ELey
Sgt A Allen
FO F N Hough
23/24 Jul
DONGES
2224-0324
SL T E Ison and crew
24/25 Jul
STUTTGART
2212-0521
FL R F Griffin and crew
25/26 Jul
STUTTGART
2200-0601
SL T E Ison and crew
28/29 Jul
HAMBURG
2251-0320
SL G C Hemmings DFC
FO S J Richards
FL T J Pye
Sgt A Green
FO P A Taylor
FO T H A Hill
PO G B Stone
5 Aug
FORET DE NIEPPE/V1 SITE
1141-1403
SL T E Ison and crew
6/7 Aug
CABOURG
2141-0025
SL T E Ison and crew
7/8 Aug
'TOTALIZE'/NORMANDY
2205-0039
SL T E Ison and crew
9/10 Aug
FORT D'ENGLOS/V1 SITE
2127-0022
FL K W Kitson
FO B J L Dodd
Sgt J G Stewart
Sgt W G Taylor
FL J Booth
Sgt R O'Donnell
FS E G Gusway
11 Aug
SOMAIN
1417-1708
SL T E Ison and crew
12/13 Aug
RUSSELSHEIM
2158-0246

FL P F Clayton DFC
FL F W Chandler
FO T G Greene
FS R J Bruce
FO P B Kettle
FO A G Lindsay
FO P O Bone
14 Aug
'TRACTABLE'/NORMANDY
1324-1625
SL T E Ison and crew
16/17 Aug
KIEL
2122-0232
SL T E Ison and crew
FS W J Connolly in (RG)
18/19 Aug
BREMEN
2140-0241
SL T E Ison and crew
25/26 Aug
RUSSELSHEIM
2106-0402
FL P F Clayton and crew
26/27 Aug
KIEL
2025-0119
FL A J Hiscock
FL J A Turk
FS J S Turner
FO W N Bingham
PO C Wilson
FO J G Cooper
PO E W C Brackett
27/28 Aug
STETTIN
2125-0600
FL H Rollin
FO W O Chmilar
Sgt S Hudson
Sgt M G Phillips
FS J W Pollock
Sgt E Murray
Sgt A W Hornsley
31 Aug
LUMBRES
1419-1653
FL K P C Doyle
FS E F Hearn
FS D K Green
PO J A Brookes
PO A Astle
FO J A Noble
FS J D Sanders
FO W A M Savill
3 Sep
EINDHOVEN A/F
1600-1856
FL H Rollin and crew
5 Sep
LE HAVRE
1719-1946
SL T E Ison and crew
(Deputy Master Bomber)
9 Sep
LE HAVRE
0700-0935
SL T E Ison and crew
(raid abandoned due to poor
visibility and storms)
10 Sep
LE HAVRE
1537-1824
SL T E Ison and crew
11 Sep
GELSENKIRCHEN
1652-2037

SL T E Ison and crew
12/13 Sep
FRANKFURT
1859-0100
FL W J Cleland
PO G J Hudson
FS A J Wilson
FS J R Watson
FO N N Wray
FS W Appleby
FS J A McGregor
14 Sep
WASSENAR
1321-1535
FL K P C Doyle and crew
WO E H Edinburgh in (MU)
(Deputy Master Bomber)
15/16 Sep
KIEL
2234-0350
FL K P C Doyle and crew
17 Sep
BOULOGNE
0740-0937
FL K P C Doyle and crew
(FL Doyle and crew failed to
return 24 September)
20 Sep
CALAIS
1520-1736
SL T E Ison and crew
(Master Bomber)
25 Sep
CALAIS
0935-1128
SL T E Ison and crew
PO T Kennedy (N)
Sgt L T Walton
Sgt F Gregory
FO N W Wray
PO A J Eley
FS W J Connolly (MU)
SL D F Allen GM BEM (RG)
26 Sep
CAP GRIS NEZ
0853-1124
SL T E Ison and crew
FO A P Willoughby (N)
Sgt L T Walton
Sgt F Gregory
FL J R Chislett
PO A J ELey
FS W J Connolly
FO J Costigan (RG)
5/6 Oct
SAARBRUCKEN
1921-0011
FO L T R H Williams
WO P A Robertson
Sgt J Burgess
Sgt B A Butterfield
FS H F Wilkinson
Sgt D Reed
Sgt R J Heatrick
7 Oct
CLEVE
1206-1540
SL T E Ison and crew
FL E T Cook
FS L T Walton
FL J Booth
PO A J Eley
FS W J Connolly
FS J W Close
14 Oct
DUISBURG
FO F D Wallace

Sgt G W Blick (N)
FS J T Barnes (WOP)
Sgt E Ogden (WOP2)
Sgt N I Davies (N2)
FO J M MacKrory RCAF
Sgt J Hayton
15 Oct
WILHELMSHAVEN
1727-2208
FO F D Wallace and crew
19 Oct
STUTTGART
1806-2354
FL A C Pope
FO L E Munro
Sgt K Autcliffe
Sgt G E Batten
Sgt E A Marlow
FO J F Aspinall
FS I W Kelly
FS R C Fletcher
23 Oct
ESSEN
1707-2115
FO A B Pelly
FO D F Sinclair
FS W G Pearce
Sgt R Morgan
FO A J McLeod
FS L Ayres
FS T S Carr
25 Oct
ESSEN
1339-1707
FL H Rollin and crew
28 Oct
COLOGNE
1409-1751
FO J H Deremore-Denver
FS S Carpenter (N)
WO J R Jones RAAF
Sgt J C Bennett (FE)
FS J L Jacobs RAAF
Sgt J V McClosky (MU)
Sgt G MacQueen (RG)
29 Oct
WALCHEREN
1045-1301
FO A B Pelly and crew
1 Nov
OBERHAUSEN
1800-2238
FO F D Wallace and crew
2 Nov
DUSSELDORF
1647-2120
FO F D Wallace and crew
4 Nov
BOCHUM
1749-2144
FO N Jackson
Sgt G W Blick
WO A Everest (WOP)
FS J S MacDonald
Sgt J Harvie
WO K K Muir
FS R R Willgoss
16 Nov
DUREN
1330-1722
FO A B Pelly and crew
Sgt J D Routledge (2N)
(FO Pelly and crew failed to
return 20 February 1945)
27 Nov
NEUSS
1759-2216

FO F D Wallace and crew
28 Nov
ESSEN
0314-0745
FO F D Wallace and crew
30 Nov
DUISBURG
1718-2154
FO F D Wallace and crew
3 Dec
HEIMBACH
0813-1206
FO F D Wallace and crew
FS F Reed in (FE)
(raid abandoned by Master
Bomber)
4 Dec
KARLSRUHE
1704-2229
FO F D Wallace and crew
PO R E Page in, FS Reed out
5 Dec
SOEST
1828-0024
FO K T Wallace DFC RCAF
FS W D Bonter RCAF (N)
FL W Walker RAAF
Sgt S B Glasper (FE)
FO C L Carlson RCAF
Sgt L A Bedford
FS E C Barnes
17 Dec
DUISBURG
0350-0830
FO J H Deremore-Denver and
crew
Sgt A M Gunton (visual mark-
er)

1945
5 Jan
HANNOVER
1935-2357
FO M T Wilson
PO L N Hill
PO E A Jackson RAAF
Sgt R P Kinn (FE)
FO B W Munden RNZAF
FL S L Hyde
Sgt J H Addison
7 Jan
MUNICH
1918-0230
FL N Jackson and crew
PO H J Scull and WO H N

Whitmore in;
Sgt Blick and FS Willgoss out
2/3 Feb
WIESBADEN
2110-0215
SL A W G Cochrane DSO
DFC** RNZAF
FO J Aaron RCAF (N)
FL R F Jenkins (WOP)
Sgt R E English (FE)
FL J R Burns RCAF
FO G K Dee RNZAF (VM)
FS E Reed
FO B W Felgate
7 Feb
GOCH
1948-0140
SL A W G Cochrane and crew
(Master Bomber)
17 Feb
WESEL
1230-1659
FL V M Todd
FO A B Walters RAAF
FO A Watson
FO J N Ashton
Sgt G S Kay
FO R L Martin
FS D J Price
FL J Jackson
20/21 Feb
DORTMUND
2215-0339
FL H G Hughes RCAF
FS F H Cripps (N)
FS P McEwen RAF(WOP)
Sgt L Moody (FE)
FO G J Smith RCAF
Sgt P G Hinton
Sgt L Jackson
21 Feb
WORMS
1720-2309
FL H G Hughes and crew
1 Mar
MANNHEIM
1227-1742
FL I G Paull RAAF
PO S F Ladner RAAF
FO K T Glaziou RAAF
Sgt T L Pearson
FS V M Walsh RAAF
Sgt D R Haywood
Sgt G W D Riley

5/6 Mar
CHEMNITZ
1725-0114
FL I G Paull and crew
PO J W Close RAAF in, Sgt
Riley out
7/8 Mar
DESSAU
1723-0154
FL W J Taylor RAAF
FS E L Crispin RAAF
FS J E Ritchie RAAF
Sgt D A McLean RAAF
FO P S Green RAAF
FS K T Jones RAAF
FS R H Bennington RAAF
8 Mar
HAMBURG
1846-2350
FL W E B Mason RCAF
FO H J Collison RCAF
FS C Saunderson
PO F V Walton
FS C S Fuller
FS H G Lee
FS A S Orchard
12 Mar
DORTMUND
1338-1828
FL I G Paull and crew
Sgt D S Greenback in PO
Close out
13 Mar
DAHL/BOCHUM
1806-2302
FL W J Taylor and crew
14 Mar
HOMBERG
1748-2243
FL W J Taylor and crew
15/16 Mar
HANNOVER
1741-0018
FL I G Paull and crew
16/17 Mar
NURNBERG
1810-0130
FL W J Taylor and crew
18/19 Mar
HANAU
0109-0727
FL W E B Mason and crew
20/21 Mar
HEIDE
0207-0710

FL H G Hughes and crew
22 Mar
HILDESHEIM
1144-1616
FL H G Hughes and crew
24 Mar
HARPENERWEG
1342-1813
SL P F Clayton DFC
SL F W Chandler
FL T G Greene
WO R J Bruce
FL J R Burns RCAF
FO J F Aspinall
FS L E Ryenolds
FO H Cornforth
31 Mar
HAMBURG
0645-1129
FL H G Hughes and crew
FL R L Thompson RNZAF
(VBA)
4/5 Apr
LUTZKENDORF
2150-0505
FL H G Hughes and crew
8/9 Apr
HAMBURG
1928-0047
FL H G Hughes and crew
9/10 Apr
KIEL
1947-0057
FL H G Hughes and crew
10 Apr
LEIPZIG
1434-2108
FO C E Light
FO B B Coles
FO R J Grefell
FS J T Thompson
Sgt G W Wilkes
Sgt D Greenbank
FS A J Greenacre
11 Apr
NURNBERG
1208-1814
FL H G Hughes and crew
13/14 Apr
KIEL
2031-0227
FL H G Hughes and crew

NE181

MIKE, THE CAPTAIN'S FANCY

The final Lancaster of the 600 produced under order No 1807, NE181 was a Mk III which came off the production line in the spring of 1944 with four Merlin 38 engines. It was assigned to No 75 (New Zealand) Squadron based at Mepal, Cambridgeshire, on 20 May and although the aircraft may have been given the well-known squadron code letters of AA (A & B Flights), it soon wore C Flight's codes of JN. NE181's individual identification letter was M and from then on the Lanc was affectionately known as 'Mike'.

Its first sortie on 21/22 May was to Duisberg, with Pilot Officer C. Crawford; the second was to Aachen one week later. Mike flew on both D-Day nights with Flight Sergeant J. Lethbridge RNZAF as skipper and he flew the Lanc on 26 trips during its first three months of operational duty. Mike's next regular captains were Flying Officer G. Cuming who took the aircraft on 14 ops, and Squadron Leader N. A. Williamson, who flew seven trips. Even the squadron CO, Wing Commander R. J. A. Leslie AFC did a couple.

On a raid in support of the imminent Arnhem operation, on the night of 16/17 September, No 75 Squadron dropped not bombs but miniature dummy parachutists near Moerdijk airfield, in order to create a diversion – NE181 being one of those which took part in this unusual op.

Paddy McElligott who was Gordon Cuming's rear-gunner on this occasion also remembers the raid to Saarbrucken on 5 October:

'I recall we flew in formation low over France and in a vic formation before climbing in darkness to our bombing height of 14,000ft. I saw one of the aircraft just behind ours dip and hit the ground in a ball of flame and smoke, as it collided with a Lancaster from No 115 Squadron.

'This operation had personal connotations for me, for over the target we were attacked by a Junkers 88. To find an enemy fighter amongst the flak in the target area was unusual. I returned its fire and felt sure I had hit the fighter from the impression given by the tracer. The mid-upper gunner opened fire after me but soon his guns automatically cut-out when his turret was rotated towards our twin tail-fins. Thinking it was a stoppage he operated the manual over-ride, whereupon I became aware of tracer shooting over my turret.

'As I continued to fire, I called for the skipper to corkscrew port, which he did, and the Ju88 went down to our starboard. I asked the mid-upper where the second fighter came from that was firing over my head and he calmly said it was him!'

Paddy McElligott also recalls a raid to Cologne on 31 October/1 November, taking the Station Commander, Group Captain A. P. Campbell with them:

'This was the second time we visited Cologne within 24 hours. The Group Captain flew with us as bomb-aimer and Syd Sewell assessed his ability on this occasion as "shows promise"! I understand the Group Captain had already completed the requisite operational flying hours allocated to him for October, so by going as bomb-aimer, he no doubt hoped his transgression would go un-noticed by Group HQ.'

On 2 November, Jack Leslie took Mike to Homberg. Tiny Humphries was the crew's navigator:

'We had the misfortune to have one bomb hang-up and Leslie then announced he was going round again, but the bomb-aimer couldn't release it manually on this second run. If anything was calculated to be fool-hardy, going round again to drop one bomb over a place which had just been done over, that was! None of us wanted to fly with him again, but that was Jack Leslie. We got back somewhat late!'

Mike hardly recorded a single mechanical problem during its first seven months of operations. Not until 6/7 January 1945 did the aircraft lose an engine, during a mining sortie off Danzig, but on the next raid on 11 January Mike lost its port outer due to a coolant leak, while the port inner gave the pilot, Squadron Leader J. M. Bailey DFC, indications of overheating.

John Bailey – or Jack as he was generally known – became Mike's regular skipper late in the Lanc's career, flying a total of 14 trips in this aircraft and ending his tour with a Bar to his DFC. Flight Lieutenant Alex Simpson also did a couple of trips in Mike, and recalls the aircraft, its nose-art painting and Jack Bailey:

'I had a lot of time for Jack Bailey. He was an Irishman of Southern Irish descent and had that delightful Irish sense of humour. He had tried very hard to get approval to fly Mike to New Zealand, being the first [and only] New Zealand heavy bomber to make 100 operations, he even went to the extent of soliciting aid from Bill Jordan, the New Zealand High Commissioner in London.

'We all knew Mike was getting near its 100th, in fact one of my arguments to Jack was that it was such a clapped-out old heap and I had my own aircraft in JN-K-King, that I didn't want to fly it.

'I do not recall the background of naming Mike "The Captain's Fancy" other than knowing that the "Captain" as depicted on Mike was Captain Reilly-Foull from the wartime *Daily Mirror* cartoon strip.'

Paddy McElligott remembers Reilly-Foull also had:

'...in his right hand a pint of beer and the inevitable dart, ready to be thrown, in the left hand. The cartoon invariably began with his expression – "Strop me...!"

'When we went to Coblenz on 6 November, my log states we flew on three engines all the way. We "cut a few corners" on the route back because of our reduced flying speed, nevertheless we got back to Mepal fairly early, called up base but got no response. We soon realised our radio was u/s so climbed above the other orbiting aircraft and waited.

'Eventually the circuit lights went out and we were left alone in the darkness. Our only recourse was to fly in low over the airfield, fire a red Very cartridge and hope someone eventually got the message, which they did. The lights went on and we finally touched down some 1 1/2 hours after everyone else!

'Sadly this was our last trip in M-Mike, though not in our tour – we still had eight to go. "She" was, in the terminology of the day, "a great old kite" – reliable and a proven operational veteran.'

There was some confusion as to when Mike did its 100th trip. According to a press release, this was flown on 29 January 1945 which is in keeping with the Form 541 details. However, there must have been an earlier count of its ops, for initially it was thought the aircraft's 100th was due on 5 January. Jack Bailey, feeling superstitious about it, talked his deputy flight commander, Alex Simpson, into flying the sortie, which he reluctantly did. Only after this had been flown was another count done, which indicated that Alex had in fact flown the 101st sortie. In the final event, that was only the aircraft's 96th. Therefore, Mike went on to do the 100th on 29 January – with Jack Bailey (it was his 47th op overall). His was a very experienced crew and as the press release also noted, the New Zealand pilot and his six English crewmen had, between them, flown a total of 294 ops. Mike managed one more sortie on 2/3 February before enough became enough for both Mike and any crews who might be assigned to the veteran Lanc.

No doubt part of the confusion of ops was due to the existence of two aircraft 'M' on No 75 Squadron, the other being AA-M (ME752). In fact in the press release, this other Mike was reported as flying the 99th trip on 22 January –whereas JN-M (NE181) did not fly that night! Undoubtedly the earlier confusion was due to someone counting up the 'M's and not taking notice of the serial numbers.

Mike was retired after this 101st trip, and as authority finally thwarted the efforts to get Mike sent to New Zealand, Alex Simpson was detailed to fly Mike to Waterbeach on 17 February. Here, after a refit, the aircraft was assigned to No 514 Squadron which operated from this base, on 19 July. No 514 was disbanded in August and on 4 September Mike went to No 5 MU where it was finally scrapped on 30 September 1947. The aircraft would have made a proud museum piece in New Zealand but, alas, bureaucracy dictated that nothing could be done to get NE181 despatched to its far-off adopted country, where it could have represented the major efforts made by its airmen – both RAF and RNZAF – in World War 2.

1944
No 75 Squadron
21/22 May
DUISBURG
2255-0320
PO C Crawford
FS E Rivers (N)
Sgt T Mason (BA)
Sgt T Feaver (WOP)
Sgt A Frost (FE)
Sgt N Heslop (MU)
Sgt R Phillips (RG)
28/29 May
AACHEN
1855-0210
FS J D Perfrement RAAF
Sgt W Hall
Sgt A Kirkham
PO J Craven
Sgt J Tomlinson
Sgt L King
FS D Trigg
2/3 Jun
WISSANT
0115-0355
FS J Lethbridge RNZAF
Sgt P Crane
FO J Dickinson
WO C Newark
Sgt A Barnes
FS A Markham RNZAF
FS J Snodgrass
RNZAF
(target not identified, did not bomb)
3/4 Jun
CALAIS
0035-0240
FS J Lethbridge and crew
5/6 Jun
OUISTREHAM
0330-0640
FS J Lethbridge and crew
6/7 Jun
LISIEUX
2359-0355
FS J Lethbridge and crew
10/11 Jun
DREUX
2310-0330
FS C Nairne RNZAF
FS L Perry RNZAF
FS D Kidby
Sgt A Stannard
Sgt R Smith
Sgt S Woodford
FS P Falkiner RNZAF
11/12 Jun
NANTES
2350-0515
FS C Nairne and crew
14/15 Jun
GELSENKIRCHEN
2340-0310
FS C Nairne and crew

15/16 Jun
VALENCIENNES
2305-0230
FS C Nairne and crew
21 Jun
DOMLEGER/V1 SITE
1810-2040
FS J Lethbridge and crew
23/24 Jun
L'HEY/V1 SITE
2305-0230
FS J Lethbridge and crew
24/25 Jun
RIMEUX/V1 SITE
2330-0200
30 Jun
VILLERS BOCAGE
1810-2120
FS J Lethbridge and crew
2 Jul
BEAUVOIR/V1 SITE
1257-1615
FS J Lethbridge and crew
5/6 Jul
WATTEN/V1 SITE
2307-0055
FS J Lethbridge and crew
7/8 Jul
VAIRES
2259-0329
FS A McKenzie
FS W Stoneham
FS T Bunce
WO J Wright
Sgt G Robertson
Sgt G McKellow
Sgt W Barker
9 Jul
LINZEUX/V1 SITE
1305-1606
PO F Timms RNZAF
Sgt W Morton
FO J Noble RNZAF
Sgt P McKerrel
Sgt R Woods
Sgt J Hindley
Sgt J Kemp
10 Jul
NUCOURT/V1 SITE
0420-0730
FS J Lethbridge and crew
12 Jul
VAIRES
1811-2146
FS J Lethbridge and crew
15/16 Jul
BOIS DES JARDINES/V1 SITE
2330-0225
FS J Lethbridge and crew
17 Jul
VAIRES
1220-1320
FS J Lethbridge and crew
(recalled)

18 Jul
CAGNY/CAEN
0440-0735
FS J Lethbridge and crew
18/19 Jul
AULNOYE
2238-0226
FS J Lethbridge and crew
20/21 Jul
HOMBURG
2345-0250
FS J Lethbridge and crew
23/24 Jul
KIEL
2244-0344
PO F Timms and crew
25/26 Jul
STUTTGART
2140-0554
FL G Gunn
FO F Smith (N)
FO A Millar (BA)
FL W Naismith (WOP)
Sgt J Bruce (FE)
FO C Robertson (MU)
FO S Haines (RG)
28/29 Jul
STUTTGART
2158-0543
FL G Gunn and crew
30 Jul
AMAYE-SUR-SEULLES
0605-1001
FS J Lethbridge and crew
1 Aug
LE NIEPPE/V1 SITE
1913-2134
FS J Lethbridge and crew
3 Aug
LISLE ADAM/V1 SITE
1156-1536
FS J Lethbridge and crew
4 Aug
BEC-D'AMBES
1335-2136
FS J Lethbridge and crew
5 Aug
BASSENES/V1 SITE
1422-2214
FS E D O'Callaghan RNZAF
FS C Busfield
Sgt J Mitchell
FS S Matheson
Sgt C Simpson
Sgt E Baines
sgt A Shepherd
7/8 Aug
MARE DE MAGNE
2157-0057
FS J Lethbridge and crew
9/10 Aug
FORT D'ANGLOS/V1 SITE
2208-0019
FS J Lethbridge and crew

11 Aug
LENS
1428-1746
FS J Lethbridge and crew
12/13 Aug
MINING/GIRONDE
2244-0541
FS J Lethbridge and crew
15 Aug
ST TROND A/F
0955-1331
FO J H Scott RNZAF
FS A Scott RNZAF (N)
FS K Anderson RNZAF
FS E Howard RNZAF (WOP)
Sgt H M Thomas (FE)
Sgt J T Beardmore (MU)
Sgt J T Boyes (RG)
(Scott and his crew killed on a raid to Solingen, 4 November)
16/17 Aug
STETTIN
2109-0516
FS J Lethbridge and crew
18/19 Aug
BREMEN
2135-0309
FS J Lethbridge and crew
26/27 Aug
KIEL
2005-0153
FS P L McCartin RAAF
Sgt J Miles (N)
FO L A Martin (BA)
Sgt P F Smith RAAF(WOP)
Sgt W J Warlow (FE)
Sgt D G A Bryer (MU)
Sgt J N Gray (RG)
29/30 Aug
MINING/DANZIG
2020-0555
SL N A Williamson RNZAF
FO J Watts
FO G Coull
Sgt S Cook
FO S Moss (FE)
Sgt R Jones
FO J Tugwell (RG)
31 Aug
PONT REMY
1630-1930
SL N A Williamson and crew
FS G Wllis RCAF in, Sgt Jones out
5 Sep
LE HAVRE
1739-2118
FS P L McCartin and crew
6 Sep
HARQUEBEC/LE HAVRE
1556-1943
FO G Cuming RNZAF
FS J G Scott (N)
FS S G Sewell (BA)

FS J D Christie (WOP)
Sgt J C Lambert (FE)
Sgt W Scott (MU)
Sgt D P McElligott (RG)
8 Sep
DOUDENEVILLE
0621-1046
FS P L McCartin and crew
(McCartin and crew missing
20 November; all but R/G
killed)
12/13 Sep
FRANKFURT
1847-0152
FO G Cuming and crew
14 Sep
WAASENAAR
1300-1536
FO K Sutherland RNZAF
FO A Thompson RNZAF
FO B G Clare RNZAF
Sgt E W Vero
Sgt D J Roberts
Sgt L Cooper
Sgt T Burnett
(FO Sutherland and crew
failed to return 6 October; he
was killed and the others all
became prisoners)
16/17 Sep
MOERDIJK
2127-0019
FO G Cuming and crew
17 Sep
EMMERICH
1933-2235
FO G Cuming and crew
20 Sep
CALAIS
1429-1808
SL N A Williamson and crew
Sgt D P McElligott in, FO
Tugwell out
23 Sep
NEUSS
1927-2318
FO G Cuming and crew
25 Sep
CALAIS
0820-1124
FO G Cuming and crew
26 Sep
CAP GRIS NEZ
1147-1411
FO G Cuming and crew
3 Oct
WESTKAPELLE DYKE
0150-1508
SL N A Williamson and crew
PO G Ellis RCAF in, Sgt
McElligott out
5 Oct
SAARBRUCKEN
1730-2231
FO G Cuming and crew
(attacked by a Ju88 night
fighter)
6 Oct
DORTMUND
1652-2258
FO G Cuming and crew
7/8 Oct
EMMERICH
2220-0614
SL N A Williamson and crew
FO Tugwell back

14 Oct
DUISBURG
0650-1114
SL N A Williamson and crew
14/15 Oct
DUISBURG
2239-0319
SL N A Williamson and crew
15/16 Oct
MINING/KATTEGAT
1827-0022
FO E Robertson RNZAF
FS A Herrold RNZAF
FO S Richmond RNZAF
FS F Tubby RNZAF
Sgt F Thompson
Sgt R Maryan
Sgt P Smith
19 Oct
STUTTGART
1737-2353
FO J McIntosh RNZAF
FS R Morgan RNZAF
FS P Newman RNZAF
FS R Boag RAAF
Sgt E Graves
Sgt C Brewer
Sgt E Cooper
21 Oct
FLUSHING
1117-1410
FL T Waugh
PO C Woonton RNZAF (N)
FS R Swetland
FS P Kidd
Sgt N Southgate
FS J Nickells RNZAF
FS D Sage RNZAF
22 Oct
NEUSS
1327-1733
FL T Waugh and crew
23 Oct
ESSEN
1316-1722
FO G Cuming and crew
Sgt J Huckle and FS A Weston
in;
Sgts Lambert and McElligott
out
25 Oct
ESSEN
FO G Cuming and crew
1316-1722
Lambert back, WO R Powell
in
26 Oct
LEVERKUSEN
1304-1716
FO G Cuming and crew
WO I Cornfield in, Powell out
28 Oct
COLOGNE
1315-1746
SL J M Bailey DFC RNZAF
FO J G Brewster (N)
FO J C Wall (BA)
FO N Bartlett (FE)
Sgt R Pickup (WOP)
FS T Gregory (MU)
PO J Bryant RAAF
30 Oct
WESSELING
0902-1335
SL J M Bailey and crew
30 Oct
COLOGNE

1803-2318
FO E Butler RNZAF
FS H Holliday RAAF
FS H Stratford RNZAF
Sgt D Brazier
Sgt C Payne
Sgt J Heaton
Sgt J Messer
31 Oct
COLOGNE
1820-2235
FO R Cuming and crew
Sgt McElligott back;
GC A P Campbell in
2 Nov
HOMBERG
1139-1544
WC R J A Leslie AFC
FS A L Humphries RNZAF
FO E Holloway RNZAF
WO F Chambers RNZAF
FS S Cowen
FO R J Scott RNZAF (MU)
FS A McDonald RNZAF
4 Nov
SOLINGEN
1129-1603
SL J M Bailey and crew
WO I Cornfield in, PO Bryant
out
5 Nov
SOLINGEN
1044-1459
FO J McIntosh and crew
Sgt G Knights in, Sgt Graves
out
6/7 Nov
COBLENZ
1659-2144
FO G Cuming and crew
8 Nov
HOMBERG
0752-1219
SL J M Bailey and crew
20 Nov
HOMBERG
1242-1720
FO J McDonald
FS C Aylott
WO E DeShaine
WO E Hughes
WO J Dunn
FS W Davies
FO H Campbell
21 Nov
HOMBERG
1254-1656
WC R J A Leslie AFC
PO C Woonton RNZAF
FO G Coull
FS P Kidd
Sgt N Southgate
FS J Nickells RNZAF
FS D Sage RNZAF
23 Nov
GELSENKIRCHEN
1249-1707
FO J McDonald and crew
27 Nov
COLOGNE
1228-1659
FO J McDonald and crew
Sgt J Messer in, FO Campbell
out
28/29 Nov
NEUSS
0258-0712

SL J M Bailey and crew
2 Dec
DORTMUND
1246-1705
FO J McDonald and crew
4 Dec
OBERHAUSEN
1217-1608
FO D Williams RNZAF
FS D Sim RNZAF
WO G Duncan
WO O Harrison RNZAF
Sgt E Round
FS I Carrington RNZAF
FS R Smith RNZAF
5 Dec
HAMM
0911-1347
SL J M Bailey and crew
(bombed secondary target)
6/7 Dec
MERSEBURG
1708-0027
FO A D Simpson RNZAF
PO R Woodhouse RNZAF
FS J Hemingway
Sgt A Dibbs
Sgt J Johnstone
FS E Thomas RNZAF
Sgt C Chippendale
8 Dec
DUISBURG
0830-1243
SL J M Bailey and crew
11 Dec
OSTERFELD
0842-1253
SL J M Bailey and crew
12 Dec
WITTEN
1112-1618
FL L W Hanan RNZAF
FS H McLeod RNZAF
FO W Brizley RNZAF
FS W Jenkins RNZAF
Sgt P Yellin
Sgt R Williams
Sgt C Wilkinson
16 Dec
STEGAN
1118-1728
SL J M Bailey and crew
FL A Creagh RNZAF in, FO
Brewster out
21 Dec
TRIER
1226-1707
PO G S Davies RNZAF
FS C C Greenhough RNZAF
Sgt H E Chalmers
Sgt T M White
Sgt I R H Evans
Sgt J J Maher
WO W Reavely
23 Dec
TRIER
1200-1617
SL J M Bailey and crew
FO Brewster back
27 Dec
RHEYDT
1210-1646
SL J M Bailey and crew
31 Dec
VOHWINKEL
1134-1629
FL L W Hanan and crew

1945
1 Jan
VOHWINKEL
1603-2153
FL L W Hanan and crew
3 Jan
DORTMUND
1232-1740
FL L W Hanan and crew
5 Jan
LUDWIGSHAVEN

1136-1723
FO A D Simpson and crew
6/7 Jan
MINING/PILAU
1618-0158
FO D Clements
FO Hewitt
FO R Cato
FS T Hepard
Sgt W Richardson
Sgt J Wildish

Sgt F Watte
(returned on three engines
from Danish coast)
11 Jan
KREFELD
1152-1647
SL J M Bailey and crew
(lost port outer but continued on and bombed)
16/17 Jan
WANNE EICKEL

2316-0417
SL J M Bailey and crew
29 Jan
KREFELD
1012-1558
SL J M Bailey and crew
2/3 Feb
WIESBADEN
2046-0231
SL J M Bailey and crew

PA995
THE VULTURE STRIKES!

Part of order No 1807/C4A, this Avro-built Mk III was produced in the spring of 1944 with four Merlin 38 engines and was assigned to No 550 Squadron at Waltham, near Grimsby, Lincolnshire, on 29 May 1944 where it became BQ-K. The aircraft carried the letter K until September after which it became V-Victor, but the V soon took on other connotations for a large vulture had been painted on the nose and the legend 'The Vulture Strikes!' was written above it.

PA995's bomb symbols began on 3/4 June following a raid to Wimereux, and at first these symbols were in neat rows of 10. After 43 bombs, the style changed to smaller, slanting bombs and the fifth row which had been started with three, was completed with 11 slanting bombs to make a row of 14. The next three rows were all of 13 (despite superstitions!) and then the original fourth row had four more bombs added. All these can be seen on a well-known photograph of PA995 when it sported 98 bomb symbols, with Flying Officer G. E. Blackler in the cockpit. Although the picture shows 98 bombs, it was taken the morning after the 100th trip. Another photograph taken at the same time shows the Lanc surrounded by Blackler's men and the squadron's air and groundcrews, plus the Squadron Commander, Wing Commander J. C. McWatters DFC.

George Blacker, who flew Victor on 27 raids, plus one abort and one recall, out of his tour of 37 ops, recalls the Vulture insignia as being black and yellow, edged in white. He also remembers on the morning after the 100th trip a Tannoy message ordered everyone in B Flight to attend the aircraft, and on doing so photographs were taken of the event. George arrived first and was thus photographed in the cockpit before the rest of the men arrived on the scene.

Flying Officer F. S. Steele RCAF had been PA995's first regular skipper back in mid-June 1944, flying at least 25 ops in the aircraft and winning the DFC. When Steele completed his tour, George Blackler took over and he too received the DFC.

Several crews that flew PA995 also flew EE139, another centenarian. One unusual pilot was an American on detachment from the US 8th Air Force, Flight Officer G. P. Fauman, who arrived on the squadron on 4 May 1944, flew at least five ops in PA995 and two in EE139, before he returned to the USAAF in mid-September.

PA995 had almost no mechanical problems of any note, although Steele did hit a tree on 18 July during the Caen breakout operation, which slightly damaged the

H2S blister under the rear fuselage. Otherwise the aircraft flew on merrily and not until 19 October did it have to turn back from a mission, following the failure of the port outer engine.

Blackler took PA995 to Chemnitz on 5/6 March 1945, recorded as the aircraft's 100th sortie, and then became tour-expired, going to No 1656 CU. His mid-upper gunner was John Nicholson, who relates:

'We carried out 27 ops in PA995. On our second in V-Victor we lost two engines on the first 100 miles or so and aborted. The aircraft was promptly overhauled, including two new engines, after which she behaved beautifully.

'We had our moments of excitement and anxieties. I shot down a Me163 rocket fighter on one sortie (I saw it explode), although I didn't know it was a '163 until after the war ended. It was never confirmed and was all but forgotten until our 1993 Squadron Reunion.'

PA995's luck ran out on its 101st sortie, a raid on Dessau, when Flying Officer C. J. Jones RCAF and crew were one of three squadron Lancasters failing to return that night. It appears that some of PA995's crew escaped, although Jones and two Canadian crewmen, John Buckmaster and Leslie Harvey, are known to have died, the latter being buried in Nederwaert Military Cemetery.

1944
No 550 Squadron
3/4 Jun
WIMEREUX
2351-0243
FO K Bowen-Bravery
Sgt L A Thompson (FE)
PO C E Thomas (N)
FS J P Fyfe (BA)
Sgt A Cleghorn (WOP)
Sgt R Blackburn (MU)
Sgt R A Thomson (RG)
6/7 Jun
ACHERES
0005-0457
SL P A Nicholas
Sgt J E Legg (FE)
FO W Dinney RCAF (N)
FS F C Wilkinson (BA)
FS C J Fuller (WOP)
Sgt W A Ansell (MU)
Sgt N S Smart (RG)
(raid aborted by Master Bomber)
9/10 Jun
FLERS A/F
0054-0527
SL P A Nicholas
Sgt W J Killick
FS A E Stebner RCAF
FO M S Merovitz RCAF
Fl A R Tippett
Sgt J A Ringrow
Sgt W A Drake
10/11 Jun
ACHERES
2304-0407
FO M L Dubois RCAF

Sgt H Tulip
FO W F Cox RCAF
FO J C Young RCAF
Sgt H Wood
Sgt R Eves
Sgt L R Haynes RCAF
12/13 Jun
GELSENKIRCHEN
2303-0311
Flt Off G P Fauman USAAF
Sgt W J Killick
FS A E Stebner RCAF
FO M S Merovitz RCAF
Fl A R Tippett
Sgt J A Ringrow
Sgt W A Drake
14/15 Jun
LE HAVRE
2050-0009
Flt Off G P Fauman and crew
16/17 Jun
STERKRADE
2307-0317
Flt Off G P Fauman and crew
22 Jun
MIMOYECQUES/V1 SITE
1405-1704
FO F S Steele RCAF
Sgt R W E Walters
FO R G Fink RCAF
Sgt B R Railton-Jones
Sgt W Merrills
Sgt R G Roberts RCAF
Sgt E Smith RCAF
23/24 Jun
SAINTES
2228-0545
FO F S Steele and crew

24/25 Jun
PAS DE CALAIS/V1 SITE
0126-0501
FO F S Steele and crew
27/28 Jun
PAS DE CALAIS/V1 SITE
0124-0511
FO F S Steele and crew
29 Jun
DOMLEGER/V1 SITE
1157-1459
Fl/Off G P Fauman and crew
30 Jun
OISEMONT/V1 SITE
0602-0944
Fl/Off G P Fauman and crew
2 Jul
PAS DE CALAIS/V1 SITE
1215-1552
FO F S Steele and crew
4/5 Jul
ORLEANS
2203-0410
FO F S Steele and crew
6 Jul
PAS DE CALAIS/V1 SITE
1846-2227
FO F S Steele and crew
7 Jul
CAEN
1930-2321
FO F S Steele and crew
12/13 Jul
REVIGNY
2127-0655
FO F S Steele and crew
(raid abandoned over target due to cloud)

14 Jul
REVIGNY
FL R P Stone
Sgt C E White
Sgt R F Ferry
FS E W Holliday RCAF
Sgt D E Norgrove
Sgt L G B Wartnaby
Sgt F Wright
(raid abandoned as target not identified; Sqn CO lost on this sortie)
18 Jul
SANNERVILLE/CAEN
0339-0729
FO F S Steele and crew
(damage to H2S blister when a tree was hit)
20 Jul
WIZERNES
1908-2239
FO F S Steele and crew
23/24 Jul
KIEL
2245-0334
FO F S Steele and crew
24/25 Jul
STUTTGART
2132-0606
PO J J W Dawson
Sgt E W Edmunds
Sgt F W Willmer
FS K P Brady
Sgt J M Farmer
Sgt J Earnshaw
Sgt W A Harkness
25/26 Jul
STUTTGART

2143-0604
FL R P Stone and crew
28/29 Jul
STUTTGART
2134-0215
FL R P Stone and crew
(aborted due to Gee going
u/s)
30 Jul
CAHAGNES
0639-1046
PO L W Hussey RCAF
Sgt E Elliott RCAF
FO M A DeGast
FO H S W Nelson RCAF
Sgt P E Binder
Sgt A G Sale
Sgt R L Holmgren RCAF
4 Aug
LE HAVRE
1807-2130
PO L W Hussey and crew
2 Aug
BELLE CROIX/V1 SITE
1848-2145
PO L W Hussey and crew
3 Aug
LE HAVRE
1701-2028
FO F S Steele and crew
4 Aug
PAUILLAC
1331-2139
Sgt G H Town
Sgt G Hope
FS D J T Slimming
FS J H Windsor
FS P D Probert
PO E C Ball
WO J Teasdale
5 Aug
PAUILLAC
1425-2227
FO F S Steele and crew
8/9 Aug
FONTENAY
2109-0043
FO F S Steele and crew
10 Aug
DUIGNY
0927-1440
FO F S Steele and crew
11 Aug
DOUAI
1314-1735
FO J Dawson and crew
15 Aug
LE COULOT A/F
1010-1330
FL F S Steele and crew
7/18 Aug
STETTIN
2059-0528
FL F S Steele and crew
18/19 Aug
GHENT TERNEUZEN
2211-0121
Flt Off G P Fauman and crew
25/26 Aug
RUSSLESHEIM
2020-0451
FL F S Steele and crew
26/27 Aug
KIEL
2020-0158
PO A Abrams RCAF
Sgt K W Nettleton RCAF

Sgt J W Brown
Sgt R F Vennese
FS P L Brooker
Sgt A P Soper
Sgt K R Salten
28 Aug
WEMARS/CAPPEL/V1 SITE
1800-2123
FL F S Steele and crew
29/30 Aug
STETTIN
2115-0618
FL F S Steele and crew
31 Aug
AGENVILLE
1258-1654
FO L W Hussey and crew
FS J Fairclough in, Sgt Binder
out
3 Sep
GILZE RIJEN A/F
1600-1926
FS R A Tapsell
Sgt F S Adley
PO D J K White
FO H Black
Sgt G W Collinson
Sgt J P Sheridan
Sgt P J Sculley
5 Sep
LE HAVRE
1626-2014
FL F S Steele and crew
6 Sep
LE HAVRE
1717-2058
FL F S Steele and crew
8 Sep
LE HAVRE
0640-1038
FL F S Steele and crew
10 Sep
LE HAVRE
1649-2049
FL F S Steele and crew
FO H A Shenker RCAF (2P)
12/13 Sep
FRANKFURT
1827-0144
SL T D Misselbrook
FO J Parr
FL E Keuffling
FL J P Blackie
FO L A Cox
FO H Yates
FO H Cornforth
16/17 Sep
STEENWIJK A/F
2159-0131
FO J Dawson and crew
20 Sep
SANGATTE
1528-1901
FO G G Kennedy
Sgt G W Soundy
Sgt H Luxton
Sgt H H Powell
Sgt R T Wesley
Sgt A Frame RCAF
Sgt J Hogg
23 Sep
NEUSS
1900-2338
FO G G Kennedy and crew
25 Sep
CALAIS
0720-1059

FO S H Hayter
Sgt L A Bassman
FO T Y Thomas
FO R R Bradshaw RNZAF
Sgt A J Pearce
Sgt Mills
Sgt F E Self
26 Sep
CALAIS
1112-1409
FO S H Hayter and crew
Sgt E M Watkins in, Sgt Mills
out
27 Sep
CALAIS
0852-1212
FO S H Hayter and crew
28 Sep
CALAIS
0803-1153
FO G G Kennedy and crew
(raid abandoned over target
due to cloud)
3 Oct
WALCHEREN
1307-1609
FL M L Dubois and crew
5 Oct
SAARBRUCKEN
1847-0051
FO G E Blackler
Sgt W R Ross (FE)
Sgt H P Nicholls (N)
FS J W Bold (BA)
Sgt E Mozley (WOP)
Sgt J Nicholson (MU)
Sgt M McCutcheon (RG)
11 Oct
FT FREDERIK HENDRIK
1458-1813
FO G W Bell RCAF
Sgt R J McElroy
FO D C R Hills RCAF
FO B H Lowen
Sgt J F Noonan
Sgt D E Hookham
Sgt R West
14/15 Oct
DUISBURG
2220-0405
PO E S Allen
Sgt E T Smith
Sgt E B Dennison
Sgt G A Maginley
Sgt R W Lundy
Sgt E G Crump
PO J B M Sherreff
19 Oct
STUTTGART
1704-2013
FO G E Blackler and crew
(sortie aborted due to port
outer engine going u/s)
23 Oct
ESSEN
1628-2154
FO G E Blackler and crew
25 Oct
1255-1800
ESSEN
FO G W Bell and crew
28 Oct
COLOGNE
1325-1827
PO R P Franklyn RAAF
Sgt S Dennis
Sgt R H Lucas

FS J A Dibley RAAF
FS K J Chester RAAF
Sgt K Sharp
Sgt P J James
30 Oct
COLOGNE
1746-2356
FO J C Adams RCAF
Sgt W P Scott
FS B Sterman
FO W R Elcoats
Sgt F Papple
PO F S Benton
Sgt K D Winstanley
31 Oct
COLOGNE
1753-2327
FO G E Blackler and crew
2 Nov
DUSSELDORF
1620-2147
FO G E Blackler and crew
4 Nov
BOCHUM
1729-2239
FO G E Blackler and crew
6 Nov
GELSENKIRCHEN
1127-1624
FO G E Blackler and crew
9 Nov
WANNE EICKEL
0727-1317
FO G E Blackler and crew
11 Nov
DORTMUND
1600-2126
FO G E Blackler and crew
16 Nov
DUREN
1257-1830
FO G E Blackler and crew
18 Nov
WANNE EICKEL
1539-2203
FO G E Blackler and crew
27 Nov
FREIBURG
1548-2245
FO G E Blackler and crew
29 Nov
DORTMUND
1119-1734
FO G E Blackler and crew
4 Dec
KARLSRUHE
1631-2305
FO J C Adams and crew
6/7 Dec
MERSEBURG
1642-0037
FO G E Blackler and crew
12 Dec
ESSEN
1623-2231
FO C J Clarke RCAF
Sgt J T Tunstall
FS H E Meill RCAF
FO A L Caldwell
Sgt L O Precieux
Sgt F W Bradley
Sgt L A Gauthier RCAF
15 Dec
LUDWIGSHAVEN
1442-2035
FO C J Clarke and crew
(FO Clarke and crew failed to

return 8 January 1945)
22 Dec
KOBLENZ
1522-2200
FO G E Blackler and crew
24 Dec
COLOGNE
1439-2058
FO G E Blackler and crew
29 Dec
SCHOLVEN-BUER
1500-2109
FO G E Blackler and crew

1945
16/17 Jan
ZEITS
1728-0127
FO G E Blackler and crew
22 Jan
DUISBURG
1659-2255
PO G E Mearns
Sgt F A Norris
FS W G Kelly RCAF

FS J F McKeown
Sgt D O'Neill
Sgt J Chambury
Sgt A G Slater
1 Feb
LUDWIGSHAVEN
1533-2239
FO G E Blackler and crew
7/8 Feb
CLEVE
1838-0034
FO G E Blackler and crew
8/9 Feb
STETTIN
1903-0356
FO G E Blackler and crew
13/14 Feb
DRESDEN
2136-0741
FL J Jarvis
Sgt I S Freeman
FO E B Hornsby
FS R W Richardson
FS P F Flux
FS Rees
FS R West RCAF

14/15 Feb
CHEMNITZ
2010-0541
FO R D Harris RCAF
Sgt K J Smith RCAF
FS D J Yemen
FO J C Nicol RCAF
Sgt G P Kelleher
Sgt W Towle
Sgt D J Hicks
20/21 Feb
DORTMUND
2108-0351
FO G E Blackler and crew
21/22 Feb
DUISBURG
1914-0131
FO G E Blackler and crew
23 Feb
PFORZHEIM
1544-2341
FO G E Blackler and crew
28 Feb
NEUSS
0816-1044
FO G E Blackler and crew

(raid abandoned on recall)
1 Mar
MANNHEIM
1137-1807
FO G E Blackler and crew
2 Mar
COLOGNE
0647-1220
FO G E Blackler and crew
5/6 Mar
CHEMNITZ
1635-0159
FO G E Blackler and crew
PO R F Wallace RNZAF (2P)
7 Mar
DESSAU
1659-
FO C J Jones RCAF
Sgt S J Webb RCAF
FO J Buckmaster
WO L W Harvey
FS F M Main
Sgt M B Smith
Sgt Pelham
(Failed to Return)

PB150

An Avro-built Mk III, this was part of order No 1807/C4A and came off the production line in the spring of 1944, with four Merlin 38 engines. Assigned to No 625 Squadron at Kelstern, Lincolnshire, the squadron code letters CV were painted on the aircraft's sides with the individual identification letter 'V'. Ops began shortly after D-Day, taking in the full range of Bomber Command targets during that summer which included V1 sites, rail centres and ammunition dumps, as well as supporting Allied troops in Normandy. These also included the Caen attack prior to the Allied breakout of the bridgehead on 18 July, and numerous oil production centres, which became another priority target in late 1944.

PB150's first regular captain was a New Zealander, Pilot Officer Trever G. Wilson who would take the aircraft on 22 sorties prior to finishing his tour in September, winning the DFC, then going to No 11 OTU as an instructor. He was followed by an Australian, Pilot Officer E. P. Twynam who went on 11 sorties, but he was then lost in PB154 on 4 November during a raid on Bochum.

Although reputed to have achieved 100 ops, the only photograph discovered of PB150 has the total number of bomb symbols obscured by the engine, although 80 are clearly visible, together with the insignia of a duck standing on a bomb. The squadron Form 541 lists 93 sorties, which include five 'Manna' food drops, depicted on the nose by five tins of Spam. It was while flying these sorties that the aircraft is supposed to have reached its century.

On 10 August 1945 it went to No 38 MU, then No 32 MU the following day, and back to No 38 MU in March 1946, before finally being Struck off Charge on 22 May 1947. Giving this Lanc the benefit of the doubt, its known ops are listed below. One problem was that Lancaster PB158 was also on the strength of No 625 Squadron and one wonders if someone misread an eight for a zero and counted a few of this Lanc's ops in PB150's total?

1944	2228-0310	25 Jun	1223-1539
No 625 Squadron	PO T G Wilson RNZAF	LIEGESCOURT/V1 SITEFL	Sgt A P Sims
16/17 Jun	Sgt T H Howie	0732-1022	Sgt F Brighton
DOMLEGER/V1 SITE	Sgt R L Baker	PO T G Wilson and crew	Sgt D M Craig RCAF
0022-0350	FS D J Rant	D W Webber DFC in, Sgt	Sgt S F Robertson
PO F Collett	Sgt H Raine	Hagues out	Sgt H Anderson
Sgt N Jones (FE)	Sgt Callas	27/28 Jun	Sgt P M E Grandmaison
FO N R Lott (BA)	Sgt C E Tennant	VAIRES	Sgt R J Farr
FO J Stephenson (N)	24 Jun	0025-0503	30/1 Jul
Sgt E Evison (WOP)	LE HAYONS/V1 SITE	PO T G Wilson and crew	VIERZON
Sgt L Naylor (MU)	1554-1919	Sgt R A Rimmer in, FL	2214-0355
FS W A Peterkin (RG)	PO T G Wilson and crew	Webber out	FO W E B Mason RCAF
22/23 Jun	Sgt G T Hagues in, Sgt Callas	29 Jun	Sgt F V Walton
REIMS	out	SIRACOURT/V1 SITE	FO J A Noble RCAF

FO H J Collison RCAF
Sgt C Saunderson
Sgt H G Lee
Sgt K Steele
2 Jul
OISEMONT/V1 SITE
1211-1525
PO F G Parker
Sgt L Brice
Sgt E C Williams
Sgt E J Coupland
PO D H Hutchinson
FO J A Knight
Sgt W Reed
4/5 Jul
ORLEANS
2221-0403
PO T G Wilson and crew
6 Jul
FORET-DU-CROC/V1 SITE
1849-2223
PO T G Wilson and crew
7 Jul
CAEN
1936-2322
PO T G Wilson and crew
12/13 Jul
TOURS
2128-0352
PO T G Wilson and crew
Sgt R T Squire in, Sgt Rimmer
out
14/15 Jul
REVIGNY
2123-0534
PO T G Wilson and crew
(raid aborted by Master
Bomber as target could not
be identified)
18 Jul
SANNEVILLE/CAEN
0339-0715
PO T G Wilson and crew
20 Jul
WIZERNES/V1 SITE
1916-2238
PO T G Wilson and crew
FS E W Sadler in, Sgt Haine
out
23/24 Jul
KIEL
2247-0336
PO T G Wilson
FO R M B Cairns (2P)
Sgt R D R Rees (FE)
Sgt R L Baker (BA)
PO P A C Ansell (N)
FS E W Sadler (WOP)
Sgt R T Squire (MU)
Sgt C E Tennant (RG)
25 Jul
ARDUVAL II
0649-1032
FO R M B Cairnes
Sgt R D R Rees
Sgt J Crew
FO R J Cann
PO P A C Ansell
Sgt R Perry
Sgt W Richie
28/29 Jul
STUTTGART
2122-0601
PO T G Wilson and crew
31/1 Aug
FORET DE NIEPPE/V1 SITE
2212-0145

FO E A Eckel RCAF
Sgt T Ord
FO K R Menzies RCAF
FS B G Little RCAF
Sgt W B Codd RCAF
Sgt F Onyski RCAF
Sgt J R Hammond RCAF
3 Aug
TROSSY ST MAXIM
1137-1602
WO R B Pattison
Sgt R T Bryer
FS D P Ross RAAF
FS E G Russell
FS F G A McConnell
Sgt A Jacques
Sgt A Murray
4 Aug
PAUILLAC
1330-2125
WO R B Pattison and crew
Sgt C Cletheroe in, Sgt
Jacques out
5 Aug
PAUILLAC
1420-2218
FL R Banks
Sgt M P Mary
FS J E Tooth
Sgt S T Kerr
Sgt R W Small
Sgt G T Hagues
Sgt W R Bates
7/8 Aug
FONTENAY
2052-0113
FO E A Eckel and crew
10 Aug
OEUF-EN-TERNOIS/V1 SITE
1058-1450
FO T G Wilson and crew
11 Aug
DOUAI
1352-1735
PO J N Harvey RAAF
Sgt W A Edwards
FO J P M Brady RAAF
FS R T Williams RAAF
FS J W Smith RAAF
FS F J Allnutt RAAF
FS H R Brady RAAF
12 Aug
BRUNSWICK
2119-2340
FO T G Wilson and crew
(sortie aborted due to engine
trouble)
14 Aug
FONTAINE LE PIN
1324-1721
FO T G Wilson and crew
15 Aug
VOLKEL A/F
1008-1328
PO J N Harvey and crew
16/17 Aug
STETTIN
2100-0455
FO T G Wilson and crew
18 Aug
VINCLEY
1915-2305
PO E P Twynam RAAF
Sgt B H Fetch
FO G W Brown RAAF
Sgt D J Lincoln (N)
Sgt D Bousefield

Sgt A West
Sgt J Jones
25/26 Aug
RUSSELSHEIM
2019-0457
FO T G Wilson and crew
26/27 Aug
KIEL
2013-0203
FO T G Wilson and crew
29/30 Aug
STETTIN
2111-0613
FO T G Wilson and crew
31 Aug
RAIMBERT
1243-1600
FO T G Wilson and crew
3 Sep
GILZE RIJEN A/F
1544-1919
FO T G Wilson and crew
FL R C Gordon DFM RNZAF
(BA) in
5 Sep
LE HAVRE
1643-2013
PO W Hutchinson
Sgt R A Greig
FL R C Gordon DFM (BA)
FS J Alcock (2BA)
PO J Ockieshaw
FS A A Soloman RAAF
Sgt J M Lehrie
6 Sep
LE HAVRE
1702-2039
PO W Hutchinson and crew
10 Sep
LE HAVRE
1647-2017
WC D D Haig DFC
Sgt A L Robotham
FO H J Binns RCAF
FS J Davies
FO D J Travis
Sgt A Jacques
PO S A L Beacroft
12/13 Sep
FRANKFURT
1830-0136
PO E P Twynam and crew
16/17 Sep
RHEINE A/F
0050-0444
PO E P Twynam and crew
FO P L S Hathaway in, Sgt
Lincoln out
17 Sep
ELKENHORST
1716-2011
PO C D Mattingley RAAF
Sgt C E Bailey
FS A Fisher RAAF
PO R R Murr RAAF
FS R G Watson RAAF
FS N Ferguson RAAF
FS A J Avery RAAF
20 Sep
CALAiS
1514-1910
PO L A Hannah RCAF
Sgt G Maynard
FS T M Baird RCAF
FS K R Strachen RCAF
Sgt J Soule

Sgt G E Way RCAF
Sgt J H LoughranRCAF
26 Sep
CALAIS
1008-1323
FO E A Eckel and crew
Sgt J E Cunliffe RCAF in, Sgt
Ord out
3 Oct
WESTKAPELLE
1234-1531
PO E P Twynam and crew
5 Oct
SAARBRUCKEN
1847-0150
PO E P Twynam and crew
7 Oct
EMMERICH
1152-1606
PO E P Twynam and crew
14 Oct
DUISBURG
0624-1100
PO E P Twynam
FO C K Cochran RCAF
Sgt T Ritchie (FE)
Sgt R Murphy (BA)
FS D J Lincoln (N)
WO R A SpencerRCAF
Sgt B A Mead
Sgt Jones
14/15 Oct
DUISBURG
0031-0539
PO R A Court RCAF
Sgt P Garrison
Sgt E J Burke RCAF
Sgt G P Wright
FS G W H Smith RAAF
Sgt J W Gilpin RCAF
Sgt W F Larson RCAF
20/21 Oct
STUTTGART
2113-0404
PO E P Twynam and own
crew
25 Oct
ESSEN
1249-1446
FO J L Lane
Sgt T H Jakes
Sgt B J Mitchell
FO P R Talboys
FS E L M Bear RAAF
Sgt K Southwood
Sgt P J Kettle
30 Oct
COLOGNE
1738-2307
PO E P Twynam and crew
31 Oct
COLOGNE
1748-2248
PO E P Twynam and crew
2 Nov
DUSSELDORF
1610-2128
PO E P Twynam and crew
4 Nov
BOCHUM
1734-2228
PO P H Allen
Sgt G K Williams
Sgt W H Bell
Sgt D S Barker
Sgt C S Cathie
Sgt E G Whiteman

Sgt M Martin
6 Nov
GELSENKIRCHEN
1158-1612
PO P H Allen and crew
9 Nov
WANNE EICKEL
0826-1241
FO A Fulbrook
Sgt E G E Philips
FS A E Napper
Sgt D W Tizard
Sgt W Magill
Sgt A R Huxtable
Sgt I S Goodman
11 Nov
DORTMUND
1558-2123
FO W J Bulman
Sgt M Miles
FO J A Cottrell RCAF
FO R L Stevenson RCAF
Sgt E G Thale
Sgt L Smith
Sgt O R Sharman RCAF
16 Nov
DUREN
1246-1733
FO D R Ward RAAF
WC J L Barker (2P)
Sgt M J Harris
FS J R Rudd
FO R H Thomson RAAF
FS R N Powell RAAF
FS C Sykes
Sgt R Murcott
18 Nov
WANNE EICKEL
1605-2120
FL A B Fry
Sgt A L Sykes
FO G G Davies
Sgt J Corrigan
FS J Soule
FO T A W Harper
FS J McCandlish
21 Nov
ASCHAFFENBURG
1550-2210
FO J S Bray RCAF
Sgt E Foakes
FO H J Taylor
FO J Stevenson
Sgt R C Turner
FS D E Fumerton
FS B Ogrodnick
27 Nov
FREIBURG
1631-2301
FO H W Hazell
Sgt J O Pulford
FO M J Shenton
Sgt S Sellers
FS J Soule
Sgt A W Hall
Sgt W J Harrison
29 Nov
DORTMUND
1245-1721
FO W L Russell
Sgt S V Inwood
FO B J K Challes
FO W J Drysdale RCAF
Sgt J Gilchrist
Sgt G M Tulk RCAF
Sgt R F Cook

4 Dec
KARLSRUHE
1635-2309
FL A B Fry and crew
FSs J N Wadsworth and D G
McHardy in;
FO Davies and FS Soule out
6 Dec
MERSEBURG
1639-0031
FO T W Alexander RCAF
Sgt C C Lear
FO F R Chapman RCAF
FO W Petrachenko
Sgt G W Morgan
Sgt R Pyett
Sgt J B Williams
12 Dec
ESSEN
1629-2200
FO W L Russell and crew
Sgt G Attenborough and FS G
A Turner in;
Sgts Inwood and Cook out
15 Dec
LUDWIGSHAVEN
1438-2121
FO W L Russell and crew
24 Dec
COLOGNE
1512-2004
PO J Jamieson
Sgt E J Cuthil! (FE)
Sgt C G Aickin (BA)
FO E W O'Reilly (N)
Sgt D E Smith (WOP)
FS H W Elliott (MU)
Sgt E Wilson (RG)
26 Dec
ST VITH
1317-1713
PO J Jamieson and crew
28 Dec
MUNCHEN GLADBACH
1533-2052
FL A B Fry and crew
FS T J McLeod in, FS McHardy
out;
FSs D O O'Malley and R R Job
in;
FO Harper and FS
McCandlish out

1945
2 Jan
NURNBERG
1436-2312
FL A B Fry and crew
FS K E Campbell in, FS
McCandlish back;
FSs O'Malley and Job out
5/6 Jan
HANNOVER
1913-0014
FL A B Fry and crew
FL J R W Orr in, FS Campbell
out
6 Jan
HANAU
1604-2206
PO J K English RCAF
Sgt J A Munday
FS T M Baird RCAF
FO H R Gottfried
Sgt M T Chalk
Sgt B A Thomas

Sgt G A Stowe
14/15 Jan
MERSEBURG
1908-0310
FL A B Fry and crew
FO T A W Harper in, FL Orr
out
16/17 Jan
ZEITZ
1733-0145
FO D R Paige RCAF
Sgt R B Bennett
FS J A Puttick
WO J P Sullivan RCAF
Sgt J Wallace
Sgt K E Campbell RCAF
FS J K McRorie RCAF
7/8 Feb
KLEVE
1915-0056
PO C R Applewhite
Sgt R H T Winpenny
FS H G O Allan
FS C Wilson
Sgt A T Kaye
Sgt R E Edwards
Sgt C E H Cole
8/9 Mar
KASSEL
1732-0103
FO B C Windrim RAAF
FS J E Platt
Sgt F S Tolley
FS W Porter
FS D E R Steen
Sgt S C Simmonds
Sgt J W Slater
11 Mar
ESSEN
1140-1706
FL J R Forsythe
Sgt J W Stevens
FO K H Gindley
FO G Baker
FS W A Hughes
FS R W Liversedge
FO C E Jones DSM
15 Mar
MISBURG
1723-0109
FL P Lennox
Sgt D Abbott
FO E J Harbord
FO M Brook
Sgt R G H Wilsdon
Sgt K Cowley
Sgt W Birkey
16/17 Mar
NURNBERG
1745-0205
FL P Lennox and crew
18 Mar
HANAU
0039-0747
FL P Lennox and crew
22 Mar
BRUCHESTRASSE
0108-0705
FL P Lennox and crew
23 Mar
BREMEN
0745-1206
FL P Lennox and crew
3 Apr
NORDHAUSEN
1328-1955

FL P Lennox and crew
4/5 Apr
LUTZKENDORF
2121-0602
FS M McDermott RAAF
Sgt E Fryer
Sgt D Johnson
Sgt G A Baillieu
Sgt C Wood
Sgt J Stock
Sgt E Stevens
9/10 Apr
KIEL
1853-0200
FL P Lennox and crew
14/15 Apr
POTSDAM
1808-0320
FL P Lennox and crew
22 Apr
BREMEN
FO J K English and crew
(sortie aborted over target by
Master Bomber due to cloud
and smoke)
25 Apr
BERCHTESGADEN
0545-1353
PO K E Fife
FS J Flower
FS J G Morgan
FS F M Thomson
WO L W Keen
Sgt L D Coiner
Sgt H F Coakwell
29 Apr
'MANNA'/THE HAGUE
1150-1520
PO K E Fife and crew
2 May
'MANNA'/ROTTERDAM
1236-1537
PO K E Fife and crew
3 May
'MANNA'/ROTTERDAM
1215-1515
PO W D Street
FS D F Gumbley
FS L C P Cheng RAAF
FS R D Given RAAF
Sgt W H G Bond
Sgt J R Jenner
Sgt J G King
7 May
MANNA'/ROTTERDAM
1245-1613
FS W J Norton
Sgt L Rickier
FO F W Wreggitt RCAF
Sgt E V Alkenbrack RCAF
Sgt T Langley
Sgt A Hutcheon
Sgt L J Clarkin
8 May
'MANNA'/THE HAGUE
1108-1440
FO H J McMonagie
Sgt D J Kegan
FS W Fowler
FS E O Butts
FS D H Irvine
Sgt J Craig
Sgt A D Williams

Appendix A

NIGHT PLANE TO BERLIN
by Sergeant Ben Frazier, *Yank* Staff Correspondent

The following account was featured in the American magazine *Yank* following Ben Frazier's trip to Berlin in ED888 on the night of 29/30 December 1944, when ED888 was V-Victor in No 576 Squadron at Elsham Wolds. The aircraft's crew on this night was Flying Officer G. S. 'Taff' Morgan, Sergeant J. R. 'Jock' Mearns, navigator, Flight Sergeant N. A. 'Digger' Lambrell RAAF, bomb-aimer, Pilot Officer E. M. Graham, flight engineer, Sergeant J. R. 'Blondy' O'Hanlon, wireless operator, Sergeant A. Newman, mid-upper gunner, Sergeant C. E. 'Bob' Shilling, rear-gunner.

England. A small village lay tucked away in the fold of a valley just below the high, windswept, bleak plateau where a Lancaster bomber station was situated. Housewives were busy in the kitchen preparing food, and the men had left their ploughing to come in for the noon-day meal. In the lichen covered Gothic Church, the minister's wife was arranging decorations, and placing on the altar freshly cut chrysanthemums that had managed to escape the north winds and were still blooming in December.

  The placidness of the village life was in sharp contrast to the bustling activity at the airfield. It seemed as remote from war as any hamlet could possibly be, although the provident farmers, living so close to an obvious military target had wisely provided themselves with shelter trenches at the edge of each ploughed field. Nevertheless, the name of this quiet, lovely village, had spread far. By borrowing it, the bomber station had made it one to strike terror into the heart of the Nazi High Command.

  At the airfield, V for Victor's crew lounged around B Flight's Office waiting to see if operations were on. They kept looking up into the sky as if trying to guess what the weather was going to be like. Some of the men chuckled. 'Papa Harris is so set on writing off the Big City that he hardly even notices the weather,' one of them said. 'The last time, there were kites stooging around all over the place. The met boobed that one.'

  It was a strange new language. What the airmen were saying was that the last time out the meteorological men had given a wrong steer on the weather, and the planes had been flying all over looking for the field on the return trip. 'Papa' Harris was Air Chief Marshal Harris, chief of Bomber Command.

V for Victor's captain came back from the operations room with the news that there would be ops. That settled the discussion. You seemed to be aware, without noticing anything in particular, of a kind of tension that gripped the men; like they were pulling in their belts a notch or two to get set for the job ahead.

And with the news, everybody got busy – the aircrews, the ground crews, the mechanics, the Waafs, the cooks. The ships already had a basic bomb and fuel load on board, and the additional loads were sent out in ammunition trailers and fuel trucks. The perimeter track lost its usually deserted appearance and looked like a well travelled highway, with trucks and trailers, buses and bicycles hurrying out to the dispersal points. It was just like the preparation at any bomber base before taking off for enemy territory – but going over the Big City was something different. These men had been there before. They knew what to expect.

In the equipment room, June, the pint-size Waaf in battledress, was an incongruous note. Over a counter as high as her chin, she flung parachutes, harnesses and Mae Wests. The crew grabbed them and lugged them out to the ships. You kept thinking they ought to be able to get somebody a little bigger for the job she was handling.

In the briefing room, the met officer gave the weather report and the forecast over enemy territory. There would be considerable cloud over the target. The men grinned. An operations officer gave a talk on the trip. The route was outlined on a large map of Germany on the front wall. It looked ominously long on the large scale map. He pointed out where the ground defences were supposed to be strong and were fighter opposition might be expected. He gave the time when the various phases should be over the target. He explained where the 'spoof' attacks were to be made and the time. He told the men what kinds of flares and other markers the Pathfinders would drop. There was the usual business of routine instructions, statistics and tactics to be used. The Group Captain gave a pep talk on the progress of the Battle of Berlin. And all the while, that tape marking the route stared you in the face and seemed to grow longer and longer.

Outside it was hazy and growing more so. But this was nothing new. The men were convinced that the weather was always at its most variable and its dampest and its haziest over their field. What could you expect? Ops would probably be scrubbed after all. Hell of a note.

In the fading light the planes were silhouetted against the sky. They looked, on the ground, slightly hunched and menacing like hawks. Seeing them there, in the half light you would never guess how easy and graceful they are in flight. Nor would you realise when you see them soaring off the runway, what an immense load they take up with them. It is only when you see the open bomb bay on the ground, that you get some idea of a Lancaster's destructive power. The open bomb bay seems like a small hangar. The 4,000lb block buster in place looks like a kitten curled up in a large bed. It is a sobering sight.

In the evening some of the men tried to catch a few winks; most of them just sat around talking. The operational meal followed. It was only a snack, but it was the last solid food any one would get until the fresh egg and bacon breakfast which has become a ritual for the proper ending of a successful mission.

As there was still some time to wait before take-off, V for Victor's crew sat around the groundcrew's hut near the dispersal point, warming themselves by the stove or chewing the rag with the groundcrew. The Wingco came around to make a last minute check-up. The medical officer looked everyone over. The engineer officer checked the engines.

The minutes crept by until at last the time came to get into the planes. The deep stillness of the night was awakened by the motors revving up, one after another until each one was lost in the general roar. The crews scrambled into the planes and took their places. The great ships were guided out of their dispersal areas by the ground crews who gave a final wave as the Lancs moved off slowly down the perimeter track. They appeared more menacing than ever creeping along in the dark with their motors roaring. One by one they turned onto the runway and noisily vanished into the night.

From now on, until they would return, the members of V for Victor's crew were a little world in themselves, alone and yet not alone. For all around them were other similar little worlds, hundreds of them with a population of seven, hurtling through space, lightlessly – huge animated ammunition dumps. For its safety, each little world depended utterly and completely on its members – and a large dash of luck.

There was not much conversation over the intercom. When you're flying without running lights on a definite course, and surrounded by several hundred other bombers, you have not time for any pleasantries. The navigator was busy checking the air speed and any possible drift. Almost everyone else kept a look out for other aircraft, both friend and foe. A friendly aircraft is almost as dangerous as an enemy plane, for if two block busters meet in mid-air, the pieces that are left are very small indeed.

Occasionally the ship jolted from the slipstream of some unseen aircraft ahead, and frequently others overhauled V for Victor, passing by to port and starboard, above and below. V for Victor gained altitude very easily for maximum ceiling. She was a veteran of over 50 ops and had the DFC painted on her port bow to celebrate the fiftieth, but she had the vitality of a youngster. Blondy, the wireless-operator, broke the silence. 'Taff, the W/T has gone u/s.'

The wireless is not used except in an emergency such as ditching, but it is nice to know it's there. We went on. Occasionally Taff, the pilot, would call into the intercom, 'Bob, are you OK?' There would be a silence for a moment while the rear-gunner fumbled to turn on his intercom, until you wondered if he had frozen back there. Then he'd sing out, 'OK, Taff.' He and the mid-upper gunner were the only two outside the heated cabin. Inside the cabin it was warm and snug. You didn't even need gloves. Jock, the navigator, wore no flying gear, just the Air Force battledress.

Up ahead the Pathfinder boys dropped the first route markers, flak shot up into the air and the men knew that V for Victor was approaching the Dutch coast. An enormous burst of flame lit up the night off to port. 'Scarecrow to starboard,' the mid-upper reported on the intercom. Jerry intended the 'scarecrow' to look like a burning plane but it did not take long to see that it was not.[1]

Jock's Scots accent came over the intercom: 'Taff, we're eleven minutes late.' 'OK, we'll increase speed.' The engineer pushed up the throttles. Everything was black again below. Occasionally there was a small burst of flak here and there.

'Plane to starboard below!'

'OK, it's a Lanc.' As V for Victor passed it you could seen the bluish flame from the exhausts lighting the aircraft below in a weird ghostly manner. It was unpleasant to realise that our own exhausts made V for Victor just as obvious as the other plane.

Away off the port bow, a glow became visible. It looked like the moon but it was the first big German searchlight belt, encompassing many cities. The beams were imprisoned under cloud.[2]

'That will be Happy Valley,' (the Ruhr) Jock said. Another route marker appeared ahead.

'Tell me when we're over it,' the navigator replied. Shortly the bomb-aimer said, 'We're bang over it now.'

'OK, Digger.'

'Taff, we're nine minutes late.' The navigator took a couple of astro sights to get a fix. From this he could determine the wind and the drift of the plane.

Another searchlight belt show up to starboard. It was enormous, running for miles and miles. It was all imprisoned under the cloud but it was an evil looking sight just the same.[3] The top of the clouds shone with millions of moving spots, like so many restless glow worms, but the impression was much more sinister – like some kind of luminous octopus. The tentacle-like beams groped about seeking some hole in the cloud, some way of clutching at you as you passed by protected by the darkness. The continuous motion of the searchlights caused a ripple effect on the clouds, giving them an agitated, angry, frustrated appearance. Once in a while one found a rift and shot its light high into the sky. Flak came up sparkling and twinkling through this luminous blanket. V for Victor jolted violently from close bursts, but was untouched. It passed another Lanc which was clearly silhouetted against the floodlit clouds.

Another leg of the trip was completed. The navigator gave the new course over the intercom and added; 'Seven minutes late.'

'OK, Jock. Mac, make it 165.'

V for Victor passed plane after plane and occasionally jolted in the slip-stream of others. A third searchlight belt showed up, this one free of cloud. It was a huge wall of light and looked far more impenetrable than a mountain. It seemed inconceivable than any plane could pass through and reach the opposite side. You thanked your lucky stars that this was not the target. To fly out of the protecting darkness into the blaze of light would be a test of courage you would rather not have to face.

Nevertheless, there were some facing it right now. The flak opened up and the searchlights waved madly about. It was a diversionary attack, the 'spoof'. You watched in a detached, remote sort of way. It seemed very far away and did not seem to concern you at all. Until suddenly, one beam which had been vertical, slanted down and started to pursue V for Victor, and you realised that it did concern you very intimately. The seconds ticked by as the beam overtook the plane. But it passed harmlessly overhead and groped impotently in the darkness beyond.

'Four minutes late,' Jock called over the intercom.

The target itself, the Big City, came into view like a luminous patch dead ahead. It was largely hidden by cloud and showed few searchlights. It seemed so much less formidable than the mountain of light just behind, that it came as a sort of anticlimax. Surely, you felt, this cannot be the Big City, the nerve-centre of Europe's evil genius.

It was quiet. There was no flak as yet, no flares, and just the handful of searchlights. You tried to imagine what it was like on the ground there. The sirens would be about to sound, the ack-ack batteries would be standing ready, the search-lights already manned. You wondered if the people were in shelters.

But it was too much of an effort. It was too remote. Your problems were flak, fighters, searchlights and whether you were on the course and on time. What happened below was an entirely different problem which had nothing to do with you. What happened below might just as well be happening on Mars. V for Victor's own little world simply hovering off this planet and leading a life of its own.

Ever so slowly V for Victor crept up on the target. The two worlds were coming inevitably together. But it still had the quality of unreality. It was like a dream where you were hurrying somewhere and yet cannot move at all. Nevertheless, Victor was passing plane after plane and jolted in somebody's slip-steam now and again. The other Lancs looked ominous bearing down on the target, breathing out blue flame as they approached.

The minute of the attack and still the target was quiet. One more minute ticked by – still quiet. The engineer opened up the throttles to maximum speed and increased the oxygen supply. Still quiet. The whole attack was a minute or two late. Winds, probably. Suddenly the whole city opened up. The flak poured up through the clouds. It came in a myriad of little lights. It poured up in a stream of red, as if shaken from a hose. It would be impossible to miss such a brilliantly marked objective. Bright flashes started going off under the clouds. That would be the cookies from the planes ahead. V for Victor started the bombing run. The bomb-aimer called the course now.

'Left, left...Steady now...Right a bit...Steady...steady...Cookie gone!' V for Victor shot upward slightly. 'Steady...Incendiaries gone...' V for Victor surged again ever so slightly.

'Stand-by, Taff,' it was the voice of Bob, the tail-gunner. 'Fighter.'

Instantly the pilot sent V for Victor over to starboard and rushed headlong downward. A stream of red tracer whipped out of the dark, past the rear turret, and on past the wing-tip, missing by what seemed inches. A second later the fighter itself shot past after the tracer, a vague dark blur against the night sky.

'Me109,' Bob said calmly.

V for Victor squirmed and corkscrewed over the sky of Berlin. You wondered how it could be possible to avoid all the other planes that were over the city. But the fighter was shaken off and V for Victor came back to a normal course again.

Down below through rifts in the cloud, you could see that Berlin was burning. The bright, white flame of the incendiaries showed up as a carpet of light, always growing. And flash after flash went off as the block busters fell. The dark, black shapes of many Lancasters could be seen all over the sky, against the brilliant clouds below. They were like small insects crawling over a great glass window. It did not seem possible that these tiny black dots could be the cause of the destruction which was going on below. The insects crawled to the edge of the light and disappeared into the darkness beyond. They had passed safely through the target, V for Victor close behind.

Shortly the course was set for the return and Berlin was visible for many miles on the port quarter. The attack was over now. It took only fifteen minutes. The ack-ack was silent. There was no flak flashing over the city, but the city was brighter than ever. The clouds were getting a reddish tinge which showed that the fires had caught hold below.

And so the capital of Nazism dropped astern, obscuring the rising moon by its flames. The Government which came into power by deliberately setting fire to its chamber of representatives, the Government which first used wholesale bombing, and boasted of it, was now perishing in fires far more devastating than any it ever devised. It was perishing to a fire music never dreamed of by Wagner.

But it was impossible to connect V for Victor with the death struggles of Berlin. There was no time for contemplation.

'Stand-by, Ju88 starboard – corkscrew,' came Bob's voice. Again with lightning speed, the pilot put V for Victor over and dived out of the way. The Ju88's tracers missed us and shot down another Lanc which had not been so fortunate.

After that the route home was uneventful. Crossing the North Sea, V for

Victor went into a gentle incline towards home base, as if by a sort of homing instinct. The searchlights of England sent out a greeting of welcome. For miles along the coast they stood almost evenly spaced, vertical sentries guarding the island. Then they started waving downwards in the direction of the nearest airfield. No doubt they were helping home a damaged bomber. How different they were from the menacing tentacles over the German cities. V for Victor arrived over the home field. The wireless-operator called base over his repaired equipment. He said simply, 'V-Victor'.

The clear voice of a girl came pleasantly over the intercom, 'V-Victor, prepare to pancake'. The short business-like message in service slang was a wonderful welcome home. V for Victor circled the field, losing altitude.

'V-Victor in funnels.'

'V-Victor, pancake,' the girl's voice said. V for Victor touched down, ran down the flarepath, and turned off on the perimeter track.

'V-Victor clear of flarepath.' The groundcrew met V for Victor and acted as a guide back into the dispersal area.

'How was it?'

'Piece of cake,' someone said. The crew got out, collected their gear, the parachutes, Mae Wests, the navigator's bag, the guns, etc, and then, as one man, lit up cigarettes. The pilot walked around the plane looking for any damage. There was one small hole through the aileron but it was too dark to see it then.

The bus arrived and the crew clambered in with all the gear and were taken back to the locker room. June was there, and gathered all the stuff over the counter and staggered away, lost from sight under a mound of yellow suits and Mae Wests.

Then back to the briefing room where a cup of hot tea with rum in it was waiting. Each captain signed his name on the board as he came in. Crew by crew, the men went into the Intelligence room, carrying their spiked tea with them. There were packages of cigarettes on the table and everyone chain-smoked, lighting up from the butt of the previous one.

The Intelligence Officer asked brief questions and the replies were brief, such as 'The heavy flak was light and the light flak heavy'. It was over in a very few minutes and you went back to the briefing room and banterred over the trip with the other crews. No trouble, any of them, but there were gaps in the list of captains chalked on the board.

'It's like that,' the Wingco remarked. 'In night flying, you usually get back intact, or you don't get back at all. If you get coned, or a fighter sees you before you see it, then very often you've had it, but if somebody else gets coned then its that much easier for you.'

You thought of the other Lancaster the Ju88 got with the same burst that missed V for Victor. And you lit another cigarette. The first signs of dawn were coming over the field now and off in the distance, on the bleak, windswept, little knoll, V for Victor stood guard over the empty dispersal points from which other men and ships had gone out a short while before. '...if somebody else gets coned then it's that much easier for you.'

[1] It was only after the war that it was discovered that the Germans did not use an explosive device to simulate an exploding bomber. What the men saw, in fact, was a fully loaded bomber exploding, having either been hit by flak or night fighter attack.

[2] The 'spoof' raids this night were to Dusseldorf and Leverkusen.

[3] This was Leipzig, where the bomber stream appeared to be heading before turning north-east for Berlin.

Appendix B

CENTURION LANCASTERS BY SQUADRON AND BY GROUP

Squadron

No 9	W4964, EE136, LL845
No 12	ME758
No 15	LL806
No 44	ED611, LL885, ND578
No 50	ED588
No 61	ED860, EE176, JB138, LL843
No 75	NE181
No 83	R5868
No 97	ED588, EE176
No 100	EE139, JB603, ND458, ND644
No 101	DV245, DV302
No 103	ED888, ED905
No 106	EE191, JB663
No 115	ME803No 153 LM550, ME812
No 156	ED860, ND875
No 166	ED905, LM550, ME746, ME812
No 189	EE136
No 405	ND709
No 463	ED611, EE191
No 467	R5868, LL843
No 550	ED905, EE139, PA995
No 576	ED888, LM227, LM594, ME801
No 622	LL885
No 625	PB150
No 635	ND709

Group

No 1	ED888, ND458, DV245, DV302, ED905, EE139, JB603, LM227, LM550, LM594, ME746, ME758, ME801, ME812, ND644, PA995, PB150
No 3	LL806, LL885, ME803, NE181

No 5	R5868, W4964, ED588, ED611, ED860, ED888, EE136, EE176, JB138, JB663, ND578, LL843
No 8	R5868, ED860, ND578, ND709

Postscript

After completing the book, items of interest continued to come in, and they are offered here rather than lose them to history.

ED588 GEORGE from Brian Holmes

'I was Ernie Berry's flight engineer, although for some reason we called him "Bill", but then I was called "Bob"! We did most of our tour in her before going our separate ways. Sadly I heard that Bill Berry returned to do a second tour and was lost on his very first operation.

'We were engaged on Ops from Skellingthorpe from February to June 1944 and completed 25 of our 33 trips in "G". I'm not sure of the number of trips that were credited to her on our final Op on 14/15 June, to Aunay-sur-Odon, but it couldn't have been far short of 100.

'As a matter of interest, on our trip to Aachen on 11 April 1944, we were hit by an incendiary bomb in the No. 2 starboard petrol tank, from one of our bombers overhead, but it didn't go off! (Lucky "G" and lucky us.) Anyway, on removal from the aircraft, the bomb was found to be still "alive" and by all accounts went off when dropped from a "test wall". We were certainly lucky that night and went on to complete our tour.'

ED888 MIKE by J. R. O'Hanlon

'I am the blond one according to relations, and yes, I was the one referred to as "Blondie" in the "Yank" magazine by Ben Frazier.

'On that trip to Berlin, the wireless receiver went off for a while, however, after a look around for the cause, it again functioned. I explained this to Flight Sergeant Molyneaux, who lives near me now, and he said no fault could be found. I still argue with him now when we meet.

'I recall another trip to Berlin, when we returned with one engine u/s to find thick fog when we arrived back at Elsham. We flew round and on our last circuit, "Bob" Shilling, our rear gunner, nearly collected the red light atop of the main hanger! We were then ordered north and eventually found clear air at Scorton. Control was asked for runway lights and when they were lit, I think everybody grabbed their parachutes, for it was like the Blackpool illuminations, very bright reds, whites and blues.

217

'We landed safely, however, but one thing I won't forget that night is that we took one of the ground crew with us. Having been diverted, we were lucky in the Mess as we only had flying clothing, so the "extra crewman" did not have to show his rank. Some time later there arrived at Elsham Sergeant's Mess, a bill for this gentleman. As far as I know it's still on the notice board.'

LM227 ITEM from George Tabner RCAF

'In July this new Lancaster arrived at Elsham Wolds, to be taken on by 576 Squadron. She landed with a very bad oil leak and repairs were immediately done by the ground crew. The Group Captain was concerned about the new aircraft so he personally air-tested her. Serviceable, she was painted with the Squadron letters "UL" and I/D "I2".

'On July 4th, with PO Stedman and crew she flew her first Op to Orleans. On the second trip, she returned to base with two engines burned out. These were replaced and she was air-tested by FL H. B. Guilfoyle.

'Liking her performance and in spite of his crew's protests, he made her their aircraft. With more speed by 5 knots, it had a better climb rate than their old "B2" for the last ten trips. Guilfoyle and crew went on to complete their tour in her, by which time she had 24 symbols on her nose, 23 for bombing and one for mine laying.'

LM550 LET'S HAVE ANOTHER ('CHARLIE') by H. William Langford

'The crew of which I was the pilot, flew on 17 operations between November 1944 and May 1945, the first being to Freiburg on 27 November. I have a photograph of "Charlie" taken just before setting out for Essen on 11/12 March 1945 (how is 12.3.45 for a date!). Since this was her 101st operation, we had "100 not out" chalked on her nose.

'I also have a record of a trip on 28 February, take-off 0840, mission abandoned and landing back at 1215, but it counted. I recall too bringing back the "cookie" from Nurnberg on 16 March, just couldn't get rid of it. That night 1 Group lost a number of aircraft, but we missed all the mayhem having to go round a few times trying to unload it, thus coming home alone some time after the rest. "Charlie" was good for us.

'As for the beer mugs, these had been removed after her change of Squadron and identity letter, but we had them restored towards the end of the war. There were 118, as far as I remember, at the end.

'I was not at all surprised when, on the last operation of the war in Europe, to Berchtesgaden, the Squadron Commander took her, and she obliged him to turn back with some engine problem, the only time it happened. Although he was a first class man in all respects, it seemed she didn't like him.'

INDEX